P9-DTR-389

ANIMAL LIFE

ANIMAL LIFE

GENERAL EDITOR

Robert Burton

New York
OXFORD UNIVERSITY PRESS
1991

CONSULTANT EDITOR
Professor Peter Haggett, University of Bristol

Jill Bailey, Oxford, UK
Central America
Elizabeth Bomford, Hereford, UK
The Nordic Countries; France and its neighbors; Southern Africa
John Burton, Saxmundham, UK
The United States; Italy and Greece; Eastern Europe
Robert Burton, Huntingdon, UK
The World of Animals
Dr Chris Dickman, University of Sydney, Australia
Australasia, Oceania and Antarctica
Dr Teresa Farino, Cantabria, Spain
Spain and Portugal
Ruth Feber, University of Oxford, UK
The British Isles
Linda Gamlin, Bath, UK
The Soviet Union; Northern Africa; Central Africa
Paul Goriup, Newbury, UK
The Middle East
Dr Emilio Herrera, Simon Bolivar University, Venezuela
South America
Carol Inskipp, Cambridge, UK
The Indian Subcontinent
Dr Andrew Laurie, Cambridge, UK
China and its neighbors
Dr Haruko Laurie, Cambridge, UK
Japan and Korea
Professor J.M. Reichholf, Munich, Germany
Central Europe
Dr Karen Burke da Silva and Dr Jack da Silva, McGill University, Canada
Canada and the Arctic
Drs Wim J.M. Verheugt, Beuningen, The Netherlands
The Low Countries

AN EQUINOX BOO'

Planned and produced by
Andromeda Oxford Limited
11-15 The Vineyard, Abingdon
Oxfordshire, England OX14 3PX

Published in the United States of America by
Oxford University Press, Inc.,
200 Madison Avenue,
New York, N.Y. 10016

Oxford is a registered trademark of
Oxford University Press

Library of Congress
Cataloging-in-Publication Data
Published by Time-Life Books Limited

Animal life / edited by Robert Burton
p. cm.
"An Equinox book"--T.p. verso
Includes bibliographical references (p.)
and index.
ISBN 0-19-520916-8 (hardcover)
1. Zoogeography. 2. Animals. I. Burton, Robert 1941-
QL 101.A55 1991
591.9--dc20 91-20117
CIP

Volume editor Victoria Egan
Designers Frankie MacMillan, Ben White
Cartographic manager Olive Pearson
Picture research manager Alison Renney
Picture researcher Linda Proud

Project editors Susan Kennedy, Candida Hunt
Art editor Steve McCurdy

ISBN 0-19-520916-8

Printing (last digit): 9 8 7 6 5 4 3 2 1

Printed in Spain by Heraclio Fournier SA, Vitoria

INTRODUCTORY PHOTOGRAPHS
Half title: *Western red colobus monkey, Gambia, western Africa (Premaphotos Wildlife/K.G. Preston-Mafham)*
Half title verso: *Cape gannets, Malgas Island, southern Africa (NHPA/Anthony Bannister)*
Title page: *Peacock butterfly, Europe, (Bruce Coleman/Mik Dakin)*
This page: *Caribou, Alaska, USA (Ardea/Martin W. Grosnick)*

Contents

PREFACE

THE EVOLUTION OF ANIMAL LIFE ON THIS PLANET STARTED AS MUCH AS 1,000 million years ago. Little is known about the development of these early forms of life but we do know that many of the main groups of animals alive today came into being over the next 500 million years. Since those almost inconceivably distant times, species, families and entire assemblies of animals have come and gone.

One of the main lessons to be gained from studying the fossil record of these early animal forms is that life is not static - animals not only evolve and take their place on Earth, they also become extinct in a natural process. Apart from a few animal groups that have survived almost unaltered, there has been a kaleidoscope of change through the ages. The pace of change has, however, been so slow that it is revealed only by reading the compacted fossil record laid down in the rock strata.

The driving stimulus to the evolution of the animal life on our planet is adaptation. Through the process of evolution by natural selection, changes can occur to an animal's anatomy, physiology or behavior that make it better able to cope with variations in its physical and natural environment. In other words it is better adapted. If the animals do not adapt to change they may die out, and proof of their existence remains only as traces in the fossil record.

Animals do not live in a particular place by chance. They are there as the result of movements of the Earth's crust, the ebb and flow of ice caps and, above all, through their ability to survive the conditions they find there. Animals have evolved adaptations to live in deserts, at the top of mountains, beneath the ground and in the sea. Some travel long distances to exploit different environments; others are only able to survive in the narrowest of habitats.

On the timescale of human experience, the animal kingdom appears to be unchanging. Only rarely can we observe the processes of evolution and adaptation, so it is easy to think that animals have reached an endpoint - it is hard, outside the realms of science fiction, to imagine how they will continue to evolve. We do not know what sort of environment they will have to adapt to. But we do know that wildlife habitats around the world are changing and disappearing very rapidly. Animal evolution can cope with change, but not at the rate now being caused by human alteration of the environment. There is simply no time for animals to evolve, and our descendants may find themselves sharing the planet with only a few species that are able to live in our modified environment.

This book is a celebration of the amazing variety of animal life to be found in every region of the world. It is hoped that, with the considerable efforts now being made to conserve species and habitats, it will not merely be a record of that variety for future generations.

Robert Burton
HUNTINGDON, UK

A lioness with her cub in Serengeti National Park, Tanzania

Land iguana on the Galapagos Islands *(overleaf)*

THE WORLD OF ANIMALS

ANIMAL DIVERSITY

ANIMALS UNDER THREAT

CONSERVING ANIMAL LIFE

Animals Today

NO ONE KNOWS EXACTLY HOW MANY different kinds of animal there are on Earth, but so far about 1.5 million species have been studied and named. Among the higher animals there are roughly 8,800 species of bird, about 4,070 mammals, more than 21,000 fish, about 4,015 amphibians and roughly 6,600 reptile species. However, these numbers represent only a tiny proportion of the world's wildlife. The overwhelming majority of living animals are invertebrates – animals without backbones – and of these more than a million species are insects. (Some estimates put the total number of insects at up to 30 million.) The numbers in which insects gather can also stagger the imagination. While we count our national populations in tens or hundreds of millions, a single swarm of Desert locusts may contain 50 billion individual insects.

Animals display enormous variety in shape and size, in their geographical distribution, and perhaps most strikingly of all in their physical and behavioral adaptations to the varied habitats in which they live. The Blue whale, the largest animal ever to have lived on Earth, still survives today; it averages 27 m (88.6 ft) in length and may weigh up to 150 tonnes, while the smallest ants are less than 1 mm (0.04 in) long and weigh a tiny fraction of a gram. The great diversity of ways in which different animals obtain the necessities of life, protect themselves from harm, and pass on their genetic characteristics to future generations provides a field of endless study and fascination. This is the world of the Barn owl swooping down on silent wings to catch a mouse in total darkness; of the diminutive dormouse spending two-thirds of its life asleep; and of the salmon migrating from its feeding grounds off Greenland to the very same upland stream in which it hatched years earlier.

The annals of natural history reveal a catalog of amazing animal lifestyles. There are bats that catch fish; an octopus that leaps from the water to catch crabs; and the Pistol shrimp that stuns its prey with shock waves. The Naked mole-rats of Africa live in colonies in a similar way to bees or ants. Only one dominant pair will breed, while the rest of the colony is divided into castes that dig burrows, gather food or guard the breeding female. Equally strange is a parasitic wasp that lays its eggs inside moth caterpillars. Each egg develops into 200 larvae, some of which become "soldiers" whose job it is to kill the larvae of other wasps that have parasitized the caterpillar. Larvae of the "soldier" caste die once their job is done, but the sacrifice is not in vain: they leave the caterpillar body free of competitors for their siblings which develop normally into adult wasps.

Animal classification

Zoologists have always been faced with the major task of having to impose some order on the plethora of animals in their collections. They began by arranging the animals into groups according to characteristics that they had in common. The basic unit of classification is the species: a particular species differs from other animals in looks and behavior and breeds only with others of its own kind. The lion and the lynx are both members of the cat family, but they are distinct and separate species. The definition of the term takes account of the fact that many of the differences and similarities between animal groups are based on inherited characteristics determined by an animal's genes. Accordingly a species is a community of animals that breed among themselves, sharing a common gene pool.

From Aristotle onward, animal species have been grouped as far as possible according to their nature rather than their superficial appearance. However, some of the older natural history books reveal some very odd bedfellows. Whales were described as "cetaceous fishes" on the basis of their streamlined bodies, their fins for swimming and their fishlike way of life. Their mammalian nature was only recognized in the late 17th century by the English naturalist John Ray, although Aristotle had pointed out some of the similarities between whales and other mammals almost 2,000 years earlier.

Today, there is a recognized hierarchy of taxonomic groupings above the species level; the main ones being: species are grouped into genera, genera into families, families into orders, orders into classes and classes into phyla. Phyla are finally grouped into kingdoms. Classification can be also carried below the species level. Subspecies are populations of a species that are geographically isolated from each other and do not normally interbreed; consequently they have evolved small differences in genetic makeup and appearance. This can be an important point for conservationists. For example, is a population of birds living in a remnant of a tropical rainforest sufficiently different from other populations of the same species to merit special conservation efforts?

Modern classification has two main functions: to provide a straightforward identification system and to show the evolutionary relationships between different groups of organisms. This grouping involves the use of features not always immediately visible – such as internal structure or physiology, or even the characteristics of the genetic material (DNA) itself. These links in relationships are often missing from the fossil record, so even today there are many species whose affinities are unclear. The Giant panda, for example, has been variously classed with the raccoons in the family Procyonidae, with bears in the family Ursidae, and in a family all of its own, the Ailuropodidae. Current views, based on DNA analysis, favor a close affinity with the bears.

A clouded yellow butterfly *(above)* about to land on a thistle head. The number of insect species exceeds that of all other groups together – one million have been identified, but there may be up to 30 million in all.

Pollination on the wing *(left)* A Buff-bellied hummingbird hovers in front of a banana flower to feed. While doing so it picks up pollen on its head, which it will deposit on the stigma of the next flower it visits, so pollinating it. There are many similar examples of partnerships between plants and animals.

Classification Lions belong to just one species *(Panthera leo)* but up to seven subspecies have been recognized, including the Cape lion of South Africa, which is now extinct. Subspecies can interbreed but do not normally do so because of geographic separation.

SCIENTIFIC NAMING

The universal system for naming both plants and animals was devised by the Swedish botanist Carl von Linné (Latinized to Carolus Linnaeus) in the 18th century. The system of using a pair of names, the first for the genus, the second for the species, enables scientists of all nationalities to refer to the same organism without confusion. The lion species, for example, has different names in different languages (Löwe in German, Simba in Swahili). However, its scientific name is *Panthera leo*, and this is recognized by scientists from all around the world.

The great advantage of the Linnaean system is that the names are arranged in a hierarchical system that demonstrates species' relationships to each other. The lion, *Panthera leo*, is closely related to the tiger, *P. tigris*, but more distantly to the puma, *Felis concolor* and the domestic cat *F. catus*.

Sometimes scientific names of animals are changed. For example, the Brown hare of Europe was first called *Lepus europaeus*. Linnaeus had earlier given the name *L. capensis* to a hare that had been found at the Cape of Good Hope. Later it was realized that this hare was not any different from the European hare. As the name *L. capensis* had been used first it had priority, and it is now accepted as the correct name for the single species incorporating both the European and African hares.

LION CLASSIFICATION

Phylum	Chordata
Subphylum	Vertebrata (vertebrates)
Class	Mammalia (mammals)
Order	Carnivora (carnivores)
Family	Felidae (cat family)
Genus	*Panthera* (big cats)
Species	*Panthera leo* (lion)

The Animal Kingdom

THE ANIMALS ARE DIVIDED INTO ABOUT 38 major groups or phyla. Many of these are very small and obscure. Traditionally all these phyla were grouped together in a single animal kingdom, but now it is customary to divide the animals into two kingdoms – the single-celled protozoans and the multicellular animals, the latter being known as the animal kingdom.

The animal kingdom is divided into two unequal groups – the vertebrates, or animals with backbones, and the invertebrates, those without backbones. Humans are vertebrates, and so are all the most familiar large animals – the birds, mammals, fish, reptiles and amphibians.

Invertebrates overwhelmingly outnumber the vertebrates, both in numbers of species and in sheer weight of organic material (biomass). They range in size from microscopic organisms to the 20-m (60 ft) long Giant squid, and include such varied groups as insects, spiders, snails, worms, jellyfish and sponges. Many of the small invertebrates are a prime source of food for other animals, and as such form one of the links in the many food chains. Others, such as earthworms and woodlice (sow bugs), perform vital roles as decomposers – breaking down organic debris so that its components can be used again.

Invertebrate mysteries

Some invertebrate groups are a puzzle even to specialists. The gnathostomulids are worm-like animals about 1 mm (0.04 in) long, discovered in the mud of the ocean floor where there is virtually no oxygen. There is no consensus among scientists about where they fit into the animal kingdom. Even well-known groups can be difficult to place if their ancestors have left insufficient fossils to show their links with other groups. The group Arthropoda was originally created to include all animals with segmented, hard-skinned bodies and jointed legs – the insects, crustaceans, spiders, mites, scorpions, millipedes and centipedes, and it had long been assumed that they had a common ancestor. However, current research suggests there are three main evolutionary lines (insects, crustaceans and spiders), none of which may be related and they are now placed in separate phyla.

Protozoans – a new kingdom

The bodies of multicellular animals show a wide range of forms. Most of those animals we see around us have complex systems of body organs – heart, lungs and blood circulation system, brain and nervous system, digestive and excretory systems, sense organs for sight, smell, taste, touch and hearing, and skeletons and muscles for supporting and moving the body. Yet in many microscopic creatures such as *Amoeba* and *Paramecium* all these functions are combined in a single cell, which has specialized regions or "organelles" instead of organ systems. These features are so significant that the protozoans and their relatives are considered to belong to a completely separate kingdom – the Protista.

Many protists are animal-like in that they have to absorb (eat) food from their surroundings, but some are plantlike. They contain the green pigment chlorophyll, through which they capture the energy of sunlight and use it to manufacture carbohydrates from carbon dioxide and water – the key characteristic of plants. The creation of the kingdom Protista avoided the problem of whether certain single-celled organisms were plants, animals or both.

DISCOVERIES AND REDISCOVERIES

A scientist was waiting for a helicopter to lift him from the top of a mountain in Venezuela, South America, when a frog hopped onto his lap. It was a new species, not known to science. The helicopter failed to arrive, and during the next ten days the scientist discovered four more new frog and three new lizard species. In 1984 another scientist was struck on the head by a bird on Gau Island in Fiji. It was the only Fiji petrel to be recorded since the first specimen was collected in 1855.

These two stories illustrate that there are still exciting finds to be made in the world of animals. New species, and specimens of those believed long-dead, turn up through a combination of luck and diligent searching. Occasionally one will hit the headlines, such as the "living fossil" coelacanth fish first discovered in 1938 and again in the Indian Ocean in 1952, but little is heard, except in scientific journals and academic papers, of the thousands of invertebrates discovered each year. Some scientists believe millions of invertebrate species remain to be discovered.

Kingdoms of the living world

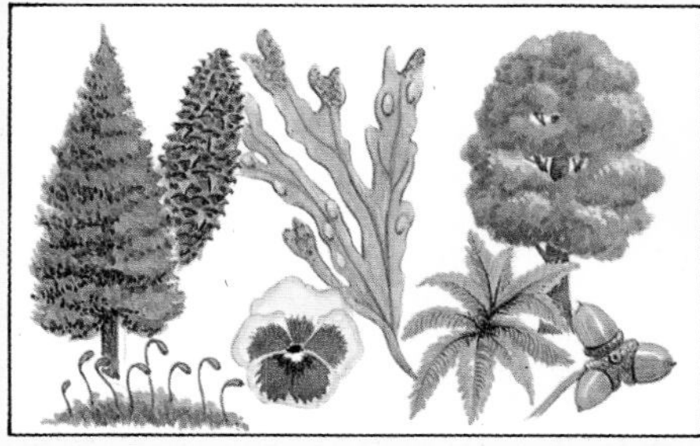

Plantae - plants Multicellular organisms that synthesize food from raw materials by photosynthesis. Body of cells with cellulose cell wall. Do not move and mostly of indeterminate growth. For example, algae, ferns, flowering plants.

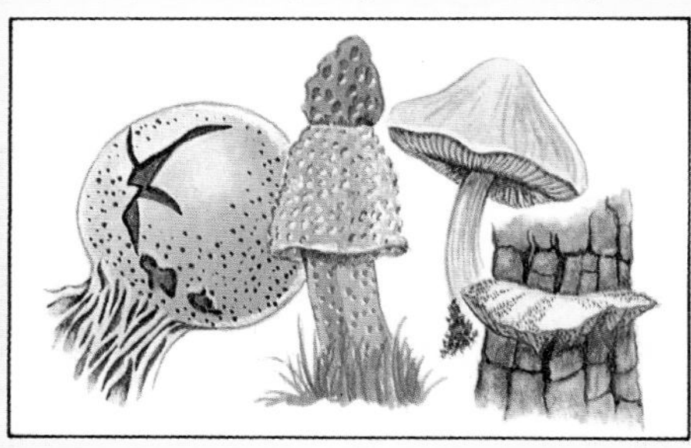

Fungi - fungi Mostly multicellular organisms (some microscopic) that digest food outside the body and absorb it. Body composed of threadlike hyphae which may mass to produce fruiting bodies (e.g. mushrooms).

Animalia - animals Multicellular organisms that ingest their food and mostly have complex body plans. They are able to move, and respond rapidly to stimuli; grow to fixed body size and shape. For example, worms, fish, mammals.

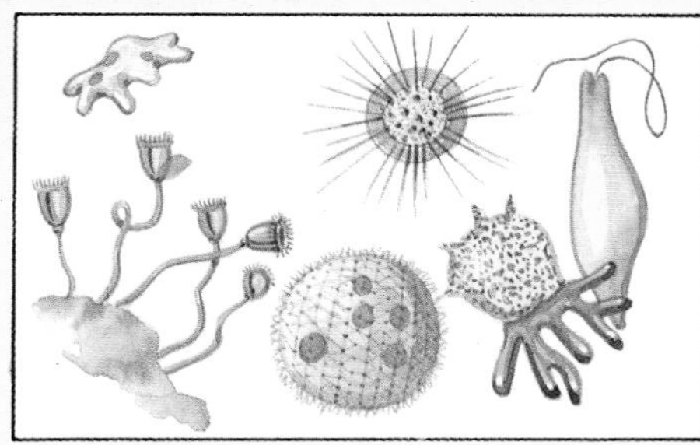

Protista - protists Mostly single-celled, but live in colonies. Protozoans are animal-like and ingest their food while unicellular algae have chloroplasts and synthesize their own food. For example, *Amoeba, Paramecium, Euglena.*

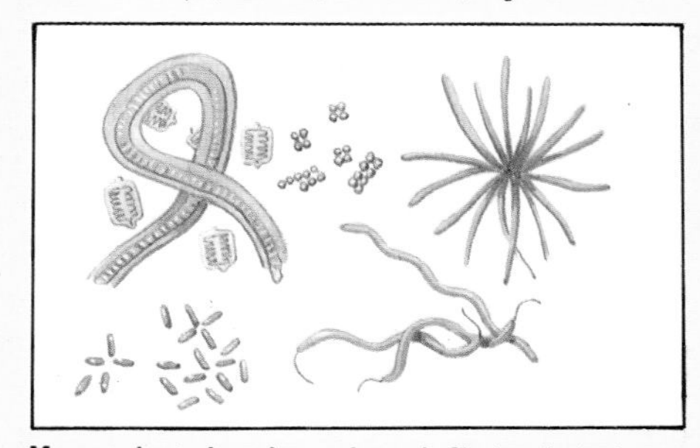

Monera - bacteria and cyanobacteria Single-celled, microscopic organisms; cells lack internal structure found in other kingdoms. Bacteria absorb nutrients from surroundings; cynobacteria (blue-green algae) are photosynthetic.

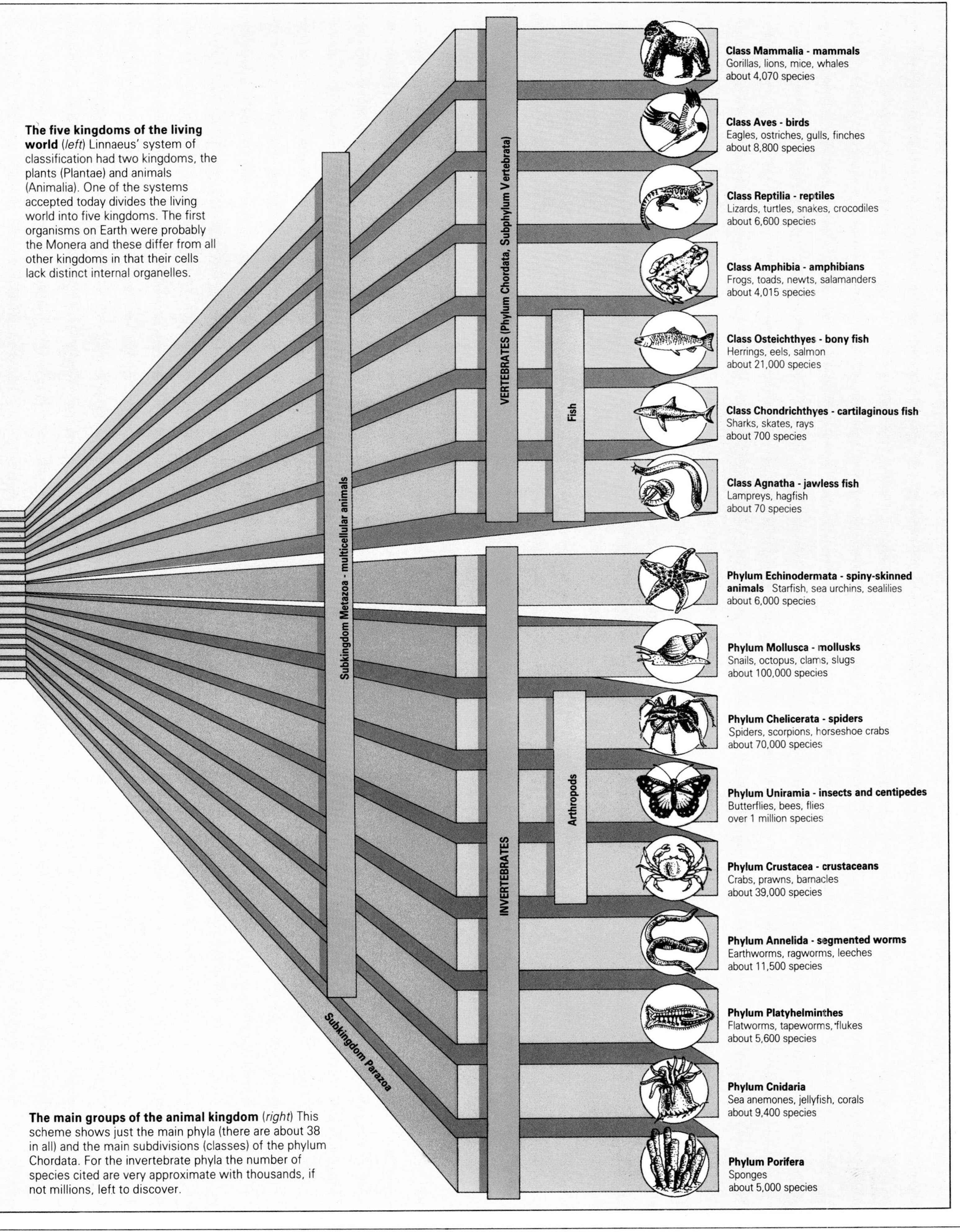

The five kingdoms of the living world (*left*) Linnaeus' system of classification had two kingdoms, the plants (Plantae) and animals (Animalia). One of the systems accepted today divides the living world into five kingdoms. The first organisms on Earth were probably the Monera and these differ from all other kingdoms in that their cells lack distinct internal organelles.

The main groups of the animal kingdom (*right*) This scheme shows just the main phyla (there are about 38 in all) and the main subdivisions (classes) of the phylum Chordata. For the invertebrate phyla the number of species cited are very approximate with thousands, if not millions, left to discover.

The Evolution of Animals

THE IDEA OF EVOLUTION DATES BACK TO the Ancient Greek philosophers. However, it was Charles Darwin and Alfred Russel Wallace who made the first scientific study of the evidence and, in 1858, they published the explanation of how evolution could take place by slow change over many generations. One year later Darwin published *"On the origin of species by means of natural selection, or the preservation of favoured races in the struggle for life"* in which he propounded the theory of evolution by natural selection.

Darwin's theory of evolution

From his observations of plants, animals and fossils during his voyage on *The Beagle*, Charles Darwin became convinced that species changed, or evolved, over a period of time. To account for a mechanism he reasoned as follows: animals produce more young than can survive to produce themselves because the environment can only hold a certain number. There was, therefore, a "struggle for existence" in which individuals compete for food, space and so on. He also noticed that individuals differ slightly from their parents and each other in the range of characteristics that they exhibit. It follows that those born with a slightly advantageous characteristic – such as speed or strength in the case of a hunter, or camouflage coloring, a toxic skin or defensive armor in the case of a potential prey species – will be more likely to survive and produce offspring than those that have inherited less advantageous characteristics. This is known as the "survival of the fittest" and zoologists now define "fitness" as the ability to survive and produce offspring which will also survive in prevailing conditions.

With each successive generation the advantageous characteristics accumulate in the population. If, at a later stage, environmental conditions change – for example if the climate becomes colder or a new predator appears – those animals with the characteristics best suited to survive new circumstances will be "selected" – that is, will survive and produce young, while the old, unsuitable, types die out. Eventually a new population with distinctive characteristics is formed that is much better adapted to the new conditions.

One problem that bothered Darwin was how the changes between parents and offspring had come about. This was eventually solved by the science of genetics. The genetic material, DNA, occasionally undergoes spontaneous, random changes (mutations) and produces new genes. These are mixed in sexual reproduction and natural selection then "weeds out" the harmful or weak ones.

Tracing the fossil record

The course of evolution between groups of animals is sometimes established by the discovery of intermediate types – often referred to as "missing links". For many years a search was made for the supposed "ape-man" that represented the halfway stage between the apes and our own species *Homo sapiens*. No such animal has been found because we are not descended directly from the apes; instead we share a common ancestor.

Nevertheless, intermediate stages have their value. The most famous is probably the fossilized body of *Archaeopteryx*, whose feathers, beautifully preserved in fine Bavarian limestone, prove it to be a bird, while its teeth, long tail and other skeletal features are undoubtedly reptilian. What is now needed is to establish the way in which birds evolved from

Scene from the prehistoric world. This artist's impression shows a range of animals that flourished in southwest North America during the last cold stage of the Quaternary period (about 30,000–20,000 years ago). Evidence of these animals comes from fossils.

LIVING FOSSILS

"Living fossils" are animals that, by surviving almost unchanged for many millions of years, provide us with an insight into life in prehistoric times. Coelacanths, large, heavily-scaled fish with fleshy muscular fins, were abundant in Cretaceous seas 135–65 million years ago and were thought to have died out about 70 million years ago. However, a live specimen, later named *Latimeria chalumnae*, was caught in the Indian Ocean in 1938. Unfortunately the specimen deteriorated before proper studies could be carried out, but another was caught in 1952. The ancient coelacanth is now known to live on the sea bed near the Comoros Islands.

Of far more ancient lineage is the Pearly nautilus, a mollusk related to the octopus, whose relatives were common in seas 500 million years ago. Although in some ways a "modern" animal, studies of the living nautilus have helped in the understanding of how these ancient ammonites used buoyancy chambers in their shells to swim at any depth.

Horseshoe crabs live off the coasts of North America and Southeast Asia where they are caught in large numbers. They have hardly changed in 300 million years. At one time they were believed to be related to the long-extinct trilobites, but their nearest relatives are probably the spiders.

It is a mystery why all these living fossils should have survived while their relatives died out millions of years ago. It may have been chance, or perhaps because they lived (and continue to live) in relatively stable environments with little competition.

Relict of a bygone age, the tuatara is the sole survivor of a group of reptiles that came into being 220 million years ago. It lives on only a few islands off New Zealand.

reptiles, that is the discovery of an earlier missing link in the story.

The origin of animal life lies some 1,000 million years ago in the seas. It is impossible to be certain because the first animals had soft bodies which left little trace in the rocks. However, fossils of corals, jellyfish and worms have been found that date back 680 million years. Fossils from the Cambrian period (570–500 million years ago) are more common; by then animals had developed hard parts that readily became fossilized. Almost all phyla were in existence by this time, but the record of their evolution is poor, so comparisons of their body plans have to be used to produce their evolutionary "tree". The pattern of evolution of any group of animals, whether phylum, family or other level, is one of radiating branches. From the ancestral type there arises a number of differing types that change in character to take up new ways of life or adapt to changing environments. Some are unsuccessful and die out. Others undergo radical changes and set up new lineages.

If this process of radiation is rapid, in geological terms, it is difficult to follow the progress because vital stages may be missing from the fossil record, but the mollusks, for instance, radiated slowly and the evolution of different mollusks - gastropods, bivalves and cephalopods - can be traced through fossils and living species

Life moves onto land

The Silurian Period (440–395 million years ago) saw a major innovation when animals emerged from the sea onto land, a move that became possible only after plants had first achieved that important transition. The first groups to emerge were arthropods, such as millipedes, and about 50 million years later wingless insects made their appearance; these were soon followed by the tetrapods (four-legged animals) in the form of amphibians. Major adaptations of body form and function were required for survival on land. These included the development of a tough waterproof skin; a strong skeleton to support the body's weight out of water; waterproof eggs – essential even in animals with internal fertilization (as in reptiles and birds); and the ability to breathe air. The mollusks and various forms of worm have been only partly successful in adapting to life on land because they dry up easily. Once on land, adapting to land-living allowed a massive radiation of new animal types to fill unoccupied space.

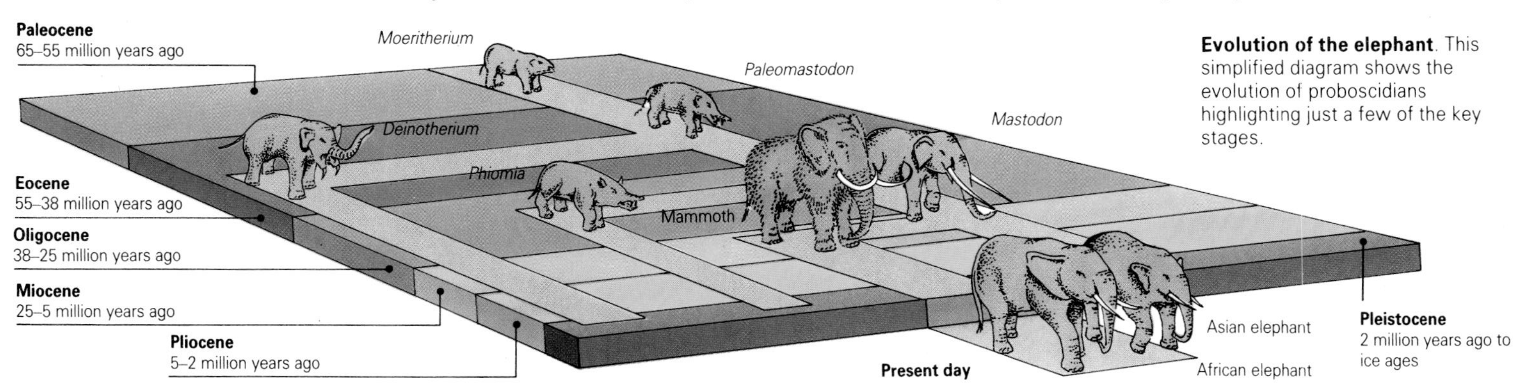

Evolution of the elephant. This simplified diagram shows the evolution of proboscidians highlighting just a few of the key stages.

Patterns of Distribution

WHEN THE AMERICAS WERE FIRST SETTLED by Europeans, the newcomers were astonished by the unfamiliarity of so much of the native wildlife. In North America there were pronghorns and gophers, skunks and raccoons, tree-climbing porcupines and a host of colorful but unknown birds. In the great landmass of South America they found yet more true wonders – armadillos, prehensile-tailed monkeys, giant anteaters and a variety of insect and bird life that staggered the imagination.

The main reason for the differences in the animal life of the Old World and the New lies in their isolation. Vast expanses of ocean lie to east and west of the Americas; only in the far northwest does North America reach out a promontory and chain of islands to within 90 km (55 mi) of Asia. This explains why the early explorers and settlers also found familiar species such as bears and weasels.

The distribution of an animal species is limited not only by the extent of its habitat – be it seashore, cloud forest, sand dune or woodland floor – but also by the geographical barriers that may stand in its way. Zoogeography seeks to explain the pattern of animal distribution. The most obvious barriers, for land animals at least, are the oceans, mountain ranges and deserts. Some species have a very wide distribution. The Barn owl, for example, is found in every continent on Earth except Antarctica. But the Devil's Hole pupfish is extremely limited in its distribution: it has been marooned for thousands of years in a single pool measuring 20 m by 3 m (65 – 10 ft) in Death Valley, California, in the United States.

The drifting continents

Each major landmass has its own typical "fauna" – an array of animal species that is characteristic of the region because it has developed more or less in isolation. Toward the end of the Permian Period, 225 million years ago, the continents were united in a single landmass called Pangaea. Since that time the continental blocks have split up and drifted apart, moving over the Earth's surface at speeds of up to 10 cm (4 in) a year. This phenomenon is known as continental drift. The animals continued to evolve in isolation after the landmasses had moved apart. North and South America were separated for a long time and the southern landmass evolved a variety of unusual mammals including marsupials, edentates (the sloths, armadillos and their relatives) and a number of strange giant hoofed animals. Continental drift then reunited the two landmasses. Central America became a land bridge, with the isthmus of Panama linking the two continents. Many North American mammals spread south over the land bridge and replaced many of the native South American species. Only a few of the South American species such as the American opossum spread north.

By contrast the island continent of Australia remained isolated. Its egg-laying monotremes – platypus and echidnas – survived and, without competition from placental mammals, its marsupials evolved into a wide range of forms.

Ice barriers

The Ice Ages, which began almost 2 million years ago and ended barely 10,000 years ago, forced animals out of large parts of their ranges. At the height of the Pleistocene Ice Age extensive ice sheets covered most of Europe, driving all but a few animals far to the south. As the climate improved some animals returned, but the rising waters from the melting ice sheets separated Britain from mainland Europe, leaving it with an impoverished wildlife. Further rises in sea level separated Ireland from mainland Britain, isolating many of the British animals: the result is that to this day Ireland has no frogs, toads, snakes or moles.

The barriers between the zoogeographic regions are not insuperable. Insects and spiders are carried over them by the wind; birds and bats fly over them. Some species have spread by sea – swept out from their native shores on rafts of vegetation by fast-flowing rivers and transported by ocean currents. Others have extended their range by overland migration. The Bering Straits were once dry land. Mammoths, bison, Musk oxen and people crossed into America from Asia before rising sea levels separated the continents some 14,000 years ago.

Since animals first attached themselves to human societies, people have either deliberately or inadvertently transported animals far beyond their natural ranges. Some have been harmless in their new homes, but others, such as the Black rat and American cockroach have become serious pests or even brought about the extinction of native species.

ENDEMICS AND RELICS

When an animal population is isolated from others of the same species it may continue to evolve and produce new forms. Thus remote oceanic islands, and even whole continents, acquire their own distinctive fauna.

After a long period of isolation a group of new species results, with a very localized distribution. They are "endemic" to that particular area and are found nowhere else. On the volcanic islands of Hawaii, lava flows have carved up the existing mountain vegetation into isolated green bands; the resident fruit flies, many of which will not fly far to mate, have evolved into hundreds of different and highly localized species. Mountain tops, particularly those of the tropics, can be "islands" of coldness surrounded by lands with a more hospitable climate. Here endemic species may evolve.

Mountain tops and other isolated habitats such as lakes and cave systems may also be refuges of animals that once had a far wider distribution and now are confined as their habitat shrank. Such species or populations are termed relics and may die out completely or be able to reinvade should the habitat expand again. These "islands" and refuges are of prime importance to conservationists.

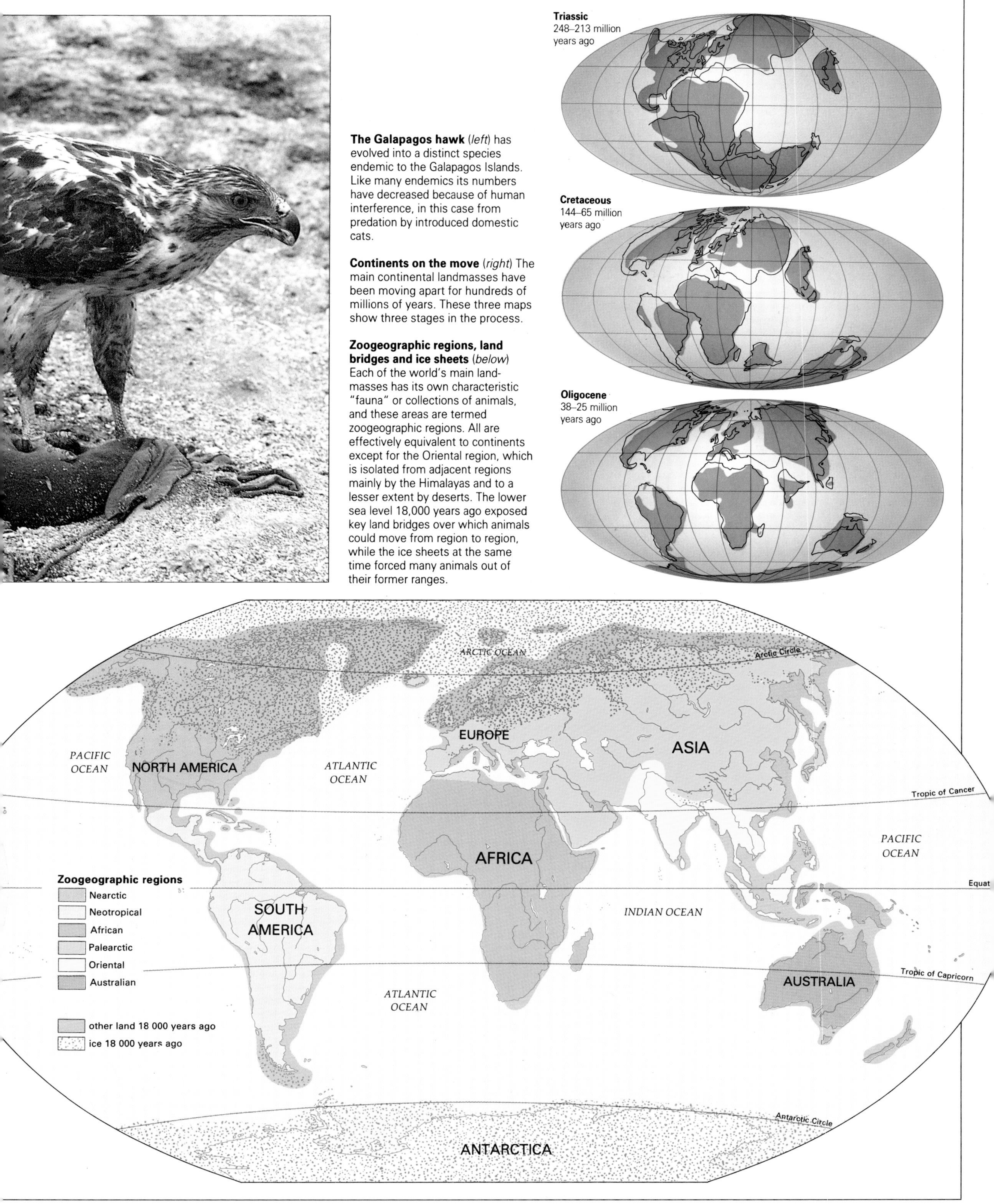

The Galapagos hawk (*left*) has evolved into a distinct species endemic to the Galapagos Islands. Like many endemics its numbers have decreased because of human interference, in this case from predation by introduced domestic cats.

Continents on the move (*right*) The main continental landmasses have been moving apart for hundreds of millions of years. These three maps show three stages in the process.

Zoogeographic regions, land bridges and ice sheets (*below*) Each of the world's main landmasses has its own characteristic "fauna" or collections of animals, and these areas are termed zoogeographic regions. All are effectively equivalent to continents except for the Oriental region, which is isolated from adjacent regions mainly by the Himalayas and to a lesser extent by deserts. The lower sea level 18,000 years ago exposed key land bridges over which animals could move from region to region, while the ice sheets at the same time forced many animals out of their former ranges.

Animals and Ecology

Ecology is the study of animals and plants in relation to their environment. It can be studied from three viewpoints. First, there is the investigation of all aspects of the life of a particular animal or plant species, which is called autecology. Next, there is the study of relationships that exist between species, such as competition for food or living space, and the relationships between predators and their prey. The third involves the study of an environment, including the physical needs of the animals; for example how gerbils and similar species survive in deserts and obtain water, or how female birds obtain the calcium they need for manufacturing eggshells.

A community of animal and plant species in their environment is called an ecosystem. It may consist of an individual tree, a whole rainforest, a coral reef or an ocean. Each ecosystem is more or less self-sustaining, and within it the plants and animals interact with each other and with their physical surroundings. Beyond that the ecosystems themselves are linked together: the coral reef with the ocean, the tree with the forest, and so on, to make up the biosphere – the totality of land, air, water and living organisms that clothes the Earth's surface.

Food chains and food webs

All the Earth's ecosystems have a number of essential processes in common. The most important of these is the food chain or food web – best visualized as the flow of energy through the system. Apart from some curious seabed communities, all living things derive their energy ultimately from the Sun. Plants harness the Sun's energy by photosynthesis – using it to synthesize carbohydrates from carbon dioxide and water. This process is made possible by the green pigment called chlorophyll, which acts as a catalyst for the chemical reaction. Plants are therefore the basic food manufacturers, so they are called "producers". Animals are "consumers", either obtaining their energy and nutrients directly from plants, in which case they are herbivores, or indirectly, as secondary consumers or carnivores; these eat the flesh of other animals.

Eventually, all organisms die, and their bodies are broken down by a host of decomposers – bacteria, fungi and scavenging animals. These release some of the energy as heat, while nutrients are returned to the soil or water and are taken up again by plants.

Energy and nutrients pass from plant to animal to animal through food chains, but as most animals have a varied diet, these simple chains become linked to form food webs. The web may be represented as a series of trophic levels (from the Greek word *trophikos* – nourishment). At the base are the producers; above them, in decreasing numbers, are the primary and secondary consumers; and lastly, on the highest level, the top predators.

At each stage in a food web, energy is lost. Of the food energy an animal takes in, most is used to keep its body warm, and to fuel its movements and bodily functions, in particular keeping it warm in the case of warm-blooded animals like mammals. Barely ten percent of the food energy is converted into new tissue and is available as food to the next consumer in the chain. This is why, in most habitats, there are large numbers of plant-eaters, smaller numbers of animals preying on the plant-eaters, and very few of the largest hunters. The population numbers at the various levels form a pyramid of numbers.

Habitat successions

Ecosystems are dynamic. Over a period of time the habitat, and the number and variety of its resident animals, will change. A new habitat, such as a bare sand dune, a recently cooled lava flow or an area of ground exposed by a retreating glacier, will be colonized by hardy pioneer plants. As they spread they modify the soil and the climate close to the ground, making conditions more suitable for plants that would have found the original habitat too inhospitable. Soon new plants take over from the pioneers, and they in turn give way to yet more new species. As the plants change, so too do the insects, birds and other animals that inhabit the area. The whole process is called a succession, and its end-product – the point at which a stable community of animals and plants is established – is called the climax. This climax will remain undisturbed until a forest fire, flood, human interference or other factor upsets the equilibrium.

Even within a climax there are local short-term changes. For example, if a forest tree dies and falls, a light-gap is created in the forest canopy. Plants that have lain dormant for years will spring to

Tiny communities Pools of rainwater in the tree tops created within the interleaving leaf bases of bromeliads are perhaps one of the smallest animal communities, but quite capable of providing most of the needs of these bromeliad frogs.

life and their characteristic animals colonize the area. Succession will run its course until the gap is closed and the forest canopy is once more reestablished.

Animals modify the environment in which they live. Rabbits and other grazing animals may destroy tree seedlings, preventing the invasion of woodland and so maintaining open grassland areas. Earthworms, moles, prairie dogs and other burrowers break up and mix the soil, while animal urine and feces enrich it with nutrients.

Population dynamics

Changes in the abundance of one animal species may have a direct impact on another if the two are linked by a predator/prey relationship. Caribbean Hawksbill turtles, for example, eat sponges. The turtles are immune to the poisons in the sponges that deter most other animals. The turtles play an important role in the life of a coral reef by controlling the blanketing growth of sponges, and allowing other species to colonize the reef. If the turtles disappeared, the reef would soon become a very different habitat, with a different composition of animal species.

In the long term ecosystems are stable, but they may have to withstand natural disturbances. These may take the form of hurricanes, droughts, fires or floods. Caribbean reefs suffer hurricane damage about once every century: the coral is smashed, and the fish, sea urchins, shrimps and other animals are left without food or shelter. Often the reef, or

KRILL – LINCHPIN OF THE ANTARCTIC FOOD WEB

The Antarctic is often described as the last great pristine wilderness. This is true for most of the ice-covered landmass but not for the surrounding seas. These have been pillaged since the 19th century for elephant seals, fur seals and whales, and the exploitation continues today with extensive fishing for krill, fish and squid.

Krill – 5 cm (2 in) long shrimps that live in huge swarms – are the principal food for many animals; those that do not eat krill usually feed on squid and fish which in turn have fed on krill. Krill feed on huge masses of tiny floating plants (phytoplankton) that thrive in the Antarctic summer, and together with various other small animals they form the base of the Antarctic marine ecosystem.

Before whaling began in the Antarctic, filter-feeding baleen whales ate over 150 million tonnes of krill each year. In the 1990's the depleted whale population probably consumes about 40 million tonnes. It is believed some of the 110-million-tonne "surplus" is being consumed by other animals. Fur seals, for example, were almost wiped out by hunters in the 19th century, but from a small colony of a few dozen surviving in the 1950s, the population has exploded to a million animals, nourished on krill. The numbers of some penguin species are also increasing. These increases could, however, hinder the recovery of the great whales, while unregulated fishing for krill, squid and fish in the Southern Ocean may already be affecting all of the Antarctic birds and mammals.

part of it will die but normality eventually returns. Increasingly throughout the world, disturbances are due to human interference. These include cattle grazing, the felling of forest to make way for agriculture, the socalled reclamation of coastal mudflats for marinas etc, and the draining of wetlands for building housing or leisure developments. Where these changes are temporary or smallscale, the habitat can recover in time: but all too often they are largescale, permanent and disastrous to wildlife.

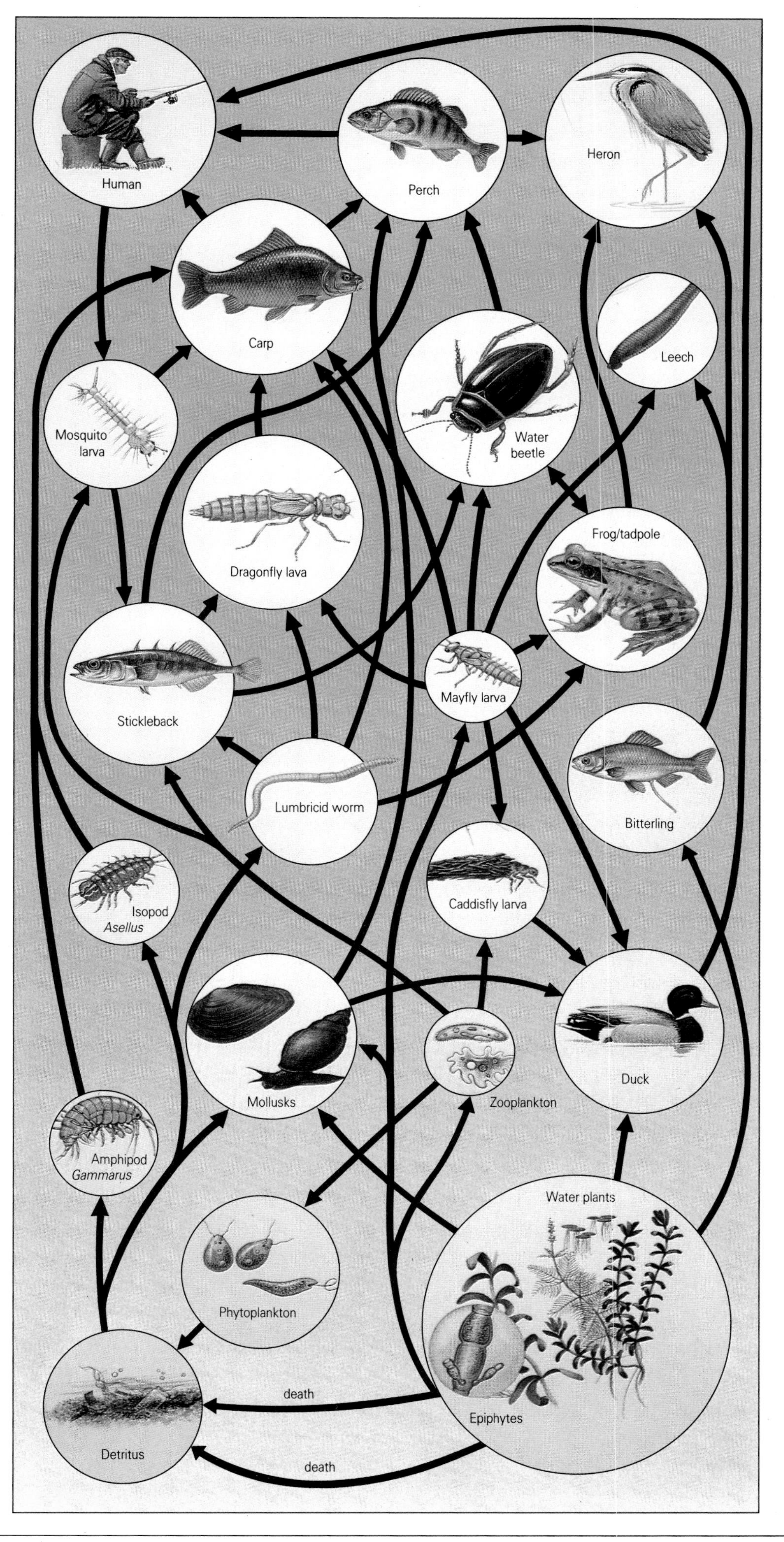

Food web in a lake community Within this fairly simple web four trophic levels are represented (plants, herbivores, carnivores that eat herbivores and carnivores that eat carnivores). The position of each species in the web will depend on its age, size and the availability of food.

Adaptation and Specialization

It has been said that nature abhors a vacuum, and this seems to be borne out by the ability of animals to live in almost any habitat on Earth. The petroleum fly lives in pools of oil around well-heads; the Emperor penguin nests in the depth of the Antarctic winter; a nematode worm lives in the nose of the stoat; and the pearlfish makes its home in the anus of a sea cucumber.

Entire animal communities live in the strangest places: in the isolation of subterranean lakes, on the underside of polar pack ice and, most amazingly of all, around plumes of hot, sulfurous water that gush from vents in the sea bed over 2.5 km (1.5 mi) below the surface of the Pacific. The only environment that have not been colonized by animals are icecaps and glaciers, and the unique dry valleys of Antarctica.

Each to its own niche

Animals have been able to set up home in these diverse and seemingly inhospitable environments because they have, by natural selection, evolved special adaptations. These characteristics enable them to cope with the physical conditions and to play a specific role in the community: each has its own ecological niche. The niche is defined by the species' own requirements and its tolerance of the environmental conditions, including its relationships with other plants and animals in the ecosystem. All animals need food and are adapted for dealing with particular foods. This can be illustrated by the teeth of mammals. Herbivorous cattle and antelopes have teeth for grinding; predatory cats have slashing and holding teeth; and the Crabeater seal has plankton-filtering teeth.

By occupying a unique niche, a species reduces competition with others. In this way the food resource of a habitat can be divided among many different species. Both kestrels and Short-eared owls hunt small mammals in open country, but the kestrel is active by day and the owl by night. However, the ecological separation is not complete since both species will compete and so suffer if small animals become scarce.

Separation of niches can be effected by quite small differences in anatomy or behavior. In the rainforests of Papua New Guinea, eight species of fruit-eating pigeons live together. This is possible because they are of different sizes: the small pigeons perch on slender branches and take the smallest fruit; the larger ones perch on bigger branches and take larger fruits. Adaptations to physical constraints also separate animals into different niches. Two species of wolf spider live in wet sphagnum bogs, but *Pirata piraticus* prefers the cool damp conditions in the deep moss near water, while *Lycosa pullata* prefers to be warm and dry and lives among the tips of moss plants.

Nature's compromise solutions

It is often said that an animal is perfectly adapted for a particular way of life: for instance that a whale is perfectly adapted for life in the sea. This is inaccurate. A whale must surface to breathe, so in this respect it is less well adapted to an aquatic life than a fish. Natural selection has to work with the material available and often produces a compromise. In the case of the whale it has transformed a land animal into an efficient marine animal by modifying its organs and body systems. The nostrils have been relocated on the top of the head, and the reproductive system has been modified for breeding at sea – something seals cannot do.

Animals also become adapted to their environments at fine levels that are not immediately apparent. Elephants face the problem of overheating because their surface area is small in relation to their bulk. Their large ears increase the surface area from which heat can be lost. However, the African elephant has adapted to life in hot open country and possesses much larger ears than the Asian species, which lives in the cooler, shady environment of forests.

Hot spots for life (*above*) Life in the hot vents on the sea floor of the Pacific Ocean is unique. Pogonophoran worms (as shown here), clams, anemones, crabs and fish are linked in a web that depends for its energy on sulfur-fixing bacteria.

Flying squirrel* *Glaucomys*

Glider *Petaurus*

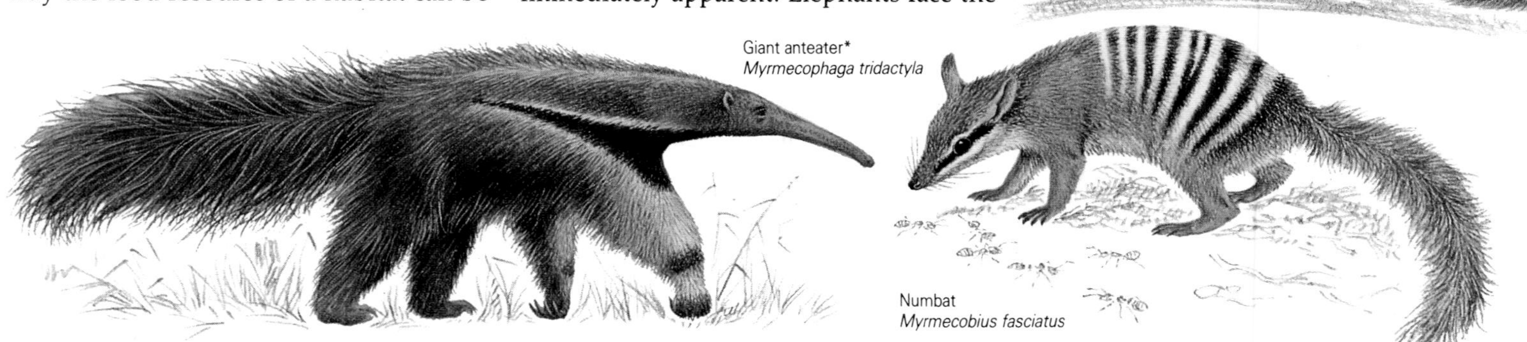

Giant anteater* *Myrmecophaga tridactyla*

Numbat *Myrmecobius fasciatus*

A classic case of convergence is that of the evolution of placental and marsupial mammals. In South America and especially Australia marsupial mammals (ones that give birth to immature young which then develop in a pouch) have occupied niches filled elsewhere on Earth by placental mammals and have evolved similar body forms. Shown here are some representative pairs of placental (identified by *) and marsupial mammals. Note, for example, how similar in appearance are the wolf (placental) and thylacine (marsupial), both being adapted to hunting. Almost the only external difference between the Broad-footed mole and Marsupial mole is their color.

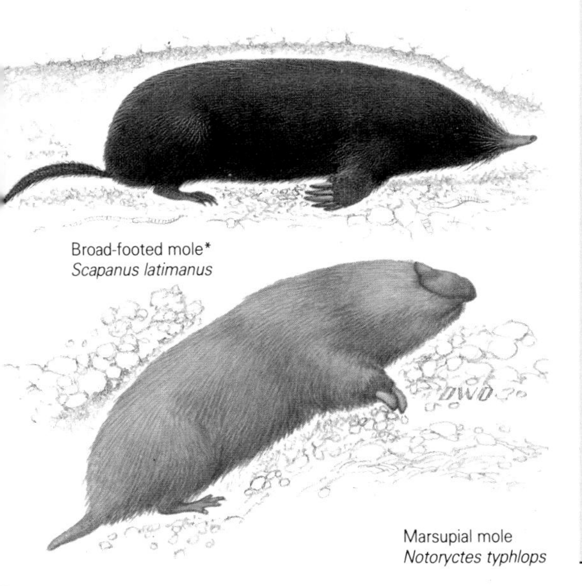

ADAPTIVE RADIATION

If a population of animals invades a new habitat that offers unoccupied niches, the stage is set for a rapid evolution of new species. It is likely, for example, that the Polar bear evolved from a population of Grizzly bears that became adapted to the icy wastes beyond the northern fringe of the grizzlies' forest home. By adopting a carnivorous rather than omnivorous diet, and by moving onto the ice floes, the new bear took over a vast territory in which there were no other large flesh-eating competitors. The evolution of the Polar bear is recent with no fossil record before 70,000 years ago.

Given time and a variety of vacant niches, isolation from competition gives opportunities for a diversity of new species to evolve from the ancestral stock. This is called adaptive radiation and was observed in detail by Charles Darwin among finches on the Galapagos Islands. When the original finch arrived there, probably blown in from the South American mainland, it was faced with a large number of vacant niches suitable for small birds. The result was the evolution of 13 species, each with a different feeding speciality.

Adaptive radiation is seen on a grand scale in Australia. As the continent drifted north from a position near the Antarctic Circle, its climate and vegetation were constantly changing, creating new niches. The marsupials, lacking competition from the placental mammals that dominate the wildlife in most other regions, diversified to occupy the huge range of new niches. The result is that today, kangaroos take the place of the antelopes so typical of other grasslands; numbats parallel the anteaters of South America; wombats have adopted the burrowing role of the North American woodchuck; and the now-extinct thylacine was a marsupial version of a wolf. In this unique collection of animals there are marsupial shrews, mice, moles and many more – all filling roles more usually taken by placental mammal species.

Perhaps the most spectacular radiation the world has ever seen was that of the many types of meat- and plant-eating dinosaurs, the flying pterosaurs, swimming plesiosaurs and ichthyosaurs and many others that contributed to the Age of Reptiles between 280 and 65 million years ago.

Centers of Diversity

Biodiversity is a relatively new term to come into popular use. It covers the enormous variety of life-forms into which living organisms have evolved while adapting to particular ecosystems. The preservation of this diversity is vitally important because it is the raw material of evolution itself. Without it, plant and animal life will not be able to adapt to future changes in the environment – whether natural or brought about by human activity. Indeed, human activity usually reduces diversity – a field of cereals has less diversity than the grassland or forest it replaced.

There are three ways of looking at animal biodiversity. *Species diversity* is the range of different animal types in an area, and even more specifically the variety of species within an individual animal group. *Genetic diversity* is the degree of variability within a species, which gives it the ability to adapt to new conditions. *Ecosystem diversity* refers to the overall variety of animals living a particular place and interacting with each other.

Global patterns of diversity

The diversity of animal life is not evenly spread over the globe. Tropical forests cover only about 7 percent of the Earth's surface, yet they hold well over half the world's total biological diversity. As a general rule, diversity is high in the tropics and diminishes toward the polar regions. It also tends to be high at sea level, dropping with increasing altitude.

Tropical forests illustrate another general feature of diversity. It seems that a year-round abundance of food and constantly equable physical conditions promote diversity. Species are less likely to be wiped out, by a drought for instance, and competition in the rich environment is increased so as to favor the evolution of specialized species that adopt narrow niches. In the forests of Panama in Central America, for example, there are 31 species of bats, including insectivores, frugivores and carnivores. Alaska in the United States, by contrast, has only a single species, an insectivore.

Clinging to the treetops a Pale-throated three-toed sloth searches for leaves. There are five species of sloth and all come from the rain-forests of Latin America. They are very successful, specializing in an arboreal leaf-eating way of life in which the effects of competitors and predators are scarcely perceptible.

Another factor affecting diversity is the age of the habitat: the Great Lakes of North America, formed since the end of the last ice age, are comparatively barren, while the lakes of the African Rift Valley, which are millions of years old, are biologically rich and support several hundreds of fish species.

Variations in these general patterns of diversity are not easy to explain. For example, penguins, petrels and shorebirds among the birds, and seals among the mammals, reverse the trend and are abundant at high latitudes. Also, papyrus swamps, which are extremely productive, harbor relatively few plants and animals. On the other hand, marine planktonic animals form a diverse fauna but their habitat is relatively uniform.

Tropical rainforests are key habitats to preserve, as they contain large numbers of species. Shown here is a scene from the treetops of the African rainforest. Redtail and Blue monkeys inhabit the rainforests in Central Africa; they generally live in harmony despite their similar diet and behavior, but the Blue monkey will supplant the smaller Redtail if there is competition. Very rarely, as in the Kibale forest, Uganda, at the edge of the range of the Blue monkey, the two species interbreed. Shown here is a confrontation over territory, between a Redtail family (1) and another family of Redtail females (2) together with a Blue-Redtail hybrid female (3). The hybrid female has more of the characteristics of the Blue monkey than the Redtail and is larger than the other females, which she tends to dominate, thereby gaining better access to food. In such confrontations, males such as this Blue, the father of the hybrid (4), tend to ignore what is going on.

Earth's biological "hot spots"

Conservationists once focused on saving endangered species, or on protecting areas of forest, wetland or seashore, but a central aim now is to conserve biodiversity, that is large sections of intact ecosystems with the genetic diversity of its plants and animals. The strategy aims to protect not only the attractive high-profile species such as elephants, pandas, gorillas and tigers, but also the many thousands of lesser creatures whose fate is often ignored by all but a few specialists. Conservationists are now faced with the challenging task of locating areas with exceptional biodiversity whose preservation will protect a large number of species and the ecosystem they live in. One such is the La Selva forest reserve in Costa Rica in Central America. The reserve is only 14 sq km (5.4 sq mi), yet contains 388 bird, 63 bat, 42 fish, and 122 reptile and amphibian species, together with more than 1,500 species of plants. This number is higher than the whole of the British Isles.

WHERE ARE THE WORLD'S RARE BIRDS?

To focus conservation efforts on areas of greatest need, the International Council for Bird Preservation (ICBP) has organized the Biodiversity Project.The aim of this project is to identify centers of genetic diversity by mapping the distribution of endemic birds with restricted ranges – species whose entire world range is estimated to be less than 50,000 sq km (19,300 sq mi). Results so far indicate that centers of endemism for birds correlate with the distribution patterns of rare amphibians, so any conclusions about where to concentrate scarce bird conservation resources will probably prove equally effective in protecting other animal groups.

When the centers of endemism have been identified, they will be matched with maps of habitats and protected areas so plans can be formulated to target scarce conservation resources. In the Philippines, for example, Mount Apo holds one quarter of the country's restricted-range endemic birds, yet although in a national park, it is being overrun by settlers. The mountains of Cameroon contain about twenty endemic bird species, of which four live only in the Bamenda Highlands. The area is heavily deforested, and the largest remaining forest block, on Mount Oku, supports the only viable populations of two of these rare endemics. Here, ICBP is running an integrated land-use project.

The Human Factor

As an omnivorous species, humans have exploited virtually every part of the animal kingdom, from great whales and sturgeons' eggs to sea cucumbers and sponges. It is often suggested that hunters and gatherers live in ecological balance with the animals they exploit. This is only likely when the people were part of the ecosystem and their numbers were regulated by their food supply. In a natural system, when food becomes scarce, predators starve, stop breeding or migrate.

Genuine hunter-gatherers, such as the bushmen of Africa, are few in number, the great majority of hunter societies have already given way under the impact of technology. Improved weapons and transportation are widely available, and human populations have increased many times over. At the same time, many prey species have been reduced by habitat destruction and overhunting – often by outside agencies. Alaskan Inuit whaling, for example, was always wasteful, but the number of whales killed was sustainable until commercial whaling by industrial nations brought the Bowhead whale to the brink of extinction. There are also records of herds of Musk oxen or bison being needlessly wiped out and even in hunting cultures animals were killed for prestige, so that rare animals would be specially sought.

The advance of technology

Over the centuries, the exploitation of animals has focused on the domestication of a limited number of species, and human populations have become increasingly divorced from the ecosystems on which they depend. The widespread destruction of animals has become a feature of technologically advanced communities, whose populations demand more than nature can provide. Indeed, replacing natural ecosystems with artificial ones has long been regarded as a worthwhile activity: uncultivated land is described as "waste", and is quickly built on or plowed for agricultural use.

Rainforests, swamps and coral reefs are highly productive environments, but human ingenuity has failed to devise a way of using them efficiently except by destroying them. The idea that rainforests could be exploited sustainably by anyone but small tribes is a recent concept that has still to be proven. Meanwhile, the great forests of Southeast Asia and the Amazon basin in South America are being felled for their timber, to make way for disastrous settlement schemes or for unsustainable raising of cattle. The soils of the new pastures are so thin and fragile that, robbed of their protective forest cover, they are reduced to sterile dust in five or six years.

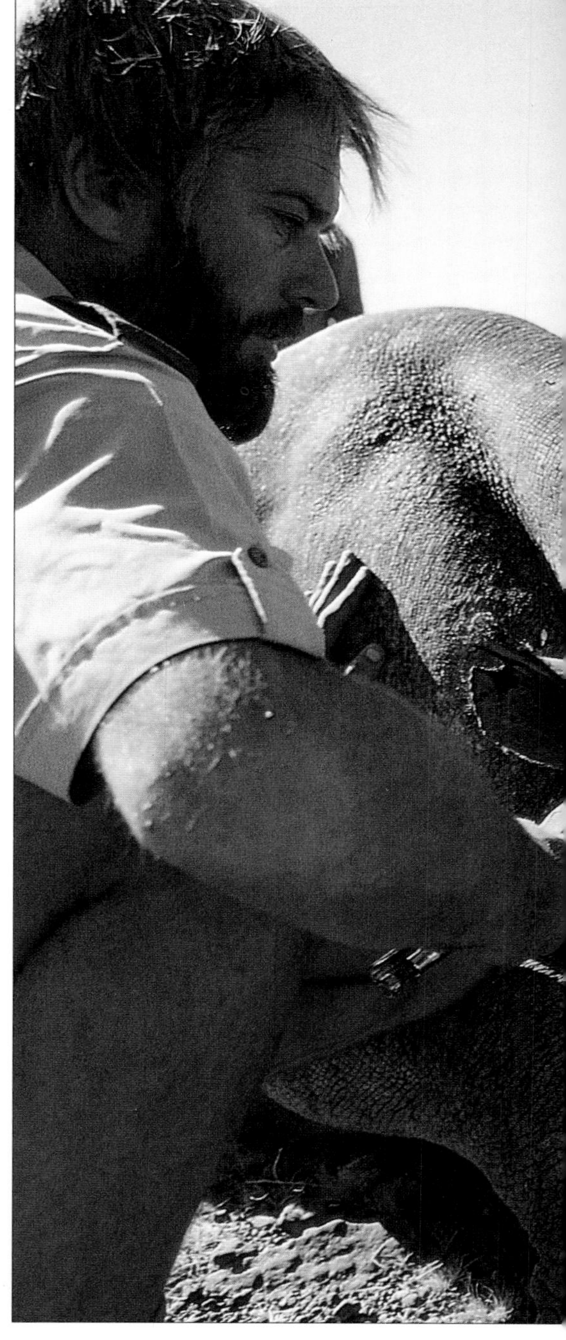

Removing the problem (*above*) Black rhinos are poached only for their horns. In an effort to prevent it, experimental dehorning has been carried out.

Cages of despair (*left*) Wild birds on sale in a street market in Djakarta, Indonesia. The trade in animals as pets is widespread around the world.

This exploitive approach to natural resources cannot be maintained. Many animals are in great danger of disappearing forever, because they have fallen victim to human greed – which includes eating bears' paws, wearing the furs of spotted cats and the skins of reptiles, carrying daggers with hilts of rhino horn

CHANGING ATTITUDES TOWARD BATS

The continued survival of some animals depends largely on public attitudes toward them. Cultural taboos protect some species, while traditional fears lead to the persecution of others. As early as the 6th century in England, wolves were treated as outlaws and were hunted down. This attitude still prevails in Norway and Sweden where wolves are shot on sight through fear, even though very few are left.

It is hard to find any animals that have suffered such undeserved antipathy in the West as bats. In the East, bats are regarded as bringers of good luck and longevity, but in the West they are traditionally associated with evil. In recent years this prejudice has been reinforced in North America by the discovery that bats can be carriers of rabies – a finding that produced an instantaneous and near-hysterical reaction against them.

In Britain and the Netherlands there has been a marked change in attitudes toward these animals over the last few years, due to a prolonged public information campaign. Whereas in the 1960s the overwhelming number of demands to local health authorities were for the destruction of bat colonies, during the 1970s most of the requests were for the safe removal of such colonies to more suitable sites. By the 1980s the situation was completely reversed, with facilities being built for attracting and observing them. From this time bats have been legally protected and public opinion favors bat conservation.

and keeping Spix macaws in cages. For the Passenger pigeon it is too late; it was destroyed in North America in only a few short decades.

The hidden losses

Destroying an animal population has far-reaching consequences. There is the "knock-on" effect linked to the animal's position in the ecosystem. It may be sad that bees are disappearing as a result of the removal of hedges and field margins and the universal use of insecticides, but it is disastrous when crops fail because there are no bees to pollinate them.

The loss of some species can have unexpected repercussions. In California, people who make their living from gathering clams and abalones would like to outlaw the Sea otter which preys on their catch. However, the otters also eat sea urchins, and research has shown that in the absence of otters, the urchins become abundant and destroy the beds of kelp and other seaweeds that not only harbor a wealth of marine life, but also represent an important crop in California.

There is now a growing public awareness of the need for conservation where not many years ago there was only a concerned minority. The world is rapidly losing many species that could be commercially valuable, provide food or health products, or simply enrich our lives. For example, the saliva of Vampire bats contains an anti-coagulant far more efficient than any used in medicine; if it can be identified and manufactured it will represent a considerable advance in heart condition treatment.

A hunter poses with his catch – a female Brown capuchin monkey and her baby. Animal populations could once cope with local subsistence hunting, but hunting with modern weapons cannot be sustained

Animal Extinctions

THE FOSSIL RECORD SHOWS THAT ANIMAL species persist on Earth for an average of 2.7 million years. For a mammal the average is about 600,000 years. (Humans, *Homo sapiens* in its modern form, have been in existence for only half a million years, so we still have some time to go unless we prematurely destroy ourselves.) In the 600 million years of the fossil record, something like 1,000 million species have lived on Earth, of which only a fraction – some 5–30 million species – are alive today.

What causes extinctions?

There are several reasons why species become extinct. Many have disappeared through the normal process of evolution – for example, when one species evolves into two new and slightly different species that replace their progenitor. Competition from new species can force an animal out of existence, as can the disappearance of a food resource or the appearance of a new predator. Species that live in small, restricted habitats, such as islands, lakes or specialized tracts of forest, are easily wiped out by natural events such as floods, fires or hurricanes.

At intervals throughout the ages there have been periods of mass extinction when entire groups of animals that were once extremely numerous have disappeared within a very short time (in geologic terms). The most famous mass extinction took place in the late Cretaceous, when the dinosaurs disappeared. For over 140 million years these reptiles had ruled the Earth. Then the two groups of dinosaurs – the ornithischians and saurischians – died out, along with the plesiosaurs, mosasaurs and pterosaurs. At the same time the ammonites – squidlike mollusks with flat, coiled shells – disappeared from the oceans, while the turtles, sponges and several other groups also lost large numbers of species.

There is some evidence that the Cretaceous mass extinction took only a few thousand years. It was certainly completed within a few million years, and even that is a staggeringly fast rate for events of such vast scale.

More than 50 explanations have been proposed for the Cretaceous mass extinction. One popular theory is that the impact of a huge meteorite upset the world's ecology by causing perturbations in the climate, perhaps because it created a blanket of dust that cut off the Sun's rays. Any explanation must, however, allow for the fact that many animals, including whole groups, survived. Birds, mammals and reptile groups such as snakes, turtles and crocodiles continued throughout the Cretaceous and survived to populate the world today. Throughout geologic time there have been regular periods when the rate of extinction has been high; it is possible that these mass extinctions occurred when extinctions in several different groups coincided.

The human threat to wildlife

Mass extinction has assumed a new importance with the realization that another episode is about to – indeed may already be – taking place. The growth of the human population during the last century, together with the power of technological advances rapidly to change the environment, will cause huge destruction of wildlife. With the exception of mass events, extinction has in the past been balanced by the evolution of new species, but the mass destruction of whole ecosystems and their genetic resources

Threatened with extinction (*above*) Green turtles on sale in the Philippines. This species is in danger of extermination throughout its worldwide range because of hunting, egg-collecting and habitat disturbance.

Extinct and endangered wildlife (*below*) This diagram highlights the better-known examples of the total destruction of wildlife. There is a high number of island species compared with mainland ones, both extinct and currently threatened. Today species with fewer than 500 individuals are in extreme danger.

1800–1900

1600–1800

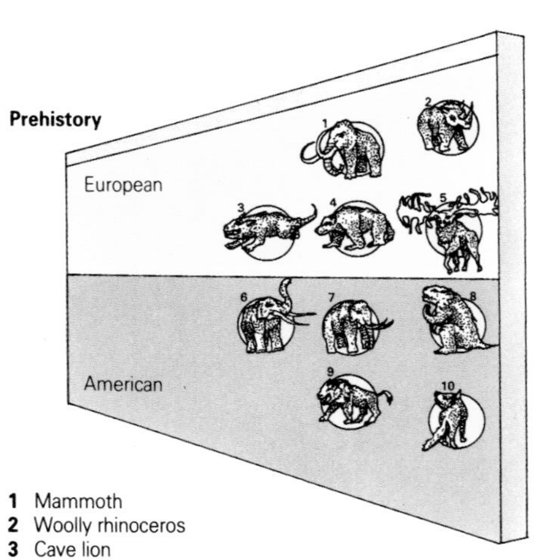

1 Mammoth
2 Woolly rhinoceros
3 Cave lion
4 Cave bear
5 Irish elk
6 American mastodon
7 Imperial mammoth
8 Giant ground sloth
9 Saber-tooth tiger
10 Dire wolf
11 Réunion solitaire
12 Dodo
13 Guadaloupe amazon
14 Elephant bird
15 Aurochs
16 Steller's sea cow
17 Blue buck
18 Hispaniolan hutia
19 Green and yellow macaw
20 Moa
21 Dwarf emu
22 Rodriguez little owl
23 Sandwich rail
24 Great auk
25 Spectacled cormorant
26 Atlas bear
27 Tarpan
28 Sea mink
29 Portuguese ibex
30 Quagga
31 Warrah
32 Palestine painted frog
33 Abingdon Island tortoise
34 Round Island boa
35 Passenger pigeon
36 Carolina parakeet
37 Pink-headed duck
38 Lord Howe Island white eye
39 Hawaiian o-o
40 Madagascar serpent eagle
41 Kauai Nukupuu
42 Greater rabbit bandicoot
43 Arizona jaguar
44 Schomburgk's deer
45 Caribbean monk seal
46 Thylacine "wolf"
47 Jamaican long-tongued bat
48 Barbary lion
49 Newfoundland white wolf
50 Bali tiger
51 Italian spade-foot toad
52 Chinese alligator
53 Central Asian cobra
54 Geometric tortoise
55 Hawksbill turtle
56 Japanese crested ibis
57 California condor
58 Red-billed curassow
59 Black robin
60 Réunion petrel
61 Abbott's booby
62 Hawaiian gallinule
63 Mauritius pink pigeon
64 Western ground parrot
65 Hawaiian crow
66 Leadbeater's opossum
67 Ghost bat
68 Woolly spider monkey
69 Mountain gorilla
70 Blue whale
71 Humpback whale
72 Indus dolphin
73 Northern kit fox
74 Baluchistan bear
75 Giant otter
76 Siberian tiger
77 Asiatic lion
78 Mediterranean monk seal
79 Grevy's zebra
80 Przwalski's horse
81 Mountain tapir
82 Great Indian rhinoceros
83 Swamp deer
84 Giant sable antelope
85 Indri
86 Orangutan
87 Sumatran rhinoceros
88 Mountain anoa

will leave a void that nature may be unable to fill.

There is nothing new about humans destroying wildlife. Even the primitive technologies of fire-lighting and weapon-making gave people the ability to alter the balance of nature, and there is ample evidence that humans caused largescale extinctions in the recent past. Numerous large North American mammals and the flightless moas of New Zealand were wiped out in this way, but never before have extinctions been so numerous as in the last 300 years. It is estimated that 63 mammal and 88 bird species have been lost since 1600 AD. Many disappeared when exotic forms such as cats, rats, pigs and mongooses were introduced to islands where the native species had no defense against the new competitors and predators. Others, such as the quagga, Great auk and Passenger pigeon, were hunted into oblivion. Recent largescale habitat destruction has resulted in increased rates of extinction.

Until very recently the world remained rich in wildlife, despite the gradual loss of species and the whittling down of animal populations. Even in Western Europe and North America, huge flocks of migrating shorebirds and waterfowl could be seen every spring and fall, and every wood had its resident birds, insects, mammals and other animals. However, wildlife is becoming increasingly confined to national parks and other protected areas. Moreover, it is not just large and spectacular animals such as bison and tigers that are being lost. Many familiar countryside species are also becoming rare.

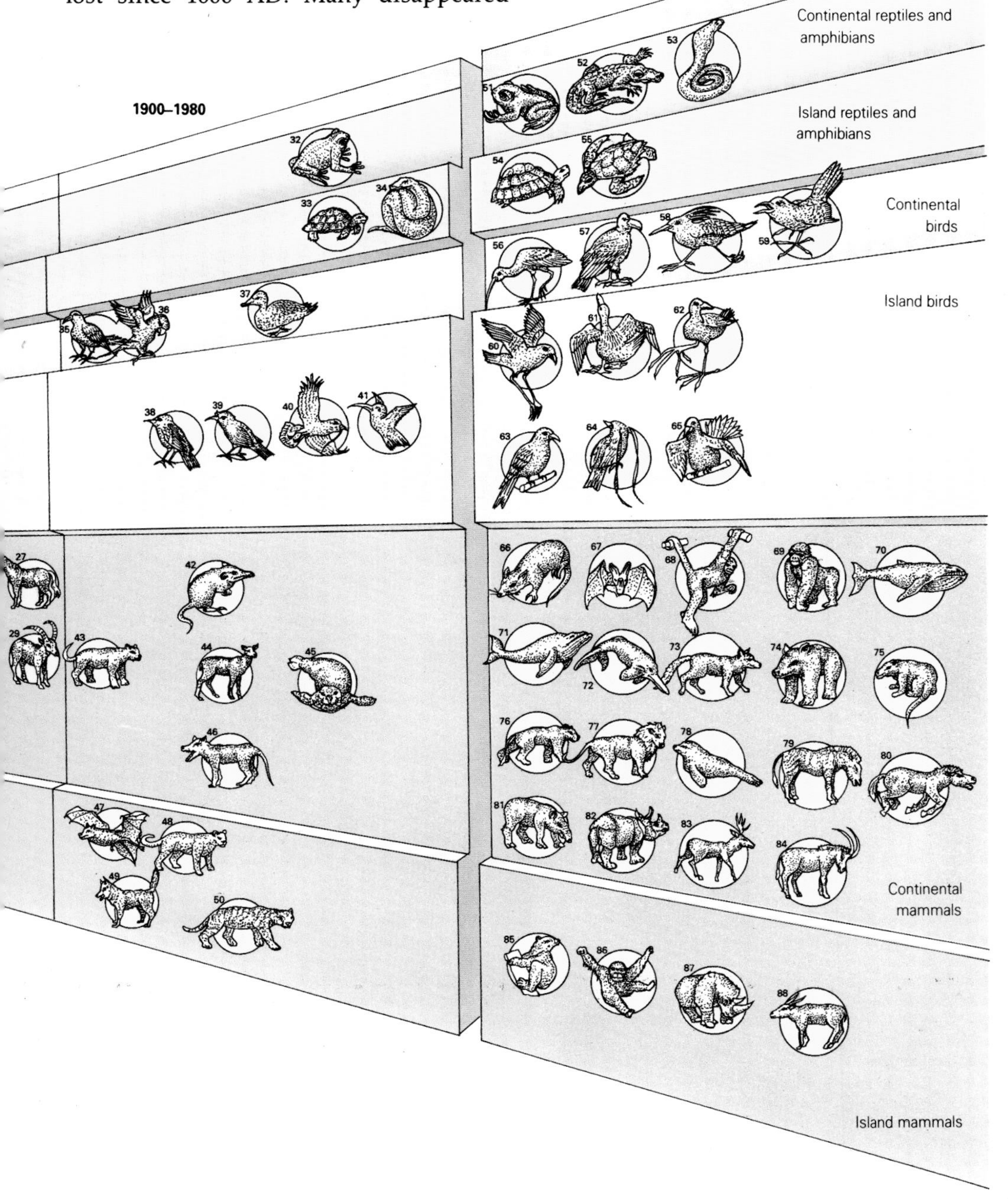

CRITICAL POPULATIONS

According to the Book of Genesis in the Holy Bible, only "two of each sort" of animal were needed in the Ark to preserve a species. In theory this is possible and species have been known to recover from very small populations. The Chatham Island black robin, for example, recovered from a population of only four, with just one fertile female, under a planned conservation program. Such cases, however, are exceptional: any species with only a tiny population is gravely at risk.

Several factors determine what constitutes a minimum viable population. One is fragmentation of the population. There are ten times more Grizzly bears in Canada than in the United States, but the former population is more at risk because the bears are widely scattered. There is also some danger of inbreeding in small populations and, because they have much reduced gene pools, this also reduces their ability to cope with disaster. The one remaining population of the Black-footed ferret in Wyoming, in the United States, for example, was almost wiped out by an outbreak of canine distemper.

Genetic studies have suggested that the minimum viable population for most vertebrates is about 500, though this takes no account of ecology or social behavior. (The demise of the Passenger pigeon was probably hastened by the fact that the birds needed the stimulation of large numbers in order to breed.) When the sex ratio, age, structure, breeding rate and normal population fluctuation of a species is taken into account, it is more likely that several thousand animals would be needed for viability. On this assumption, conservation of many of our most familiar and best loved animals might as well be abandoned, except it is contrary to the human spirit to abandon life while there is hope that it might finally be preserved.

Hunters and Collectors

WHETHER OR NOT OUR ANCESTORS CAUSED wholesale extinctions of the animals they hunted is a moot point, but there is no doubt that human hunters over the past three centuries have caused numerous extinctions, and greatly reduced the populations of many species. This covers the period of European expansion which has seen human populations expand worldwide, and the demand for worldly goods increase. As a result there has been a huge increase in the exploitation of animals to satisfy these needs.

Hunting for food and sport

Hunting is an important part of economic life for rural people in the developing world, where bushmeat in the form of rodents, monkeys, small hoofed animals, birds, turtles and many other species provides an essential supplement to the meager yield of food crops. The threat to wildlife here lies in the increased human population making excessive demands on animal populations. Similar effects are seen in industrialized countries where hunting is still a popular pastime. Some hunting, such as of duck and deer in North America, are well regulated with proper licensing, quotas and closed seasons. However, the wholesale slaughter of migrant birds as they pass through Mediterranean countries on their way to and from Africa is now having a serious impact on even the most common species.

The trade in wildlife

Exploitation of wildlife quickly gets out of hand when local demand for animal products is boosted by demands from more affluent countries. Then, target species are easily driven into local, or even total, extinction. The devastating impact of the millinery trade's demand for feathers at the turn of the century prompted the first major conservation campaigns. Half a century later the killing of cats, seals and other mammals for their furs led to an international outcry. Today there is deep concern for the African elephant as its numbers have been halved in only 10 years to satisfy the demand for ivory.

There is still a huge trade in live animals and their skins, shells, horns and feathers. Ornaments made of turtle shell, cowboy boots made of pangolin-skin and elephant-foot waste paper baskets fill market stalls, along with bizarre tourist souvenirs such as baby alligators dressed in dolls' clothes and toucans that have been stuffed and made into table lamps. Giant spiders are sold as pets; monkeys, apes and even leeches are bought for medical research and rare fish fill specialist aquaria. It is a billion-dollar business, much of it illegal. Payments to the hunters are often paltry, yet even these are worthwhile to people with few alternative ways of earning a living. Thousands of animals also die every year during capture and transportation.

NONTARGET SPECIES

Hunters often catch animals other than those they want, and this leads to unnecessarily wasted lives. The "incidental catch" can be a major problem, especially if the nontarget species already happens to be rare. Giant pandas, for example, are occasionally caught in snares set for forest deer in China. However, even this is a minor tragedy compared with the loss of life caused by commercial fishing.

Modern nets measuring several kilometers long form "walls of death" in the oceans, depleting not only the fish, squid or shrimp that are the target species of the fishermen, but also many noncommercial fish species, seals, turtles, diving seabirds and whales. Some of these animals are caught simply because they are in the way, like the endangered Atlantic Ridley's turtle, which feeds on the sea bed and is often swept into trawl nets. More often, however, the nontarget species is feeding on the target species and is caught while pursuing its prey.

The Japanese salmon fishery in the North Pacific accounts for 250,000 to 750,000 birds every year (mainly auks and shearwaters) while the California halibut fishery kills an estimated 10,000 birds a month during the summer season. In the eastern Pacific, whole schools of dolphins are often trapped and killed because they alert deep sea fishing boats to the presence of shoals of tuna. The fishermen search for dolphins and, when a school is sighted, pass a purse seine net around them and draw it in, capturing the tuna and drowning the dolphins. It has been shown that a temporary slackening of the net will allow many of the dolphins to escape by swimming over the edge, but even this humane practice is often ignored in the pursuit of the maximum catch and thousands of dolphins die every year.

Commercial exploitation

Three factors contribute to the decline of animals through over-hunting. One is summed up in the phrase "the tragedy of the commons." No one owns or controls the animals so everyone aims to take as large a share as possible before someone else does. In Chesapeake Bay on the east coast of the United States, Blue crabs were a communal resource and were fished almost out of existence. However, farther along the coast another fishery thrived because families parceled out the shore and each fished its own section carefully, to preserve its future.

Secondly, a steady farming of resources is often less profitable in the short term than destructive policies of sudden and massive exploitation in which the maintenance of the animal stock is not of importance to the exploiter. The oceans, lying outside the jurisdiction of national laws, have been the main sites of such methods. The story of whaling consists of hunting first one population and then another to the brink of economic, if not biologic extinction. Whaling expeditions involve a huge capital investment, and to operate anywhere for below maximum return makes no economic sense.

The third factor is to tip a species from economic to biologic extinction. It is worth continuing to hunt the few remaining specimens of a particularly prized animal if the price escalates as they become rarer. With Spix macaws worth US $80,000 in 1989, and the owners of the few captive birds unwilling to join in captive breeding programs, the species is on the very brink of extinction.

Trade in animal products (*above*) It is often now illegal to export or import items such as these snakeskin boots. However, so long as people in developed countries desire such souvenirs and possessions, illegal trade will continue to thrive.

A sports hunter (*left*) returns home with the product of his day's sport. Some sportsmen contribute to the conservation of animals, while others have a serious impact even on common birds, as with the slaughter of European migrants.

Introduced Animals

UNDER NATURAL CONDITIONS ANIMALS AND plants live in well-balanced communities, in which interactions between species help to regulate numbers. This balance is upset by agricultural and forestry practices, by animal husbandry and by the introduction of exotic animal species. The problem with introduced species is that as aliens they do not fit into the local ecosystem, and are not subject to the constraints imposed on native species. The newcomer may prey on local species and wipe them out, or it may multiply rapidly and outcompete them. Large parts of Australia, for example, have been devastated by rabbits. Originally they were introduced for food, but having escaped from farms they bred rapidly, causing extensive damage to sheep-grazing land; they also competed with native animals. At one time it became more profitable to hunt rabbits for their skins rather than to raise sheep. The "plague" of rabbits was virtually wiped out through the introduction in 1950 of the disease myxomatosis.

Systems out of balance

Introduced animals are a problem all over the world. Some, such as the commensal species that accompany humans, have spread all over the globe. Rats and mice, House sparrows and cats now have worldwide distributions. Settlers on the Seychelles accidentally brought in rats and mice, so subsequently they introduced Barn owls and cats to control them. This proved to be a terrible mistake. The hunters found the native birds far easier prey than the rodents. More recently the Indian house crow has spread around the Indian Ocean, from Durban and Djibouti to Australia. It damages crops, takes the eggs and chicks of other birds, and is a public health hazard because it is a carrier of disease. Like so many alien introductions it is so successful that eradication is now virtually impossible.

Equally damaging are smaller animals that have been introduced unwittingly. While fruit growers in California suffer from introduced fruit flies, British farmers keep a constant watch for the Colorado beetle, a North American species that attacks potatoes and has colonized much of continental Europe. Britain was less fortunate with the American Slipper limpet and American Oyster drill, both of which have become serious pests in commercial oyster beds.

For the love of animals

In various places around the world people have introduced animals simply for the pleasure of seeing familiar or exotic species. New Zealanders, for example, imported a wide variety of European and other species – including deer, llamas, weasels, hedgehogs, peafowl, Turtle doves, chamois, hares and many others. North America has been colonized by European starlings, Ring-necked pheasants, skylarks and House sparrows; and parts of Florida now support budgerigars, Red-whiskered bulbuls, Java sparrows, Hill mynahs and Muscovy ducks.

Both North America and Australia are inhabited by European Red foxes, introduced so they could be hunted for sport with hounds. In Britain, American Gray squirrels, Canada geese and European Fat dormice, were released because they were regarded as ornamental and harmless exotics. The success of these species depended on them having few competitors or predators. Gray squirrels arrived in Britain when predators such as martens, polecats and birds of prey were low in numbers and consequently they easily outcompeted the native Red squirrels, which have now become rare.

The impact on native wildlife

From the conservation point of view the most serious effect of introductions is the destruction of native wildlife. In Europe there was great concern that American mink, which escaped from fur farms, would destroy much of the native wildlife since it is a versatile, aggressive predator that competes with, for example, otters. However, assessing the scale of the damage has proved very difficult.

Introductions have had their most serious impact on islands, where the native wildlife is often made up of endemic species with small populations. Many of these have evolved in isolation, in the complete absence of mammalian predators. The newcomers have preyed on the island species or have outcompeted them for food. Common rats, for example, have wiped out groundnesting seabirds and pipits (the world's most southerly songbirds) from large parts of the subantarctic island of South Georgia and mongooses, introduced to the West Indies to combat snakes and rats, found the local birdlife easy prey. Goats are another example: released onto islands as a food resource for visiting sailing ships, they devastated the natural vegetation of many islands, to the detriment of local wildlife.

One of the most dramatic cases of the damage done to native wildlife occured in 1894 when a single cat, taken by the lighthouse keeper to Stephens Island off the New Zealand coast, exterminated a number of endemic bird species, including, it is believed, the Stephens Island wren: the bird was discovered and identified solely from the corpses brought in by the cat.

Introduced species can also be a problem because they dilute the gene pools of native animals. The British Isles, for example, were once the home of a distinct race of the Red squirrel. Even before Gray squirrels were introduced from America it had been swamped by introduced

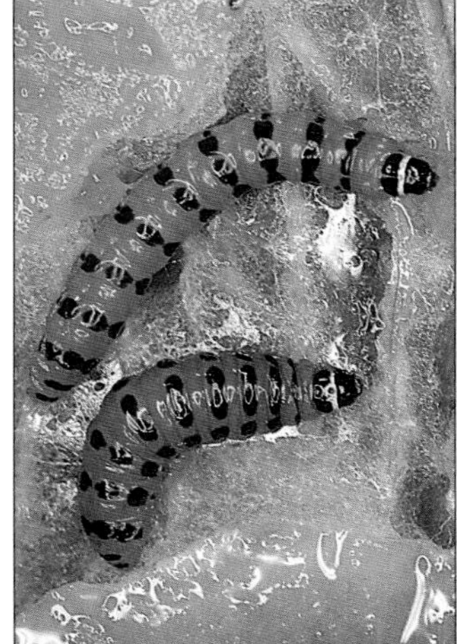

Biological control gone wrong (*above*): the Cane toad disaster. In the 1930s Cane toads were introduced to Australia to control a cane beetle; however, they prefered the surrounding habitat where they have reached near plague proportions.

Biological control wins (*left*) *Cactoblastis* moth caterpillars eating prickly pear cacti. This cactus, which was originally introduced to Australia as an ornamental plant, came to dominate extensive tracts of Queensland and New South Wales. *Cactoblastis* was introduced in 1926 and by 1933 had eaten its way through 24 million hectares (60 million acres) of cactus.

A bark-stripping pest The Gray squirrel was introduced to Britain in 1876 from the deciduous woodlands of North America, where it occasionally causes damage by stripping bark from sugar maples. In Britain, Gray squirrels damage mainly sycamore (which is also a maple), beech and oak, although other trees are sometimes attacked. Because squirrel damage is costly it discourages the planting of native beech and oak woods. Control measures, including shooting, trapping and poisoning, have not been effective.

continental Red squirrels. In New England in the United States the introduced European Red fox interbreeds with the native race. This is a factor to be borne in mind by those who favor bolstering low populations of species and subspecies by introducing stock from other parts of the animal's range. Domestic animals, too, can pose a threat. Dogs, for example, will breed with coyotes and wolves, and domestic asses with wild asses.

BIOLOGICAL CONTROL

The positive side to introductions of animals is that they can be used to control pest species. Physical control methods are costly in manpower and equipment, and chemical controls often have undesirable environmental impacts. A better solution would seem to be to set an animal to catch the pests or eat the weeds. In theory only a few would be released: they would breed, and continue to attack the pest until it had been wiped out and then die out.

In practice, biological control is not nearly so simple, for the reasons that attend any introduction of exotic species. Disaster overtook the beautiful endemic *Partula* snails of Moorea in French Polynesia when, in the 1930s, the giant African snail *Achatina fulica* was introduced to the South Pacific as a food. The giant snail started eating crops so the carnivorous snail *Euglandia rosea* was brought in to control it. Unfortunately, *Euglandia* preferred the endemic island snails to *Achatina*, and six of the seven *Partula* species on Moorea are now extinct in the wild.

Biological control is successful only after the most rigorous screening to ensure that a controlling species will attack only the target pest. When biological control does work, it is extremely efficient. The floating water fern *Salvinia molesta* of Brazil has been introduced to many tropical countries where its unchecked growth has caused problems. In the Sepik river floodplain of Papua New Guinea a weevil was introduced that reduced mats of *Salvinia* to virtually nothing in two years.

The Pollution Hazard

MAN HAS POLLUTED THE ENVIRONMENT FOR centuries, but only in the past few decades has pollution had such an obvious and disastrous effect on animal life. Newspapers describe major catastrophes such as escapes of radioactive material from nuclear power stations or oil spills from wrecked tankers or ruptured pipelines; but of no less concern is the lower-level pollution of water, air and land that is continuous and universal.

The specter of a silent spring

The problem of pollution was brought to public attention by Rachel Carson in her book *Silent Spring*, published in 1962. In it she focused on the side-effects of the widespread use of insecticides and herbicides such as DDT and 2,4-D, quoting from a letter lamenting the loss of familiar American birdlife: "I put up a feeder and had a steady stream of cardinals, chickadees, downies and nuthatches all winter, and the cardinals and chickadees brought their young ones in the summer. After several years of DDT spraying [on elm trees] the town is almost devoid of robins and starlings; chickadees have not been on my shelf for two years, and this year the cardinals have gone too; the nesting population in the neighborhood seems to consist of one dove pair and perhaps one catbird family." The birds had eaten poisoned insects. They were not target species themselves, but simply innocent victims of pest control measures.

The impact on nontarget species is highlighted by the disastrous effect of timber treatment on bats. Woodworm is a serious pest in house timbers, and spraying the roof space might seem unlikely to harm the environment at large. However, when colonies of bats take up residence in roof spaces, they can be wiped out by the chemicals. A similar problem was found with the use of TBT to keep the hulls of boats free of fouling organisms. TBT is so toxic that a few parts per million in the sea can prove lethal to fish and ruin the spawning beds of fish and shellfish. It could be argued that ridding a pest is worth the destruction of non-target species, but this can backfire. In 1956 DDT was used by the US Forest Service to control Spruce budworm. Next year the trees were devastated. The DDT had wiped out the insect predators of the Red spider mite, which had bred unchecked and defoliated the trees – never before had it been a pest.

In the last 30 years the problem of pesticide pollution has, in theory, improved. Birds such as the Peregrine falcon, badly hit by DDT in the 1960s, have recovered well since DDT was replaced by less persistent pesticides, and improved delivery systems have made lower dose-rates possible. However, these changes have not always resulted in any improvement in the environment. Recent research in Britain has shown that fields are now sprayed more frequently than they were 20 years ago. Harmless insects of field margins are not only being killed directly, but are also losing their food plants. The result is a dramatic drop in the populations of Gray partridges and other insectivorous birds. Meanwhile, pest control programs in the developing world are using chemicals that have been banned in the western world.

The search for alternatives

Safe alternatives are not easy to find. An outright ban on pesticides would cause world food production to plummet and disease to become rampant. Biological control is helpful, but can cause major problems if the predator turns on non-target species. Pure stands of single-species crops are particularly susceptible to diseases and pests and experiments on mixed cropping are under way in many countries. However, such farming systems are unlikely to suit the largescale mechanized farming that provides much of the world's food. Genetic engineering offers hope of producing new strains of crop plants with built-in protection against pests and diseases, but such developments are still a thing of the future with much research yet to do.

The recent abandonment of crop rotation in favor of supplementing the soil with artificial fertilizers has led to accumulations of nitrates in rivers, seas and groundwater. The nitrates cause algal blooms which rob the water of its oxygen (a process called eutrophication) and this leads to suffocation of fish and other aquatic animals. Nevertheless, a simple ban on nitrate fertilizers is not the answer. The complexities of soil chemistry make it difficult to reduce the nitrates in effuents and soil run-off.

Pollution on a global scale

Concern has recently been raised about global pollution and its effects on animal life, including human species. Acid rain,

A Sea otter (*left*) coated in oil spilled from the wreck of the Exxon Valdez in 1989 off the coast of Alaska. Many otters died in this incident.

Aftermath of Chernobyl (*below*): reindeer feeding within the Arctic circle. In 1986 radiation released during the Chernobyl nuclear disaster spread across northern Europe, contaminating the forage of herbivores such as reindeer and sheep. This does not appear to have threatened the animals themselves, but there was concern that human health would be affected by eating contaminated meat.

the greenhouse effect and the depletion of the ozone layer are worldwide problems. Emissions of sulfur and nitrogen oxides by factories and vehicle engines rise into the atmosphere and become dissolved in water to produce acid rain. This has wiped out fish populations in lakes in Scandinavia and North America, and has had a profound effect on forest wildlife by killing trees and even whole forests.

The Earth's atmosphere is warming up – the so-called "greenhouse effect" – which is primarily the result of a buildup of carbon dioxide (from the burning of fossil fuels and large areas of tropical forest) and methane (from the rice paddies of the tropics and the millions of cattle around the world). Scientists predict global changes in climate and habitat distribution, on a time-scale of centuries or even decades. Animals will find it very difficult to adapt or move to new areas fast enough. The melting of polar ice caps will have profound implications: rising sea level will inundate vast areas of coastal lowland, including some very densely populated areas, and thousands of islands will drown.

POLLUTION HITS SEAL POPULATIONS

When pollution kills outright and the media can show scenes littered with dead animals, the impact is immediate and dramatic. Pollution does not always kill, yet it may still destroy an animal population. Concern that the Gray seal population of the Baltic was dwindling began in the 1970s. The death rate did not appear to have risen significantly, yet the seal's numbers were falling. The cause was found to be a marked drop in birth rate: there were far fewer pups to be seen and the incidence of pregnancy in Ringed seal populations in the Gulf of Bothnia, a northern arm of the Baltic, had dropped alarmingly. Gray seals and Harbor seals were similarly affected and postmortem examinations showed a high incidence of abnormalities in the females' reproductive organs. Furthermore, chemical analysis showed that those seals that did bear pups were less heavily contaminated with DDT and PCBs (polychlorinated biphenyls) than those that failed to produce young.

At about the same time, the Harbor seal population in the Waddenzee off the mainland of the Netherlands declined rapidly and was believed to be in danger of extinction. Once again individual seals appeared in good condition but the birth rate had slumped. Pollution with PCBs appeared to be the cause, but a direct link with breeding failure was hard to prove. To examine the role of pollutants Dutch scientists fed captive groups of seals on diets of either polluted fish from the Wadden Sea or clean fish from the Atlantic. Those fed on Wadden Sea fish produced significantly fewer pups. Analysis of blood samples showed that PCBs from the polluted fish had upset the hormone balance in female seals and caused premature termination of their pregnancies through failure of the embryo to implant in the lining of the uterus.

Habitat Loss

THE DAMAGING IMPACTS ON ANIMAL LIFE OF exploitation, pollution and introduction of alien species are serious, but often containable and even reversible – given the support of appropriate legislation and sufficient funding. Yet the main cause of animal depletion today is the wholesale destruction of habitats, which affects all species, not just the high-profile animals that attract the attention of conservationists and the general public.

Without somewhere to live, animals are doomed. The problem is not new, but its scale has escalated over the past three or four decades. Forests, heathlands and wetlands are disappearing throughout Europe, and in the United States the prairies are almost gone: Illinois once boasted 145,000 sq km (60,000 sq mi) of wild grassland, but there are now just a few scattered patches totalling about 10 sq km (less than 4 sq mi). Tropical rainforest losses are such that some countries will probably lose all their virgin rainforest within the next 100 years. The animal life of entire ecosystems will vanish forever or be reduced to a few protected remnants.

Changing environments

One problem facing biologists and conservationists seeking to draw attention to the effects of habitat loss is that the environment has never been constant. Climate and other factors change and affect the distribution and character of habitats. Indeed, change itself is one of the causes of animal diversity.

This creates the difficulty of distinguishing changes due to human activities from natural phenomena. Coral reefs were once believed to be an example of a habitat, supporting a very stable community in which the diversity of species is held in a harmonious network of checks and balances. When it was discovered that damselfishes were creating gardens of algae by killing coral some experts thought this was due to overfishing of large species that preyed on the damselfishes. It is, in fact, a natural phenomena not a man-made environmental problem.

Changes in a habitat are not necessarily deleterious, except that they open debates as to their cause, which may delay realization of a problem where humans are to blame. Change is a natural part of life, but changes caused by humans cannot be ignored. Coral reefs recover after devastation by hurricanes provided that they are not disturbed further. The wholesale destruction of coral reefs worldwide leaves no chance for recovery.

The progressive loss of habitat makes it difficult for natural changes to be accommodated. When areas of bamboo forest in Sichuan province in China flower and die, as they do on a regular longterm cycle, Giant pandas used to be able to move to another area and find alternative food supplies. The bamboo forests are now so much reduced and fragmented that the few remaining wild pandas are under serious threat of extinction because they cannot move from area to area.

STAGING POSTS

Many animals exploit two or more food sources by migrating between them. Birds are the best-known migrants, but whales, wildebeest, bats, turtles, eels and butterflies are among the many animals that undertake such regular journeys. Any disruption of their migration routes and the staging posts where they feed can be disastrous.

In arctic North America, herds of caribou follow traditional routes between their summer home on the tundra and their winter quarters in the shelter of the coniferous forests. Traditional hunters knew these routes and lay in wait for the herds. The caribou now face a much more serious threat from the new roads and oil pipelines that have opened up the region. Careful planning is helping to minimize the disruption, but the same cannot be said of the migrating antelope herds in Botswana in southern Africa. There, many of the migration routes are completely barred by fences designed to restrict the spread of foot-and-mouth disease among cattle. The wild herds cannot reach the watering places they rely on in the dry season and millions have died.

For birds the problem is the loss of critical resting and feeding places. In Europe and North America, shorebirds are known to be affected by loss of estuaries through reclamation of land for agriculture, industry, recreation and building. In North America the number of breeding ducks and geese has dropped from 45 million to 31 million over the last 15 years. The Atlantic flyway has lost half its coastal wetlands in just 35 years.

Creating new habitats (*above*) Humans do not clear animal habitat without a purpose; here, in the tropical forest of Malaysia, it is for agricultural reasons. The man-made habitat is totally new and therefore excludes most of those animals from the original habitat.

Survival against the odds (*right*) Animals have always had to cope with natural disasters. A typical disaster occurred on 18 May 1980 when Mount St Helens in the United States erupted, the blast flattening huge tracts of forest. Given time and no interference from people, this habitat will recover naturally.

The pressure of people

The burgeoning populations of Third World countries are placing pressure on many habitats. With more mouths to feed, even unsuitable areas are being plowed up and forests cleared to grow crops and graze livestock. The soils are often unable to sustain such uses, and as the vegetation cover recedes, erosion takes over. In dry areas the result is the progressive advance of the deserts. The expansion of logging and mining to meet the demands of home and international markets drives access roads through wilderness areas, fragmenting habitats and populations and opening them up to settlement. The only hope for many of these beleaguered habitats is either that they will be left alone or that rational, controlled, multiple-use exploitation will replace the current short-sighted rush for profit. In some areas opening up wilderness to tourism could offer a new source of income to local people, but unless it is planned and controlled on ecologically sound principles it could lead to yet more priceless habitats being reduced to pitiful scattered remnants.

Wildlife Research

THE MODERN CONSERVATION MOVEMENT grew out of the realization that excessive hunting was damaging animal populations. Some of the first conservationists were, in fact, hunters concerned for the future of their quarry. Their aim was protection, which meant simply creating reserves where hunting was prohibited. The ecological requirements of these species, or the wildlife in general, were not understood, and in some cases the practices seem outlandish in the light of current knowledge. For example, in Africa, lions, hyenas and other predators were shot as vermin in an effort to protect antelopes, zebras and other "game" species. It is now known that a balance of predators and prey is essential to the health of an ecosystem, and the predators are now valued in their own right.

The essential database

Costly mistakes in terms of money and effort, and unnecessary loss of animal life, can be avoided only by research. The most basic need is to investigate the status of a species or population. How many of them are there? What is the age structure? Is the population increasing, declining or steady? A census provides a baseline against which to judge the effectiveness of conservation actions, and also gives a measure of the severity of any problems. The rarity of a population may not become apparent until a census has been made; however wild populations are seldom easy to count and results may be controversial. In the case of the orangutan, for example, estimates of the population rose steadily from a figure of 4,000 in 1963. This was due entirely to improved census techniques: the population was actually declining.

The difficulty of establishing the true status of an animal is illustrated by the Papuan turtle. Its rarity seemed confirmed when offers of a reward brought in very few specimens. It then transpired that the turtle was considered to be good to eat and that the reward was not large enough to compensate for the loss of a good meal. Scientists also seriously underestimated the numbers of Bowhead whales migrating along the coast of Alaska. It is easy to count whales when they surface to blow, but local people pointed out that some travel under the ice where they cannot be seen. The scientists did not believe them until they listened with hydrophones, and had to revise their estimates.

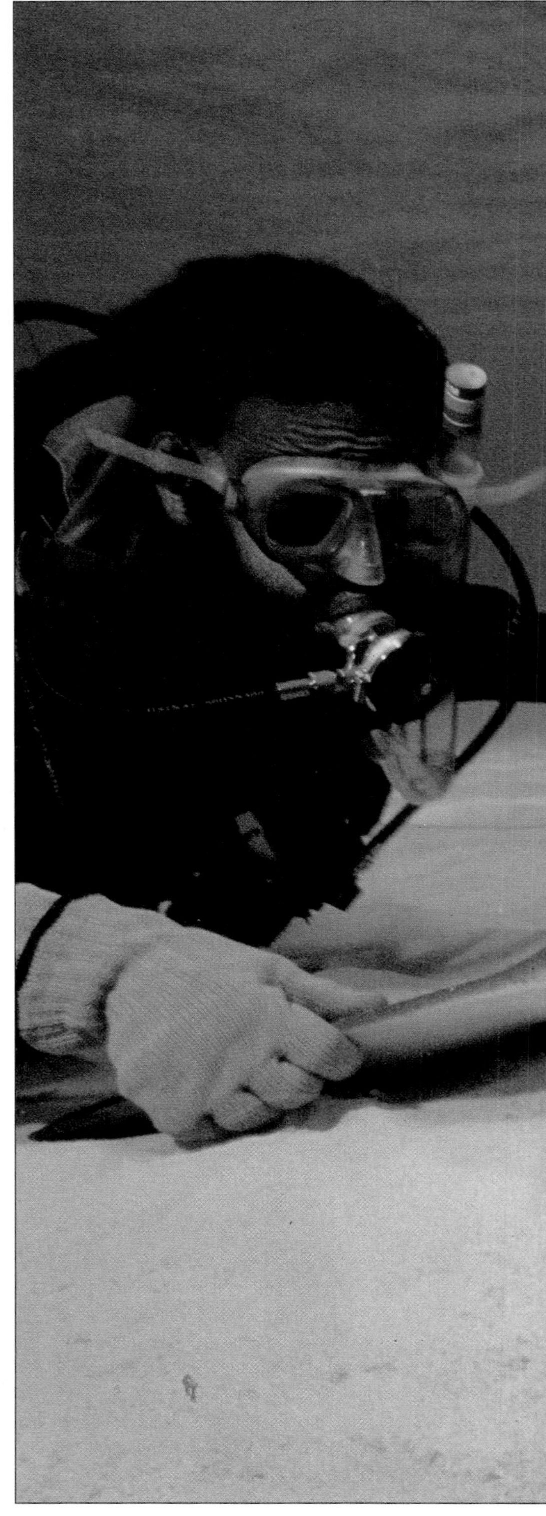

Dangerous liaisons in the deep (*right*) Scientists conducting an experiment with a live Lemon shark. As well as studies of the ecology and behavior, a full understanding of an animal cannot be achieved without an appreciation of its internal biology.

Fieldworkers counting penguin nests in Antarctica (*below*) Knowledge of breeding populations and their success at raising chicks is key to assessing the status of bird populations. However, such counts must be accurate to be useful.

The need for a holistic approach

Understanding how ecosystems work has resulted in the concept of species management. In simple terms there is little point in expending resources to protect an animal if its food source has been destroyed or it has nowhere to breed. It is no longer enough to place a defensive barrier, be it physical or metaphorical, around a species. It is essential also to ensure that its food supply, shelter, social life and other factors remain intact. Conservation can sometimes then be effected quite simply. The very rare

CAPTIVE BREEDING FOR THE FUTURE

Attempts have been made to improve the fortunes of critically small animal populations by breeding animals in captivity and returning them to the wild. There have been some notable successes, such as the return of the Arabian oryx to the deserts of Oman after its extinction in the wild. However, there is more to reintroduction than simply opening the cage door. Captive breeding must be backed by research into ecology and behavior.

The nene, or Hawaiian goose, was saved by captive breeding, despite the fact that of the 1,600 birds released initially, only four survived. One reason was that the birds were released in the wrong place and could not find enough food. The last wild nenes had been found in desolate mountain lava fields, but their natural home was probably in the lusher lowlands, from which they had been driven by persecution. Further introductions were extremely successful.

Attempts to reintroduce orangutans failed because the youngsters were released into the forest alone. Orangutans also need to learn from their relatives about the leaves, fruits, shoots and other items that make up their diet. So, to survive, they must integrate into orangutan society.

Without constant monitoring and research, such problems would not have been appreciated or overcome. Yet even when reintroduction has not been possible, captive breeding has great value. Stud books are kept of many captive populations and animals are exchanged between zoos to reduce inbreeding. In this way it is hoped that the breeding vigor of the species will be maintained until either the problems of reintroduction can be overcome or, if not, viable populations can be maintained in captivity.

cahow, or Bermudan petrel, was rescued from oblivion by modification of its nest burrows to keep out the tropicbirds that were usurping them. In South Africa and elsewhere, vultures have been helped by putting out carcasses at vulture "restaurants" to compensate for the scarcity of natural carrion in livestock-raising areas.

Realization of the need for sound ecological knowledge has not made the task of conservation any easier. The workings of ecosystems are so complex that it is not always easy to determine the reasons for fluctuations in an animal's numbers. Over the last 30 years, efforts have been made to control the plagues of coral-eating Crown-of-thorns starfish that have devastated large sections of coral reef, especially along the Great Barrier Reef of Australia. Some outbreaks were attributed to human interference in removing tritons – large mollusks that were believed to control the numbers of starfish – but it was found that under natural conditions the starfish usually escaped by forfeiting an arm. The conclusion is that plagues of Crown-of-thorns starfish are natural.

Great debate has been aroused by the plight of African elephants in the national parks of eastern Africa. Where the surrounding country has become increasingly populated and/or turned over to agriculture, the elephants have become confined to the parks and population density has risen. Unable to continue their traditional wanderings, the large numbers of elephants have damaged the habitat, to the detriment of themselves and other animals living there. The result has been highly controversial programs of culling elephants to reduce numbers.

Habitat Preservation

THERE WAS A TIME WHEN CONSERVATION WAS largely a matter of preventing animals being killed. Now, finding somewhere for them to live is probably the most important necessity as natural and semi-natural habitats disappear. Although animal life of all kinds is suffering, the problem is particularly acute for large animals such as elephants and rhinoceroses. While being killed at a rate in excess of their capacity to reproduce, they are also losing their habitats.

Habitat fragmentation

The fragmentation of habitats has very serious repercussions, not least the loss of animal diversity. In South Africa, for example, the remnants of forests that are left are too small to support leopards and Crowned eagles and, partly as a result of the loss of these predators, Vervet monkeys have become more common. At the same time sunbirds and other small birds have decreased because the monkeys raid their nests. The limited and fragmented habitat left to elephants often does not allow them to roam freely, and so they degrade the vegetation on which they and many other species depend. The problem for rhinoceroses is that they live at an extremely low population density, and it is difficult to set aside areas large enough to support viable populations.

This is not only a problem for larger animals. Kirtland's warbler is confined to a few areas in Michigan that still have Jack pine forests. The bird's requirements are very specific: it will nest only in trees that are 5-20 years old, with branches that reach low enough to conceal its ground-level nest. In the past fires caused by lightning cleared patches of forest, maintaining a mosaic in which there were always trees of the right age. However, the introduction of improved forestry fire control soon left the warbler with no breeding habitat. Its survival now depends solely on a few small patches cleared by controlled burning.

The problems caused by having isolated areas of habitat can be mitigated to some extent by leaving corridors of intact habitat between the surviving larger areas, so that small, fragmented populations can link up and interbreed. Without this, the genetic variability of small populations may decline leading to reduced fertility.

The Rufous hare wallaby of Australia has become rare because its habitat is being destroyed by fire. The Aborigines used controlled smallscale fires for a variety of purposes. This practice has now died out and, when the overgrown vegetation is set on fire by lightning, the resulting fires are much hotter and destroy the vegetation over large areas.

As with species conservation, habitats need more than just simple protection: management is essential, and that raises its own problems. A natural ecosystem is a mixture of ever-changing habitats, each with its own wildlife, so for any conservation area decisions must be made about which animals are to be the focus of concern and how best to manage the habitat for their benefit. For example, a pond left untouched will gradually fill in and become covered with trees. If the requirement is to conserve fish, or fish-eating birds, the pond must be kept open, even though this is detrimental to other wildlife occupying the pond.

National parks face the same problem on a much larger scale. Chobe National Park in Botswana in southern Africa formerly had a mosaic of grassland and acacia woodland supporting game animals and predators: precisely the variety of animals that foreign tourists want to see. However, the woodlands are disappearing due to excessive browsing by elephants and other species, and the diversity of wildlife is threatened. These woodlands developed 100 years ago when drought and an outbreak of rinderpest reduced the number of browsers and allowed the trees to grow. The problem now is to maintain a balance between the trees and the browsers.

Man and the biosphere

To preserve the maximum number of animal species we need to conserve sufficiently large areas of all the world's habitats. Such a scheme is already being put into practice under the United Nations' Man and the Biosphere Program, which aims to designate Biosphere Reserves in every major habitat. In these reserves, fully protected core areas will be surrounded by "buffer zones" in which controlled and sustainable habitat utilization by local people will be allowed, thus ensuring their involvement and support. The Man and the Biosphere Program is receiving support from both industrialized and Third World countries, and is a major step forward in preserving the Earth's wild animals.

Tourism and conservation (*above*) Tourism plays an invaluable part in East African wildlife parks. Income is applied to the management programs and public awareness is raised by first-hand experience.

Hawaiian monk seals (*left*) only breed regularly on seven atolls of the Leeward Hawaiian Chain. With no more than 1,500 individuals surviving, the species is endangered. Most breeding sites are included in the Hawaiian Islands National Wildlife Refuge.

Watch out 'roos' crossing ahead (*below*) In most countries, thousands of animals are killed by motor vehicles each year. This can have a significant impact on animal populations, particularly those under threat.

GRAZING FOR VARIETY

Grazing animals can do irreparable damage to a habitat. The removal of cattle, sheep, goats, rabbits and other aliens is usually a priority for reserve managers, especially where such aliens compete directly for food with the native wildlife. However, where the wild animals have already gone, domestic animals can have an important role in conserving a habitat.

Sheep created the habitat for the Large blue butterfly in southern England, but this was not realized until the species was almost extinct. Its caterpillars depend on ants, which carry them into their nests where they feed on the ant grubs. The main ant host prefers heavily grazed chalk grassland. As sheep disappeared with changing agricultural practices, so did the ants and the butterflies.

Keoladeo National Park in India is a famous refuge for Siberian cranes and other waterbirds. When cattle and Water buffalo were expelled by the park authorities, the birds declined. Without the pressure of grazing, grasses had grown unchecked and choked the open pools on which the birds depended.

Animal Protection

Human populations and their needs often conflict with animal conservation. Some exploitation of animals is traditional, and to most people acceptable, but increasing pressures now make some animal management very essential.

Legal protection is an important means of conserving animals. Limits can be set on the numbers that may be killed and protection can be given to animals during their breeding seasons. Historically, protection was a matter of preserving game animals so that the privileged classes could hunt them. Without game reserves the Alpine ibex and Père David's deer would be extinct. Increasingly, however, legislation has been needed to protect the growing numbers of animals that are becoming endangered. However, enforcing such laws has proved very difficult in the face of pressures from tradition, economic need or outright commercial greed. It is far easier to poach an animal than to guard it. Elephants in Zimbabwe in southern Africa, for example, would require one guard for every 50 sq km (19 sq mi), at an annual cost of $10,000 (in 1990).

Protection is even more difficult when animals cross national boundaries or live in the sea outside national or international jurisdiction. Songbirds, which are valued and protected by law in northwestern Europe, are shot in millions as they migrate across southern Europe. Whaling illustrates the problem of exploitation of animals in international waters. There is no binding legislation preventing a nation from whaling if it decides to do so and, even if it is a member of the International Whaling

CITES: REDUCING THE TRADE IN LIVES

Animals and animal products worth billions of dollars are traded around the world every year, causing severe depletion of some species. Moreover, as an animal becomes rarer its value soars, and efforts to catch specimens are intensified. With rhino horn fetching $10,000/kg in the Far East, and an ocelot coat $40,000 in Europe, hunting pressure is higher than ever. One dreadful aspect of this trade is that for every live animal offered for sale, countless more have died during capture attempts or in transportation. Up to 80 percent of parrots die before they are purchased.

Since 1973 the trade in animals and plants has been controlled by the Convention on International Trade in Endangered Species of Wild Fauna and Flora (CITES). The basis of CITES is a list of species requiring protection. APPENDIX II lists species that will soon be threatened with extinction if trade is not regulated. APPENDIX I lists species already under threat, and which cannot be traded in any form without a permit. Records allow monitoring of trade so that controls can be introduced when necessary. However, with such rich markets at stake, smuggling and falsification of documents are inevitable. Perhaps the greatest shortcoming is that CITES does not cover trade within a country. Furthermore, any country may declare a "reservation" on a particular species, which frees it from CITES controls. Adverse publicity can help to minimize the number of reservations but this agreement is only as strong as the commitment and will of its voluntary signatories.

CITES PARTIES

Countries that have joined,and when they joined.

1975	Brazil, Chile, Costa Rica, Cyprus, Ecuador, Madagascar, Mauritius, Nepal, Niger, Nigeria, Peru, South Africa, Sweden, Switzerland, Tunisia, United Arab Emirates, USA, Uruguay
1976	Australia, Finland,Germany, Ghana, India, Iran, Norway, Pakistan, Papua New Guinea, USSR, UK, Zaire
1977	Denmark, Gambia, Nicaragua, Paraguay, Senegal, Seychelles
1978	Botswana, Egypt, France, Malaysia, Monaco, Panama, Venezuela
1979	Bahamas, Bolivia, Indonesia, Italy, Jordan, Kenya, Sri Lanka
1980	Central African Republic, Israel, Japan, Liechtenstein, Tanzania
1981	Argentina, Belize, Cameroon, China, Colombia, Guinea, Liberia, Mozambique, Philippines, Portugal, Rwanda, Surinam, Zambia, Zimbabwe
1982	Austria, Bangladesh, Malawi
1983	Congo, Saint Lucia, Sudan, Thailand
1984	Algeria, Belgium, Benin, Luxembourg, Netherlands, Trinidad and Tobago
1985	Honduras, Hungary
1986	Afghanistan, Somalia, Spain
1987	Dominican Republic, El Salvador, Singapore
1988	Burundi
1989	Chad, Ethiopia, Gabon, Malta, New Zealand, Saint Vincent and the Grenadines, Vanuatu
1990	Brunei, Burkina, Cuba, Guinea-Bissau, Poland
1991	Namibia

Commission, it can simply leave if it disagrees with its decisions.

The need for popular support

Animal protection can be effective only if it has popular support. Moral arguments are increasingly important in conservation efforts. For example, the wearing of furs has become socially unacceptable in many countries. Nevertheless, popular outrage must be felt by everyone if social sanctions are to have any force. There is no point in a majority of people in the United States favoring conservation of the nation's dwindling population of wolves, or of legislation being enacted for their preservation, if livestock owners see them as a threat to their livelihood and exterminate them. Arguments in favor of animal protection are even less likely to succeed in developing countries where any consideration other than survival is a luxury, and complete protection of an animal under law is an alien concept. If conservation is to be effective here, people must see some real advantage to themselves, such as an increase in tourism or an income from regulated culling of an animal resource. Experience shows that the poaching rate of Black rhinoceroses is directly related to the money spent on conservation. Large sums of money can be raised on licenses for foreign hunters to kill many fewer rhinos than are poached. Protection therefore becomes a more viable proposition when supported by education programs that spell out local economic advantages.

Mountain gorilla infants at play (*left*) in the Volcanoes National Park of Rwanda. Conservation measures often focus on such "flagship species", but all the animals and plants in the reserve benefit.

Killing for conservation (*below*) In some protected areas of Zimbabwe, elephant populations grew to levels that the ecosystem could not sustain; consequently a controversial culling program was introduced. Scientific research is needed to ensure that the program is effective, without disrupting the elephant society.

Flagship species

The IUCN (the International Union for the Conservation of Nature and Natural Resources) has listed over 4,500 animal species as threatened, but this underestimates the real figure. With so many animals at risk, a useful strategy is to focus on flagship species: animals that are important culturally or economically and whose need for protection will be seen by politicians, officials and the general public. For example, the Giant panda and Indian tiger attracted resources that would not have been made available for less familiar, yet equally deserving, species. However, the latter can benefit by the preservation of their habitat for the flagship species.

The kagu, a flightless bird of New Caledonia in the western Pacific, is a flagship species. It is the sole member of the family Rhynochetidae. It is also the island's national bird. One of the major threats to its survival comes from the destruction of its rainforest habitat. Focusing on the plight of the kagu is therefore the most effective way of conserving not only a unique bird, but also a fragment of primeval rainforest and the thousands of animal and plant species that it supports.

REGIONS OF THE WORLD

CANADA AND THE ARCTIC
Canada, Greenland

THE UNITED STATES
United States of America

CENTRAL AMERICA AND THE CARIBBEAN
Antigua and Barbuda, Bahamas, Barbados, Belize, Bermuda, Costa Rica, Cuba, Dominica, Dominican Republic, El Salvador, Grenada, Guatemala, Haiti, Honduras, Jamaica, Mexico, Nicaragua, Panama, St Kitts-Nevis, St Lucia, St Vincent and the Grenadines, Trinidad and Tobago

SOUTH AMERICA
Argentina, Bolivia, Brazil, Chile, Colombia, Ecuador, Guyana, Paraguay, Peru, Uruguay, Surinam, Venezuela

THE NORDIC COUNTRIES
Denmark, Finland, Iceland, Norway, Sweden

THE BRITISH ISLES
Ireland, United Kingdom

FRANCE AND ITS NEIGHBORS
Andorra, France, Monaco

THE LOW COUNTRIES
Belgium, Luxembourg, Netherlands

SPAIN AND PORTUGAL
Portugal, Spain

ITALY AND GREECE
Cyprus, Greece, Italy, Malta, San Marino, Vatican City

CENTRAL EUROPE
Austria, Germany, Liechtenstein, Switzerland

EASTERN EUROPE
Albania, Bulgaria, Czechoslovakia, Hungary, Poland, Romania, Yugoslavia

THE SOVIET UNION
Mongolia, Union of Soviet Socialist Republics

THE MIDDLE EAST
Afghanistan, Bahrain, Iran, Iraq, Israel, Jordan, Kuwait, Lebanon, Oman, Qatar, Saudi Arabia, Syria, Turkey, United Arab Emirates, Yemen

NORTHERN AFRICA
Algeria, Chad, Djibouti, Egypt, Ethiopia, Libya, Mali, Mauritania, Morocco, Niger, Somalia, Sudan, Tunisia

CENTRAL AFRICA
Benin, Burkina, Burundi, Cameroon, Cape Verde, Central African Republic, Congo, Equatorial Guinea, Gabon, Gambia, Ghana, Guinea, Guinea-Bissau, Ivory Coast, Kenya, Liberia, Nigeria, Rwanda, São Tomé and Príncipe, Senegal, Seychelles, Sierra Leone, Tanzania, Togo, Uganda, Zaire

SOUTHERN AFRICA
Angola, Botswana, Comoros, Lesotho, Madagascar, Malawi, Mauritius, Mozambique, Namibia, South Africa, Swaziland, Zambia, Zimbabwe

THE INDIAN SUBCONTINENT
Bangladesh, Bhutan, India, Maldives, Nepal, Pakistan, Sri Lanka

CHINA AND ITS NEIGHBORS
China, Taiwan

SOUTHEAST ASIA
Brunei, Burma, Cambodia, Indonesia, Laos, Malaysia, Philippines, Singapore, Thailand, Vietnam

JAPAN AND KOREA
Japan, North Korea, South Korea

AUSTRALASIA, OCEANIA AND ANTARCTICA
Antarctica, Australia, Fiji, Kiribati, Nauru, New Zealand, Papua New Guinea, Solomon Islands, Tonga, Tuvalu, Vanuatu, Western Samoa

North America

CANADA AND THE ARCTIC
THE UNITED STATES
CENTRAL AMERICA AND THE CARIBBEAN
SOUTH AMERICA

Central and South America

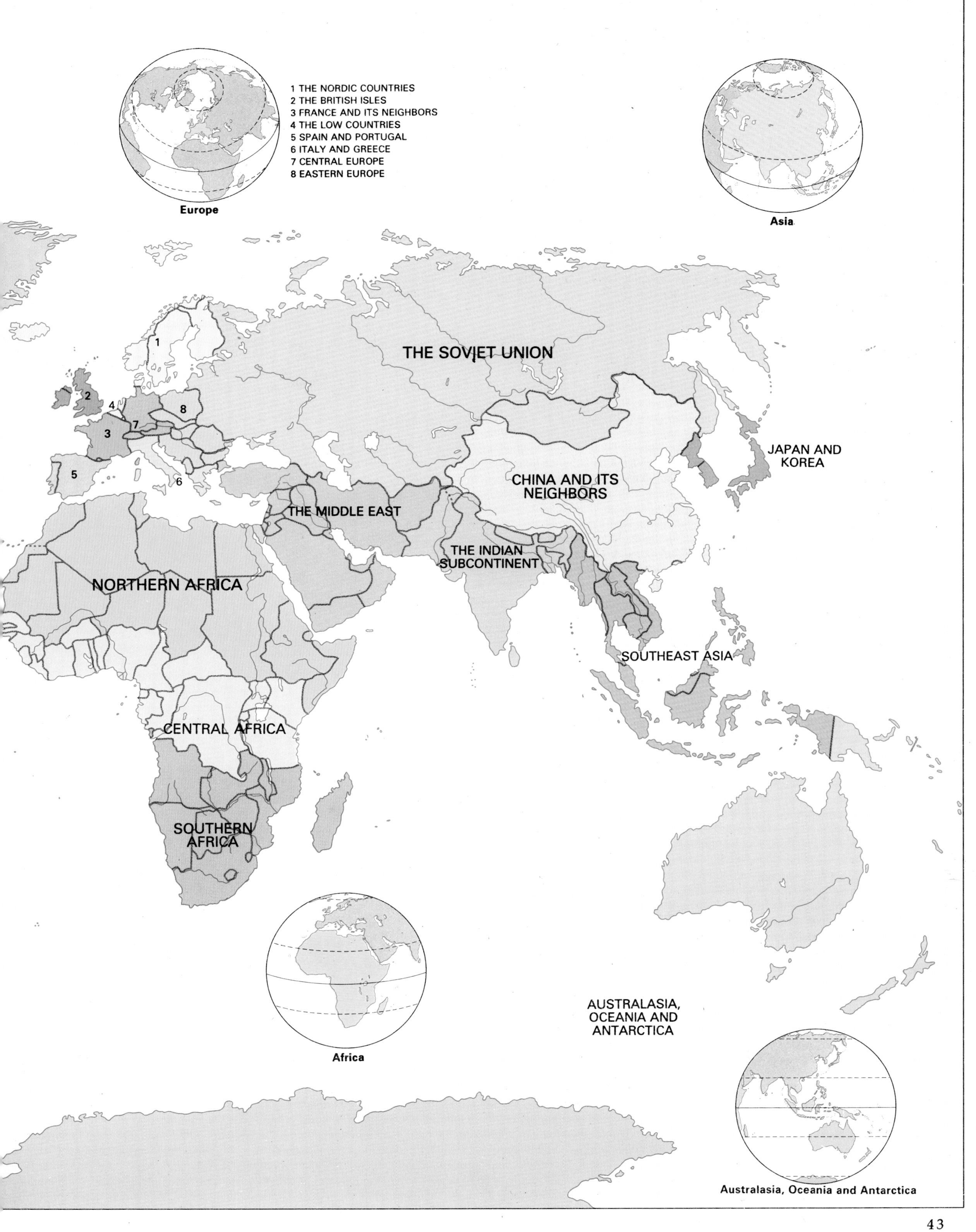

Europe
1 THE NORDIC COUNTRIES
2 THE BRITISH ISLES
3 FRANCE AND ITS NEIGHBORS
4 THE LOW COUNTRIES
5 SPAIN AND PORTUGAL
6 ITALY AND GREECE
7 CENTRAL EUROPE
8 EASTERN EUROPE
Asia
THE SOVIET UNION
JAPAN AND KOREA
CHINA AND ITS NEIGHBORS
THE MIDDLE EAST
THE INDIAN SUBCONTINENT
NORTHERN AFRICA
SOUTHEAST ASIA
CENTRAL AFRICA
SOUTHERN AFRICA
Africa
AUSTRALASIA, OCEANIA AND ANTARCTICA
Australasia, Oceania and Antarctica

SPACE FOR WILDLIFE

THE ARCTIC CHALLENGE · SECRETS OF SURVIVAL · BACK FROM THE BRINK

The varied animal life of Canada reflects its vast wildernesses, the seasonal contrasts of its extreme continental climate, and its diverse habitats. Rattlesnakes live in the most arid areas, great herds of caribou cross the tundra, and countless migrant birds visit the region for part of the year. Many of the species that live in the endless tracts of tundra and the boreal forest are common to the European landmass: they crossed into Canada during the last ice age over a land bridge where the Bering Strait now exists. Seals, walruses and Polar bears inhabit the offshore islands, and whales swim off the northern seaboard, feeding on the rich fish stocks of the Arctic Ocean. With its low human population density, Canada remains one of the last reserves capable of supporting large and healthy populations of wildlife.

COUNTRIES IN THE REGION

Canada

ENDEMISM AND DIVERSITY

Diversity Low
Endemism Low

SPECIES

	Total	Threatened	Extinct†
Mammals	136	15	0
Birds	450*	5	2
Others	unknown	28	2

† *species extinct since 1600 - Great auk* (Alca impennis), *Labrador duck* (Camptorhynchus labradorius)
* *breeding and regular non-breeding species*

NOTABLE THREATENED ENDEMIC SPECIES

Mammals Vancouver Island marmot *(Marmota vancouverensis)*
Birds None
Others Lake lamprey *(Lampetra macrostoma)*, Copper redhorse fish *(Moxostoma hubbsi)*, Periodical cicaca *(Magicicauda septendecim)*

NOTABLE THREATENED NON-ENDEMIC SPECIES

Mammals Gray wolf *(Canis lupus)*, wolverine *(Gulo gulo)*, Polar bear *(Ursus maritimus)*, Bowhead whale *(Balaena mysticetus)*, Northern right whale *(Eubalaena glacialis)*
Birds Whooping crane *(Grus americanus)*, Piping plover *(Charadrius melodius)*, Eskimo curlew *(Numenius borealis)*, Spotted owl *(Strix occidentalis)*, Kirtland's warbler *(Dendroica kirtlandii)*
Others Shortnose sturgeon *(Accipenser brevirostris)*, Atlantic whitefish *(Coregonus canadensis)*, Dakota skipper *(Hesperia dacotae)*

DOMESTICATED ANIMALS (originating in region)

none

Well concealed The Ruddy turnstone migrates from the sea shore to the tundra to nest and rear its young. The tundra is an exposed habitat that lacks cover, so the bird is well camouflaged to protect it from predators such as falcons, jaegers and Arctic foxes.

THE ARCTIC CHALLENGE

An animal must possess special qualities to withstand the extreme weather conditions of the Arctic: an area that encompasses about one-third of Canada, almost all of Greenland and the whole of the Arctic Ocean. There are far fewer permanent inhabitants than in more temperate areas: resident species include Arctic lemmings and Arctic foxes. Many of the region's animals are also found in northern North America, northern Europe and the north of the Soviet Union. This is because until relatively recent times the Eurasian and American landmasses were connected by land bridges.

After the spring melt, when the climate is more hospitable to wildlife, many species of animals are attracted to the Arctic tundra to give birth and feed their young on the abundant supply of food. Thousands upon thousands of birds migrate north every year to find nesting sites and mates, and to take advantage of the bountiful food supply provided by the grasses and wild flowers that carpet the tundra in summer, and the insects that breed among the foliage and in the pools that dot the landscape. The visitors include many of the world's long-distance migrants: Canada geese fly in from the southern United States, Ruddy turnstones arrive from Argentina, and for the Arctic tern the summer visit is just a brief pause on its year-round migration between the Arctic shores and the edge of the Antarctic pack ice. Lesser golden plovers come from South America and Northern wheatears migrate here from Africa.

The empty north

The majority of Canadians live within 160 km (100 mi) of the United States border, leaving the north relatively empty of people. Although pressure from the south – in the form of mining, damming and forestry – competes with the animals for use of the land, a great deal of the north is inaccessible to, or free from, industry. This leaves ample space for events such as the annual mass migration of caribou from their winter refuge in the coniferous forests to their summer home in the Arctic tundra. Herds of tens of thousands of animals migrate northward to their traditional calving grounds. Groups of females and yearlings arrive first and give birth in the spring: the males then join them to spend the summer grazing. In September, thousands of closely spaced bands return across ice fields, mountains and marshes to their wintering grounds. Many will have done over a thousand kilometers in their annual circuit.

The Arctic seaboard also supports a great number of migrants. In summer plankton thrive in the surface waters, supporting dense shoals of shrimplike invertebrates. These in turn are preyed on by fish, which attract seabirds and Harp, Ringed and Harbor seals. The seals are preyed on not only by Polar bears, but also by people; they are an important food source for the few remaining Canadian and Greenland Inuit who still live by hunting: the pelts of the seals are an important part of their livelihood.

There has been worldwide controversy over the commercial hunting of Harp seal pups in Canada. Annually every February and March female Harp seals gather on the ice to give birth and then mate: the perfect opportunity for hunters to harvest large numbers of newborn pups. Now, after 200 years, the clubbing has finally been brought to an end and the Harp seal population continues to thrive.

Teeming seas

The northern Atlantic and Pacific oceans support a very large quantity of marine life. For centuries fishermen from Europe, and more recently from Canada, have exploited the huge shoals of cod found along the Grand Banks of Newfoundland in the Atlantic. Lobster were also so abundant here a century ago that they were considered a "poor man's dish". The Pacific is the ocean of the giants: the Giant octopus, whose outstretched arms may reach 2 m (6 ft) across; the goeduck (pronounced gooweyduck), a clam that has a feeding siphon over 1 m (3 ft) long; and the halibut, a flat fish that can easily weigh as much as 150 kg (330 lb).

The world's largest carnivore A Polar bear roams the sea ice in search of seals. Superbly adapted to Arctic conditions, Polar bears have been known to reach Iceland and Norway over the ice. Canada has several important denning areas for Polar bears, who give birth in snow holes in midwinter.

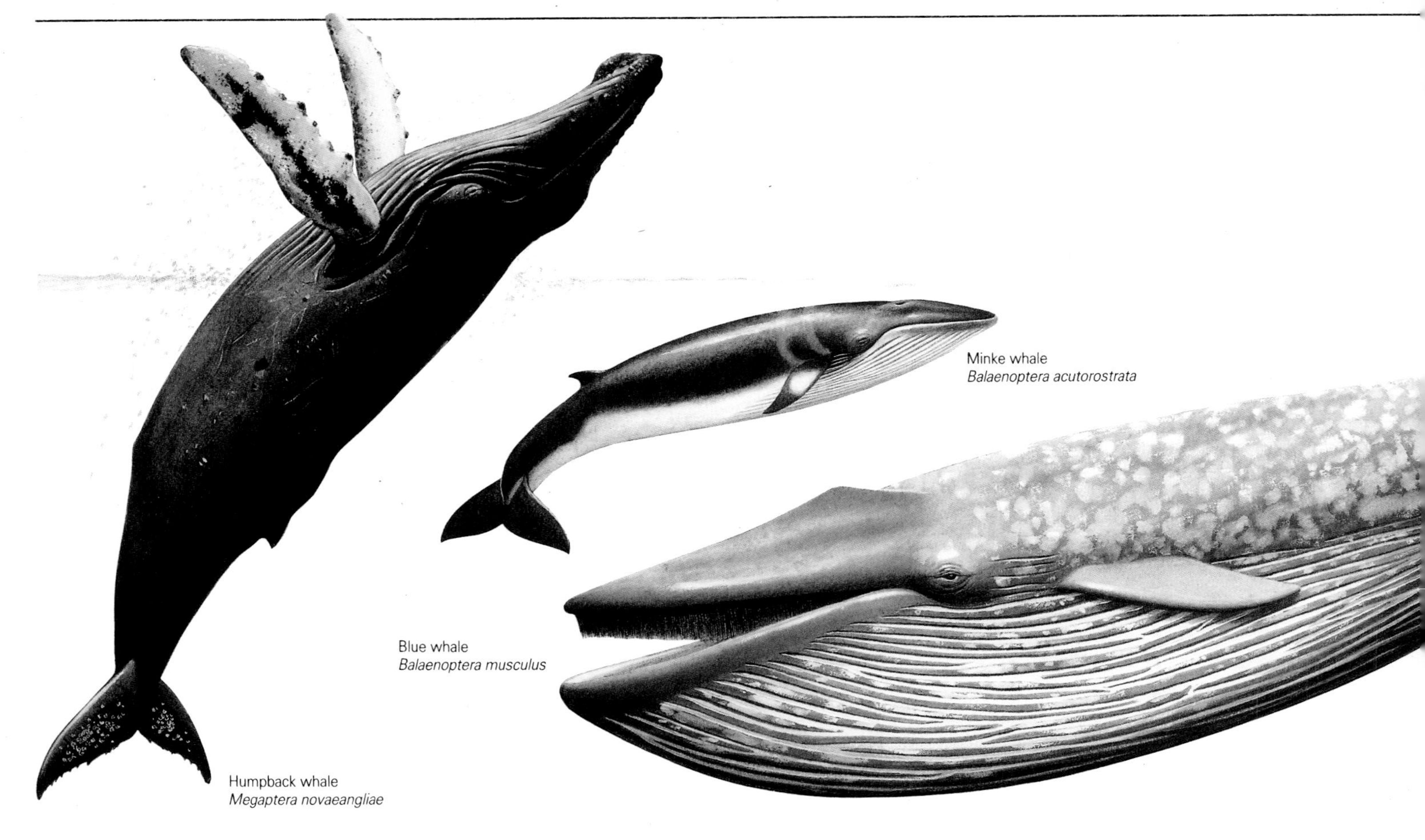

SECRETS OF SURVIVAL

The animals of the Arctic have had to adapt to withstand the severest of winter weather conditions. Insulation is an essential means of heat conservation. The Polar bear, the Arctic fox and caribou, for example, have a thick fur coat and dense underfur to keep them warm and protect them from the penetrating winds. Animals with a low surface area to volume ratio are able to conserve heat for longer; short, thick limbs and small ears, such as those of the Arctic fox, are adaptations for reducing heat loss.

Coping with snow

A typical adaptation of animals in the Arctic is the matching of coat color to the seasonal changes that affect the landscape. Camouflage aids both the hunter and the hunted on the bare tundra. The Rock ptarmigan, for example, has mottled brown plumage in summer that blends with the foliage. As winter approaches this is replaced by dense white feathers that provide insulation as well as camouflage. In severe weather it burrows in the snow for warmth. However, despite its protective coloring, it often falls prey to Arctic fox and ermine, both of which also turn white in winter.

Snow poses problems for forest species too: both the Snowshoe hare and its main predator, the lynx, have broad feet that enable them to walk on snow. The moose (elk) has another approach; its long legs and high-stepping gait enable it to move swiftly through deep snow to escape from pursuing wolves, which must plow their way through.

One way in which animals deal with cold conditions is to avoid them by hibernating. An animal in hibernation becomes completely inactive; it enters a prolonged period during which its body temperature and metabolic rate are reduced. Torpor is akin to short-term hibernation; the animal's body temperature is not reduced to the same extent, and the animal may become active at frequent intervals so that it can feed. Hibernation and torpor are both means of conserving energy when food is scarce and when conditions are too severe for normal activity. Northern bears usually enter a state of torpor and are easily wakened; chipmunks, by contrast, enter a deep and prolonged sleep.

Seals are well adapted to the cold waters in which they live for much of the time. They have developed various adaptations that enable them to swim efficiently, and to conserve heat in the cold, heat-draining waters. These include a torpedolike body shape to improve streamlining, the modification of limbs into flippers and the ability to hold their breath for long periods while diving for food. Adaptations to minimize heat loss include a thick layer of blubber, which

SCOURGE OF THE SEA LAMPREY

The Sea lamprey is a primitive eel-like fish that spends much of its adult life at sea, living as a parasite on other fish. Clinging to its host with a suckerlike mouth armed with sharp, horny teeth, it feeds on the host's blood. This seriously weakens the fish, especially if several lampreys attach themselves. The lamprey's skin has glands that secrete a poisonous mucus, which is thought to deter predatory fish. Lampreys travel hundreds of kilometers upriver in order to spawn. The males build nests for the eggs by making a shallow depression in the gravel. After the eggs have hatched the strange filter-feeding larvae remain in fresh water for several years.

The American Sea lamprey used to migrate along the St. Lawrence seaway to Lake Ontario to spawn, but was prevented by the Niagara Falls from reaching the other Great Lakes. In 1828, however, the building of the Welland Canal,which links Lake Ontario to Lake Erie, let in the Sea lamprey. The lamprey modified its ways to spend its whole life in fresh water. Its depredations devastated a once thriving fishing industry; on Lake Michigan the catch of lake trout fell from 3,000 tonnes (3,000 tons), in 1944 to 16 kg (35 lb) in 1955. There have been many attempts to get rid of the lamprey, ranging from traps to electrical barriers and poison, but despite all the money spent, it still survives – at the expense of the trout.

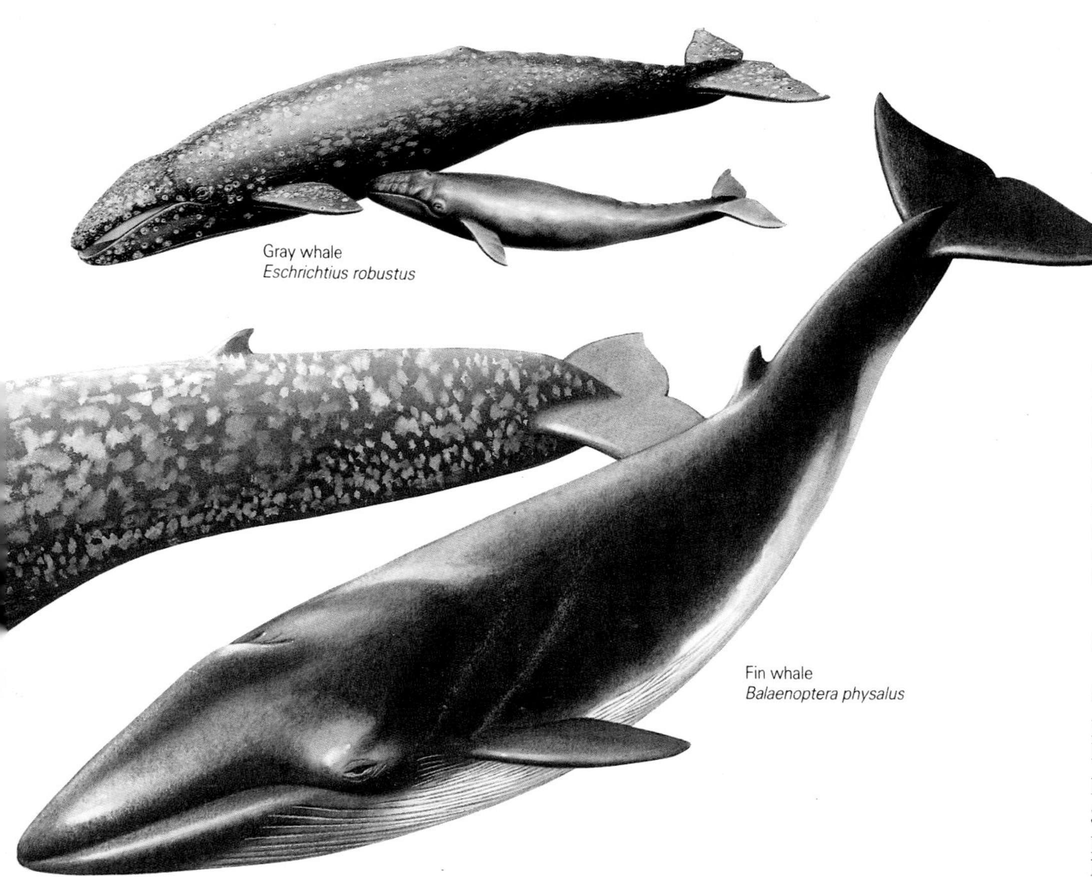

Whales of the far north (*left*) The Humpback whale is renowned for the variety of its underwater calls. The Minke whale is one of the smallest, but has been extensively hunted since stocks of the larger species were depleted. The Blue whale is the largest living creature on earth, up to 27 m (88 ft) long. Despite the worldwide ban on whaling, native Arctic peoples are still allowed to hunt the Fin whale. The Gray whale is confined to the Pacific Ocean, where it is often seen near the coast.

acts as insulation, shortened limbs – to reduce surface area – and an efficient circulation system that can reduce blood flow to the flippers while redirecting blood from the body surface to the core of the animal's body in order to conserve heat. It also directs oxygenated blood to vital organs such as the brain. In the case of the fur seal, further insulation against the cold is provided by the animal's dense pelt with its thick layer of softer underfur.

Life on the open plains

On the open Canadian prairies the lack of cover can pose problems for both predators and prey. A social organization is one adaptation to these conditions: wolves, for example, live in packs, allowing them to hunt large prey and to defend kills from other wolves and predators. In response to such pack hunters, and to being easily spotted on the grasslands, prey such as bison and Pronghorn antelopes live in large herds. This may offer advantages to the individual as a herd is more likely to detect the approach of a predator than a single grazing animal.

In addition to living in herds, the Pronghorn antelope exhibits other antipredator adaptations. For its size, the Pronghorn has relatively large eyes that probably enable it to see predators at great distances. Pronghorns also react to predators by raising two white patches of hair on their rump, signaling danger to the rest of the herd. The herd then flees as one, which reduces the danger to individual animals as they make their dash for safety. These small antelopes can attain a speed of 86 kph (53 mph), comparable to that of the cheetah, which is reputed to be the fastest land animal in the world.

Dolphins of cool waters (*below*) The Bottle-nosed dolphin is a common coastal species. The beluga or White whale has the popular name "the canary of the sea" because of its wide repertoire of underwater calls. The male narwhal has a single tusk up to 1 m (3 ft) long; its function is unknown.

BACK FROM THE BRINK

Canada's prosperity was founded on the fur trade of the early 17th century, and the tradition of hunting continues to this day. Hares, lynx, beavers and raccoons were trapped for their pelts, and wolves were killed because they, like humans, were deer hunters. A bounty was imposed on wolves to reduce their numbers, thus preserving larger deer populations for the human hunters. The irony is that humans are far more likely than wolves to be the cause of low deer numbers. The best catch for a hunter is the largest buck with the biggest antlers: a mature breeding male. This leaves younger or less healthy males to do the breeding, which may weaken the deer population. Wolves, on the other hand, tend to hunt old or sick animals (those easiest to capture), thus removing from the population individuals that are likely to die from starvation or exposure to the elements anyway.

Rescue and reintroduction

One of the best-known stories of near-extinction is that of the North American bison or buffalo. The ancestor of the two subspecies (the Plains and the Wood bison) moved into the region approximately 100,000 years ago across the land bridge that once joined North America to the Eurasian landmass. Bison spread throughout the continent wherever grasslands produced the food they required, reaching a population of 60 million or more. When European settlers reached the prairies the animals were slaughtered indiscriminately for their meat and hides, and by 1900 were practically extinct. In the safety of government reserves, however, the remaining animals survived and bred, and they are now numbered in their thousands.

The North American bison is extremely large, its massive head and shoulders indicating its great strength. The bulls can be very dangerous during the rut, when they compete with aggression to establish dominance. Ranchers have considered raising bison and selling the meat for profit, but the animal does not take well to captivity and can jump over 2 m (6 ft) fences. They have had greater success with a hybrid between beef cattle and buffalo, the "beefalo".

The stately Whooping crane has made an even more remarkable recovery than the bison. In 1941 there were only 23 Whooping cranes left in the world, two of which were being kept in captivity. Their decrease in numbers was perhaps not unexpected, considering what an easy target for hunters the great white birds made flying over the prairie: they stand close to 2 m (6 ft) tall and have a 2.5 m (7 ft) wing span. The Whooping crane migrates from the Northwest Territories (its breeding grounds are in Wood Buffalo National Park) to its wintering grounds in Texas on the coast of the Gulf of Mexico. Once it was recognized as being close to extinction public concern was quickly aroused, and publicity campaigns to save the bird were mounted along its migration route.

Whooping cranes usually lay two eggs, but raise only one young to fledgling age. In an attempt to boost the present population of about 200 birds in Canada, conservationists are using fostering techniques. One egg will be taken from a Whooping crane nest in Wood Buffalo National Park and placed in the nest of a pair whose own eggs have been lost through accident or predation. Another method is to fly the egg to Grays Lake National Park in Idaho in the northwest of the United States, where it is placed in the nest of a Sandhill crane (a closely related species). It was hoped that this would extend the range of the bird in North America as well as increase its numbers, but though the 16 Idaho-raised Whooping cranes migrate regularly, not one has so far reproduced.

The gregarious walrus *(above)* Ungainly on land, in the sea the walruses' fat provides buoyancy and streamlining. They feed on clams and other invertebrates on the seabed, squirting high pressure jets of water to uncover them. Their sensitive whiskers help them locate their food in murky waters during winter or at depths where light barely penetrates. The powerful tusks are mainly for display, but are occasionally used when hauling themselves onto the ice or for defending their young.

Exposed to danger *(left)* A vulnerable female Long-tailed duck incubates her eggs alone on the tundra for 23 days: duck eggs are a favorite prey of Arctic foxes. These ducks migrate from the open sea to the tundra to take advantage of the long summer days for feeding their young on the abundant mollusks and crustaceans in the Arctic lakes. The ducklings are well developed by the time they hatch, and are able to fly within 5 or 6 weeks.

Snow geese (*above*) migrate in their thousands to the tundra to breed in spring. They feed only on plant life, taking advantage of the flush of new growth prompted by the melting snow. The goslings are much better camouflaged than their parents.

LIFE UNDER THE ICE: THE HARP SEAL'S SONG

Female Harp seals congregate on sea ice in February and have their pups in late February or early March. Mating, which takes place underwater, follows in about mid-March. Much of what is known about the behavior of these seals under the ice comes from studies of their underwater calls: eerie chirping and clicking sounds comprising sixteen call types. An increase in calling in mid-March coincides with the onset of courtship and mating. It is thought that the calls have two main functions: to draw other seals to the breeding herd, and to attract mates. Natural selection may favor females that produce loud calls, which attract more males to the breeding area. This results in a larger pool of suitors from which a female can choose a mate. Quieter calls may be made by males that are trying to attract a female or to establish territory by warning other males to stay away from a breathing hole in the ice.

Harp seals sing in dialects. Although the calls of the seals that live in the Gulf of St. Lawrence have not changed for over 15 years, their calls are noticeably different from those of Harp seals breeding off the coast of Norway. In this respect Harp seals resemble birds, which are well known to have different song dialects in different locations. The song dialects of Harp seals may reflect their dependence on a vocal means of communication. The visual and chemical signals used by many land mammals would be ineffective underwater because they do not carry over even moderate distances: sound, on the other hand, travels considerable distances through water.

Public awareness has not produced such successful results with other species, such as the Swift fox, whose population was greatly reduced by hunting and poisoning, and became extinct in Canada. A breeding program in Alberta has led to the reintroduction of the Swift fox into parts of its former range in the dry areas of the southern prairie, but its numbers are still extremely low.

The Vancouver Island marmot has a little-known history, perhaps because its numbers have never been high. It survives only in an alpine habitat, and is a slow breeder; the young remain with their parents for three years. These two circumstances are often related to species' extinction. However, if its habitat is preserved, and if there is minimal interference from humans, it should continue to survive and perhaps even gradually increase its numbers.

The threat from humans

A grave threat to many of Canada's animals comes from habitat loss and pollution. The unique temperate rainforests of the Pacific coast – home of the Bald eagle, the Grizzly bear and Pacific salmon – are being felled at an alarming rate, and the Great Lakes are so polluted that in many places the fish are unsafe for human consumption. Dam projects threaten many wetland bird refuges, including part of Wood Buffalo National Park, home to North America's northernmost pelican colony; and moose, muskrats and many other species face the inundation of meadows and muskegs.

Hunted for their fur

Since earliest human times people have hunted animals not only for food but also for their fur, to make clothes for warmth and adornment. Until the comparatively recent development of artificial fibers that provide thermal insulation, people commonly wore animal skins to protect them from the snow and bitter winds. Animal furs have also long been valued for the status of wealth, and sometimes dignity, they confer upon the wearer. This is exemplified by the symbolic importance given to the wearing of ermine (the white winter coat of the stoat) in the courts of medieval and renaissance Europe. Ermine was originally used to border the hems and cuffs of costly robes, and came to signify royal or noble birth. It is still used on some ceremonial occasions. The coronation in 1937 of Britain's King George VI (1895–1952) is said to have accounted for 50,000 ermine, needed to trim the robes of the British peerage.

So great was the appetite of 17th century Europe for furs (particularly beaver fur, which was used to make men's hats) that the fur trade was one of the main forces in the opening up of Canada, with its vast wealth of fur-bearing mammals of all sizes. In 1670 a British trading organization, the Hudson's Bay Company, was formed so that the British could trade with the Amerindians for furs. Trading posts gradually spread across the country. Even today, although it has many other interests, furs remain an important part of the company's activities.

The colder the conditions in which an animal lives, the greater is its need for insulation, and the denser the fur it grows. The many fine hairs of its pelt enable it to retain body heat by trapping air in a layer next to the skin. Arctic foxes have such thick winter coats that they are able to sleep on the snow at temperatures of −40°C (−40°F) without coming to harm.

Grooming fur is essential for maintaining its insulatory capacity. The beaver has a unique split toenail used specifically for cleaning and oiling its fur: it spends much of its time in the water, and has developed a waterproof inner coat of short fur. The mink, another water-loving species, has one of the thickest and most luxurious furs. Fur seals, which spend even more time underwater, also have a thick layer of fat (blubber) below the skin for insulation.

Victims of fashion

Over the centuries many millions of animals in Canada have died in traps and snares set by humans, originally to provide the necessities for survival in a harsh environment, more recently to supply the whims of fashion. As an example of the numbers involved, one winter's total taking in North America might include 400,000 Red foxes, 250,000 Gray foxes and 37,000 Arctic foxes.

A few fur-bearing animals, such as Polar bears, are now protected, but many more are vulnerable to hunters who can trap them for their fur without legal restriction. Wolverines, martens, fishers, Arctic foxes, bobcats and lynx all have desirable coats. Among the sea mammals, both Harp seals and Hooded seals are still widely hunted for their fur. Beavers and wild mink are trapped in or near their freshwater habitats.

Since at least the turn of the century the market for fur has been met partly by fur farms, which raise animals such as mink

The coat of Arctic foxes occurs in two color forms, each with a different winter and summer coloration. A White, or Polar, form in winter coat attacks a ptarmigan. A Blue form, really a gray to brown color, looks for carrion in winter. A Blue fox vixen "helper" in brown summer coat with her year-younger siblings: the light gray cubs have the summer coat of the Polar form, while the dark gray cub is a variant of the Blue's summer coat. The extent of snow cover influences the predominant color form of the population.

Alone on the ice (*above*) A Harp seal pup waits for its mother to nurse it. It will be weaned in only 10–12 days. The killing of Harp seal pups for their pelts by clubbing them to death in front of their mothers has caused international uproar.

Fierce carnivore (*right*) The fisher has a unique method of catching a porcupine, repeatedly darting in to bite its face until it rolls over and the fisher can attack its soft underbelly. On this occasion it has caught a less well-defended chipmunk.

and foxes. Although this reduces pressure on the wild populations, farmed furs rarely reach the quality of the best wild furs, and the latter are still sought by connoisseurs. Nevertheless, in recent years there has been a considerable shift away from the wearing of natural furs by many people – a result of international campaigning by animal protectionists.

FROM ALLIGATORS TO POLAR BEARS

WAVES OF IMMIGRATION · THE LIVING DESERTS · ADAPTING TO FIRE · WETLAND WILDLIFE
ANIMALS OF THE SEAS · THE HUMAN IMPACT

A diverse array of animals inhabit the vast region of the United States: the caribou and the Polar bear of Alaska, the alligators, turtles and pelicans of subtropical Florida, desert rattlesnakes and nimble-footed Rocky Mountain sheep. Among the coral reefs and mangroves that fringe its southern shores live a wealth of invertebrates and fish. To the east and west the open oceans of the Atlantic and the Pacific are home to whales and sharks, while seals and walruses throng Alaska's Arctic seas. Animals from Asia, Europe and South America crossed ancient land bridges to make their home here. Many new species have arisen, too, in isolated canyons and on mountaintops, in desert pools and on offshore islands – none more famous than the unique animals of the Hawaiian Islands far out in the Pacific.

COUNTRIES IN THE REGION

United States of America

ENDEMISM AND DIVERSITY

Diversity Medium
Endemism Medium to high on mainland (many species shared with Mexico or Canada), very high on Hawaii

SPECIES

	Total	Threatened	Extinct†
Mammals	330	22	0
Birds	1,090*	42	20
Others	unknown	970	99

† species extinct since 1600
** breeding and regular non-breeding species*

NOTABLE THREATENED ENDEMIC SPECIES

Mammals Black-footed ferret *(Mustela nigripes)*, Gray bat *(Myotis grisescens)*, Stephens' kangaroo rat *(Dipodomys stephensi)*, Red wolf *(Canis rufus)*, Hawaiian monk seal *(Monachus schauinslandi)*
Birds California condor *(Gymnogyps californianus)*, nene *(Branta sandvicensis)*, kamao *(Myadestes myadestinus)*, Oahu creeper *(Paroreomyza maculata)*, Hawaiian crow *(Corvus hawaiiensis)*
Others San Joaquin leopard lizard *(Gambelia silus)*, Pallid sturgeon *(Scaphirhynchus albus)*, Devil's Hole pupfish *(Cyprinodon diabolis)*, Acorn pearly mussel *(Epioblasma haysiana)*, Little agate shells *(Achatinella spp.)*, No-eyed big-eyed wolf spider *(Adelocosa anops)*, San Francisco forktail damselfly *(Ischnura gemina)*

NOTABLE THREATENED NON-ENDEMIC SPECIES

Mammals Townsend's big-eared bat *(Plecotus townsendii)*, wolverine *(Gulo gulo)*, Gray wolf *(Canis lupus)*, jaguar *(Panthera onca)*
Birds Whooping crane *(Grus americana)*, Kirtland's warbler *(Dendroica kirtlandii)*, Black-capped vireo *(Vireo atricapillus)*
Others American crocodile *(Crocodylus acutus)*, Shortnose cisco *(Coregonus reighardi)*

DOMESTICATED ANIMALS (originating in region)

turkey *(Meleagris gallopavo)*

WAVES OF IMMIGRATION

Most of the major groups of North American animals share common ancestors with animals in Europe and Asia. This is because many millions of years ago North America and Eurasia were part of a giant northern supercontinent. During the most recent ice age, when the mass of ice locked in the polar regions lowered sea levels, animals such as mastodons and mammoths (now extinct), bison, caribou, moose (known as elk in Europe) and wapiti were able to cross by land from Siberia to Alaska. Animals could probably also reach North America from Eurasia using an intermittent link via Greenland. The Polar bear still travels this route across the sea ice.

The Arctic regions of the Old World and the New World thus have many mammal species in common. The range of some of them extends far south along the Rocky Mountains. Although the Arctic fox and Polar bear are confined to polar regions, the wolverine ranges as far as California; all three species are also found throughout the Soviet Union and northern Europe. Pikas (animals related to rabbits, and resembling Guinea pigs) are diverse and widespread in Siberia; two closely related species of pika are found in the United States. One of these species is confined to Alaska and parts of Canada, while the other's range extends throughout much of the Rockies at altitudes of 2,725–4,000 m (8,000–13,500 ft).

The immigration of Old World animal species continues to the present day. In Alaska several species of birds, such as the Yellow wagtail, bluethroat, wheatear and Arctic warbler, are very recent colonists. More mobile than mammals, these birds can migrate across the Bering Strait, and in winter they still return to Asia, rather than following the Water pipits and most of the Alaskan songbirds south through Canada and the United States. On the other side of the continent, a much greater journey has been undertaken by the Cattle egret, which in recent decades has crossed the Atlantic Ocean from Africa to South America, from where it has colonized parts of Florida.

Tropical invaders

About 3 million years ago a land bridge began to form between North and South America, permitting animals such as the Virginia opossum, armadillo and porcupine to invade the United States. Many bird colonists also arrived from the south. Although there is a tendency to think of migrant songbirds as "belonging" to the wildlife of North America, in reality they spend most of their lives farther south – many species going to the tropical rainforests where ruthless forest exploitation has seriously affected their populations. As conservation awareness spreads, there is a dawning realization that the destruction of the tropical rainforest has a direct bearing on the future of these songbirds.

Tropical songbirds visit their United States breeding grounds for only a few brief summer months to raise a brood or two of young. Around 60 species of warbler breed in the United States, the majority disappearing completely during the winter. Their distribution and migration routes have been modified by the features of the landscape. Many of the eastern species, for example, migrate

down the Florida peninsula and then island-hop across the Caribbean to their final destination in South America; the entire population of one of the United States rarest birds – Kirtland's warbler – spends the winter in the Bahamas.

While many animals are comparatively recent arrivals in the region, others that evolved in North America have subsequently spread far and wide. Perhaps the most familiar among these are the horses, which are believed to have evolved in the Americas, and went on to colonize the Old World by traveling across the northern land bridge, later becoming extinct in the New World. Camels (for example, llamas and vicunas) and tapirs also evolved in the New World, spreading to South America once the southern land bridge had formed.

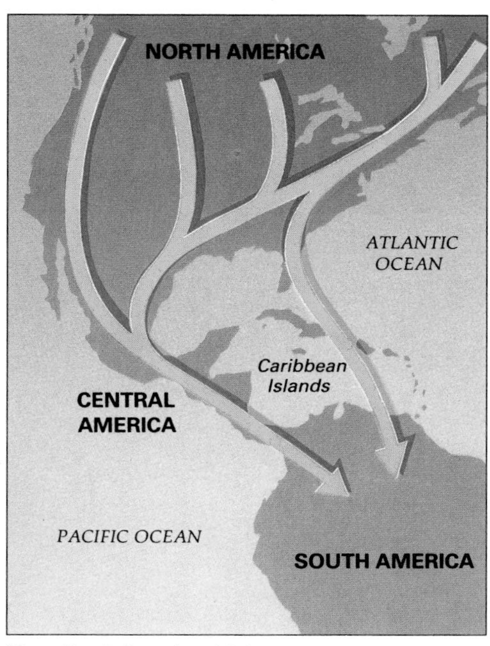

Many North American birds overwinter in South America, returning in spring to breed. They follow four main routes down through North America , and two farther south – one across Central America, the other using the Caribbean Islands as stepping stones.

The Bald eagle (*left*) is a skilled fisherman. Its toes have pronounced lumps on their undersides to help the bird grip its slippery prey. Numbers of Bald eagles have declined in recent years due to shooting and poisoning by pesticides, but breeding programs in conservation areas are helping to reverse this trend.

A versatile predator (*below*) The Mountain lion – or puma or cougar – is equally at home in snowy mountains, rainforests or desert canyons. It is found throughout most of North and South America, and feeds on a wide variety of prey.

THE LIVING DESERTS

The United States' southwest is the heartland of one of the world's great desert regions, extending from Nevada and Utah to Mexico. Parts of these deserts contain surprisingly lush vegetation and a rich animal life; they are classified as deserts only because their annual rainfall is low. The prairies of the Midwest are also relatively arid for most of the year, and their wildlife shares many of the adaptations of desert-dwelling species.

Many mammals of prairies and deserts have large ears, an adaptation that allows heat to be lost through the increased surface area. The Antelope jack rabbit, for example, has ears that are 14–19 cm (5.5–7.5 in) long on a body of only 62 cm (2 ft). With these proportions a human would have 62 cm (2 ft) ears. At night, when the temperature drops considerably, the blood vessels in the ears become constricted, and the ears are laid close to the body to maintain warmth. Kit foxes, several species of bat, the Desert woodrat and a number of other rodent species also have relatively large ears.

Most desert mammals, snakes and many desert invertebrates are nocturnal. They shelter during the day in the cool of a burrow or in crevices between the rocks, where the still air helps them to conserve moisture. Snakes, like the small mammals on which many of them prey, are generally nocturnal, but many lizards are active by day, except during the hottest time after noon. Tortoises are not only often nocturnal, but may enter estivation – the summer equivalent of hibernation – deep in an underground burrow during the hottest months.

Underground specialists

Beneath the surface of the prairies live burrowing rodents such as prairie dogs: ground-dwelling squirrels that live in underground colonies known as "towns". It is said that in 1905 a single prairie dog town in Texas, occupying an area of around 64,000 sq km (25,000 sq mi), contained 400 million prairie dogs. Today, however, these animals, long regarded as pests, have been reduced to a mere 10 percent of their former numbers.

Prairie dog burrows old and new give shelter to other animals. These include snakes, the Burrowing owl and the Black-footed ferret, whose lithe body is adapted to hunting prairie dogs underground. With the loss of its prey the Black-footed ferret has come close to extinction; only a captive breeding operation has saved it.

Saving water

A widespread adaptation found among mammals living in arid areas, particularly rodents and hares, is their ability to metabolize water from their food and live their entire life without drinking. When the food is oxidized inside the body, 1 gram (0.0035 oz) of starch produces about 0.6 grams (0.021 oz) of water, and 1 gram of fat more than 1 gram of water.

Desert birds – whether insectivorous (such as the mockingbirds, thrashers and flickers), birds of prey (for example, the American kestrel, the lizard-killing roadrunner and the hawks) or scavengers (the raven and the Black and Turkey vultures) – generally obtain their fluids from their prey. Liquids are also obtained from the nectar of flowers – for example by hummingbirds – as well as from berries and fruit. Birds can travel in search of water. Although they cannot burrow, desert plants provide them with shade and shelter. The Elf owl, the world's smallest owl, lives in holes in Saguaro cacti.

Many desert mammals spend the day in a cool humid burrow, but large mammals such as Bighorn sheep, coyotes and bobcats cannot avoid the heat so readily. Even when resting in the shade they need to regulate their temperature by sweating or panting; the evaporation causes cooling, but it does use up some water.

Perhaps the most unlikely animals to be found in the desert are amphibians. Spadefoot toads make the most of temporary pools, remaining buried in the mud for two years or more, protected by a hard cocoon of skin. They can withstand a body-weight loss of over 50 percent. When water does arrive, their eggs hatch in just 13 hours, and the tadpoles become frogs 10 days later. Slender salamanders of the Californian deserts manage to do without water for their tadpoles, which develop within the egg and hatch later into miniatures of the adults. In some species the female even broods the eggs to maintain humidity.

Tent caterpillars (*right*) are moth larvae that feed on tree foliage, eventually denuding the tree completely. The young spin a tentlike shelter of silk, from which they venture out to feed. When they are fully grown they spin cocoons in sheltered crevices. The adults emerge in midsummer, mate and lay strips of eggs around the twigs. These will not hatch until the following spring.

FISH IN THE DESERT

A desert is hardly the place one would associate with fish. Yet a remarkable diversity of fish is to be found in the southwestern deserts of the United States. Isolated by the surrounding desert, many species are restricted to a single river, and in some cases even to a single waterhole or hot spring.

This provided the circumstances for new species to evolve, but today many of these desert fish are under threat. The Devil's Hole pupfish, for example, is confined to a single natural well in Death Valley, California, where its only known habitat is threatened by the drilling of boreholes nearby, which will lower the water table.

Elsewhere introduced fish such as bass and mosquitofish prey on the desert fish or compete with them for food. Their extinction would be a great loss as many are so unusual. For example, some of the springs that fish such as the Desert dace inhabit are at temperatures of over 50°C (122°F), almost hot enough to boil an egg.

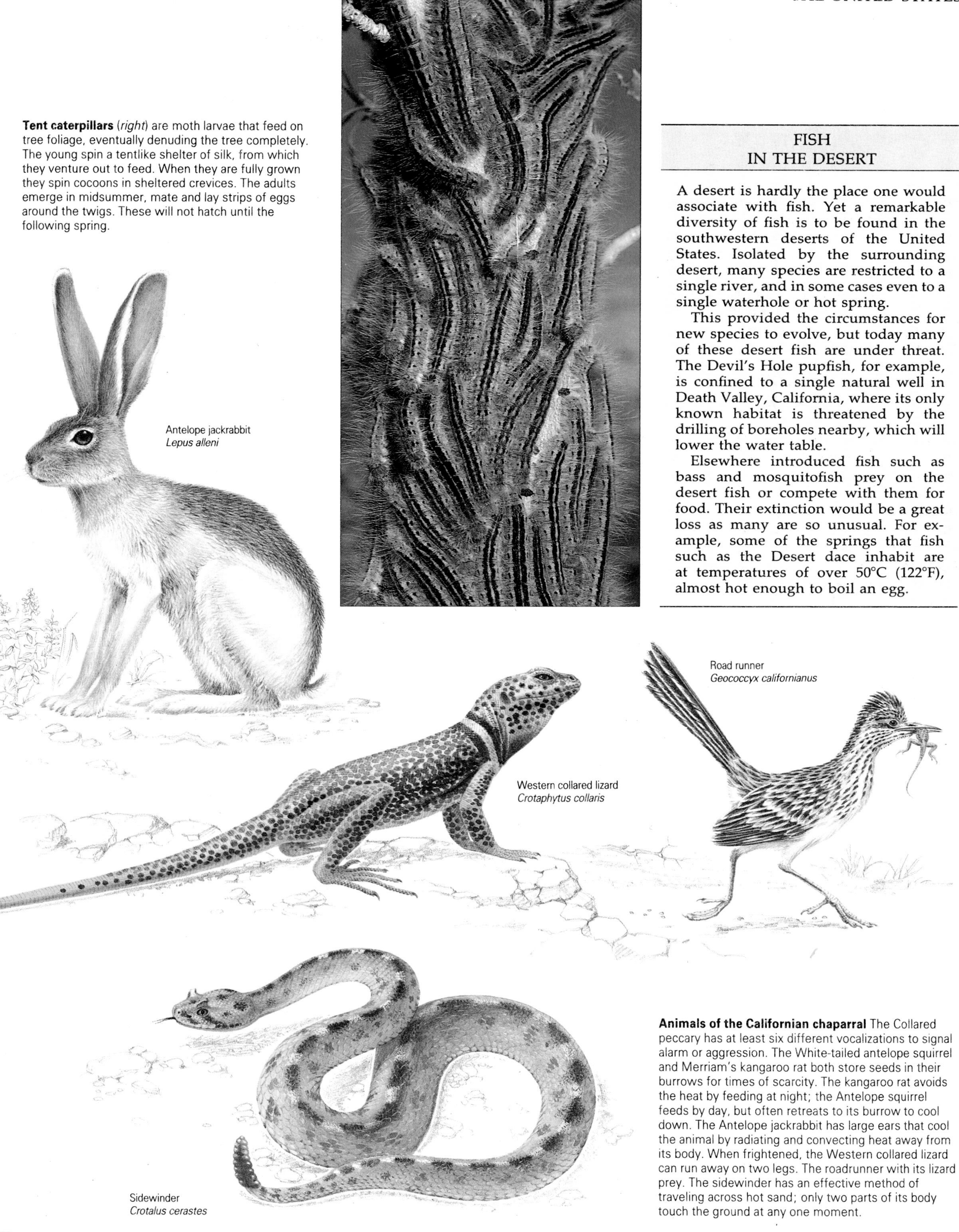

Animals of the Californian chaparral The Collared peccary has at least six different vocalizations to signal alarm or aggression. The White-tailed antelope squirrel and Merriam's kangaroo rat both store seeds in their burrows for times of scarcity. The kangaroo rat avoids the heat by feeding at night; the Antelope squirrel feeds by day, but often retreats to its burrow to cool down. The Antelope jackrabbit has large ears that cool the animal by radiating and convecting heat away from its body. When frightened, the Western collared lizard can run away on two legs. The roadrunner with its lizard prey. The sidewinder has an effective method of traveling across hot sand; only two parts of its body touch the ground at any one moment.

ADAPTING TO FIRE

Many of the animals and plants of the United States are well adapted to the fires that, ignited by the lightning strikes of late summer storms, regularly swept over the prairies until controlled by human intervention. Elsewhere forest fires could take hold and rage for weeks or months.

Birds have the advantage of being able to fly to avoid fire, but mammals (with the exception of bats), amphibians and reptiles have to adopt different methods. Plains mammals tend to be extremely fleet of foot so they can run ahead of the fire. One of the fastest land mammals known, the Pronghorn antelope, can sprint at 115 kph (70 mph) and cruise at 48 kph (30 mph). The Swift fox can reach 40 kph (25 mph), and the coyote 65 kph (40 mph).

Beneath the prairie

The burrowing mammals of the plains are safe beneath the ground provided the fire does not last too long and is not too intense. During the drier part of the year – when the risk of wildfires is intensified, evaporation rates are greater and food is scarce – many reptile and amphibian species estivate. Gopher tortoises dig a tunnel up to 9 m (30 ft) long and 5.5 m (18 ft) deep, ending in a small chamber, and these burrows provide shelter for over 40 other animals, including the Burrowing owl, Gray fox, Diamondback rattlesnake, Indigo snake and opossum, and various toads and rodents.

Franklin's ground squirrel and several related species hibernate as soon as they have accumulated sufficient reserves of fat, sometimes as early as the end of July. They may sleep for eight months, oblivious to fires raging above.

Among the most remarkable adaptations to life in arid grasslands is the ability of females to breed without mating (parthenogenesis). Although common in insects and many other invertebrates, it is rare in vertebrates. It does, however, occur in some species of whiptail lizard from arid habitats. The widespread Chihuahuan spotted whiptail is exclusively parthenogenetic – males are unknown – but the Checkered whiptail, another parthenogenetic species, does occasionally produce males.

The benefit of this extraordinary behavior is that it allows a population to build up rapidly – an obvious advantage where an environment and a population are periodically devastated by disasters such as wildfires. For a parthenogenetic species, only a single female needs to survive for the population to reestablish itself. All her offspring will also be females; and even if each female can lay just six eggs, then theoretically the fifth generation of the species could have grown to an impressive 7,500 individuals (whereas that of a normal lizard would be fewer than 500 individuals).

Plains pocket gopher
Geomys bursarius

Although there are no parthenogenetic rodents, some are extremely fecund and can increase in number fast. The Prairie vole, which normally lives for less than a year, can produce several litters of up to seven young in its short lifetime.

People and fires

The indigenous Amerindians learned to harness fire as an aid for hunting, using it to drive herds of bison and other game. The European colonists who succeeded them as farmers on the plains suppressed the natural fires, with devastating effect on the landscape and wildlife. Wildfires prevent a litter of dead leaves and twigs from accumulating. The fires sweep through the grasslands quickly, and most animals are able to escape underground. However, when a wildfire erupts after years of suppression, the accumulated layers of litter act as tinder, making the fire burn much more intensely causing great damage beneath the ground.

In recent years it has become apparent to the managers of national parks, wildlife refuges and other nature reserves that controlled fires are often essential to maintain the natural environment.

Pocket gophers (*above*) A Plains pocket gopher returning from a foraging expedition with its cheek pouches full of food. A Valley pocket gopher showing in detail the cheeks that give pocket gophers their name. A male Valley pocket gopher of a different color making a mound of soil. A second color variant of the Valley pocket gopher, a female, somewhat smaller than the male. Pocket gophers are specialized for digging: they have a large skull and a very short neck, stocky and powerful legs and a short tail. As in moles, the skin is loose, allowing the gophers to turn around in tight spaces. Most species use their large claws for digging, aided by their incisors. The incisors and claws have a fast growth rate to compensate for excessive wear.

Prairie grazer (*above*) Prairie dogs live in social groups called "coteries". They feed in the immediate vicinity of their burrows, removing tall plants to obtain a better view. They often stand on their haunches, especially when on guard duty at the burrow entrance. If danger is sighted, a loud whistle alerts all the other animals in the surrounding area.

The mighty American bison (*left*) may weigh over 900 kg (1,985 lb); an adult bull stands about 2 m (6.5 ft) tall at the shoulder. In winter the bison grows a thick coat of woolly hair for insulation. In wooded areas, Bison often graze beneath the trees as snow is not so deep here and it is easier to reach the grass underneath.

THE BISON AND THE PRAIRIE DOG

Until a comparatively short time ago animals lived on the prairies in huge numbers. Before the arrival of European settlers, bison numbered more than 30–70 million, while prairie dogs, living in their underground "towns", were even more numerous.

Bison and prairie dogs were often found living in close proximity to each other. Prairie dogs preferred to site their colonies in areas grazed by bison, where the grasses were shorter, while the bison liked to take their dust-baths near the prairie dog burrows, where the grass had been cropped even closer.

As settlers moved into the prairies they destroyed vast numbers of both animals because they competed for grazing with their newly introduced cattle: 25 prairie dogs can eat as much grass as a cow in a day. The slow, lumbering bison were reduced to fewer than 1,000 individuals in just 30 years, saved from extinction only by the efforts of a few conservationists. The prairie dogs were smaller and less obvious targets. With their rapid breeding rate they were able to recover more quickly, but neither species will ever reach its former numbers, as most of the prairies have now been plowed up and the grassland habitats destroyed.

Hawaii's vulnerable birds

The Nene goose (*left*) is found in the wild only on Hawaii. Fewer than 50 individuals remained by 1950, but a captive breeding program in Great Britain, combined with the protection of its remaining habitat, has boosted the population to a safer level.

Hawaiian finches (*right*) These colorful birds occupy the forests and shrublands of the islands, feeding on a wide range of insects, fruit and nectar; the different species have adapted to a wide variety of feeding niches. The Grosbeak finch is a seed eater. The Kauai akialoa probes for insects in thick mosses or deep cracks. The akiapolaau uses its lower mandible to chip wood and its upper mandible to probe for insects. The ula-ai-hawana eats fruit and seeds. The Maui parrotbill uses its lower bill to chisel into branches and its upper bill to extract beetles and their grubs. The Crested honeycreeper is an insect and nectar eater. The iiwi has a bill shaped for reaching inside a flower and a tubular tongue for sipping nectar.

When the British explorer Captain James Cook (1728–1779) visited the Hawaiian Islands (which he named the Sandwich Islands) in 1778, the naturalists accompanying his epoch-making voyage of scientific discovery encountered a dazzling array of birds that had never been seen before by Europeans. They also observed a unique range of plants, insects, mollusks and other wildlife.

The volcanic Hawaiian Islands – relatively new in geological terms (some are less than a million years old) – provide a striking example of the way a few colonizing species, carried to the islands by freak winds or on sea currents, have adapted without competition from other species to fill a wide range of ecological niches. They did this by becoming specialist feeders, or developing some particular type of behavior. Scientists call this process adaptive radiation.

The Hawaiian Islands consequently contain a remarkably high number of endemic species – those that are found nowhere else in the world. They are known to possess more than 4,000 endemic insects and 1,000 endemic snails, as well as freshwater fish and shrimps, and thousands of plants. It is Hawaii's endemic birds, however, that have most fascinated scientists, who have traced among them the same pattern of species evolution that Charles Darwin (1809–1882) discovered in his famous study of the finches of the Galapagos islands, off the coast of South America.

The 70 or so known native Hawaiian land birds all originated from as few as 15 ancestral forms. Of these the most spectacular are the Hawaiian finches, also known as the Hawaiian honeycreepers and honeyeaters. The 28 species of honeycreepers, which all evolved from a single finchlike species from 3,000 km away, demonstrate more clearly than any other group in the world the process of adaptive radiation. Some feed on nectar, others on insects, yet others on snails, fruit and seedpods, and their beaks have developed different forms to exploit these food sources.

Species in decline

The honeycreepers and honeyeaters also demonstrate the susceptibility of isolated island populations to any form of interference: many are now extinct. They have been exterminated by a combination of factors – destruction of habitat, introduced disease and predation – stemming from human occupation of the islands. The first people to colonize the islands reached them from Polynesia about 1,500 years ago. They started to clear the forests for agriculture, depriving many specialized animals of their habitat, so that several species had already disappeared before the Europeans arrived.

The process has been immeasurably speeded up since then – by the 1950s less

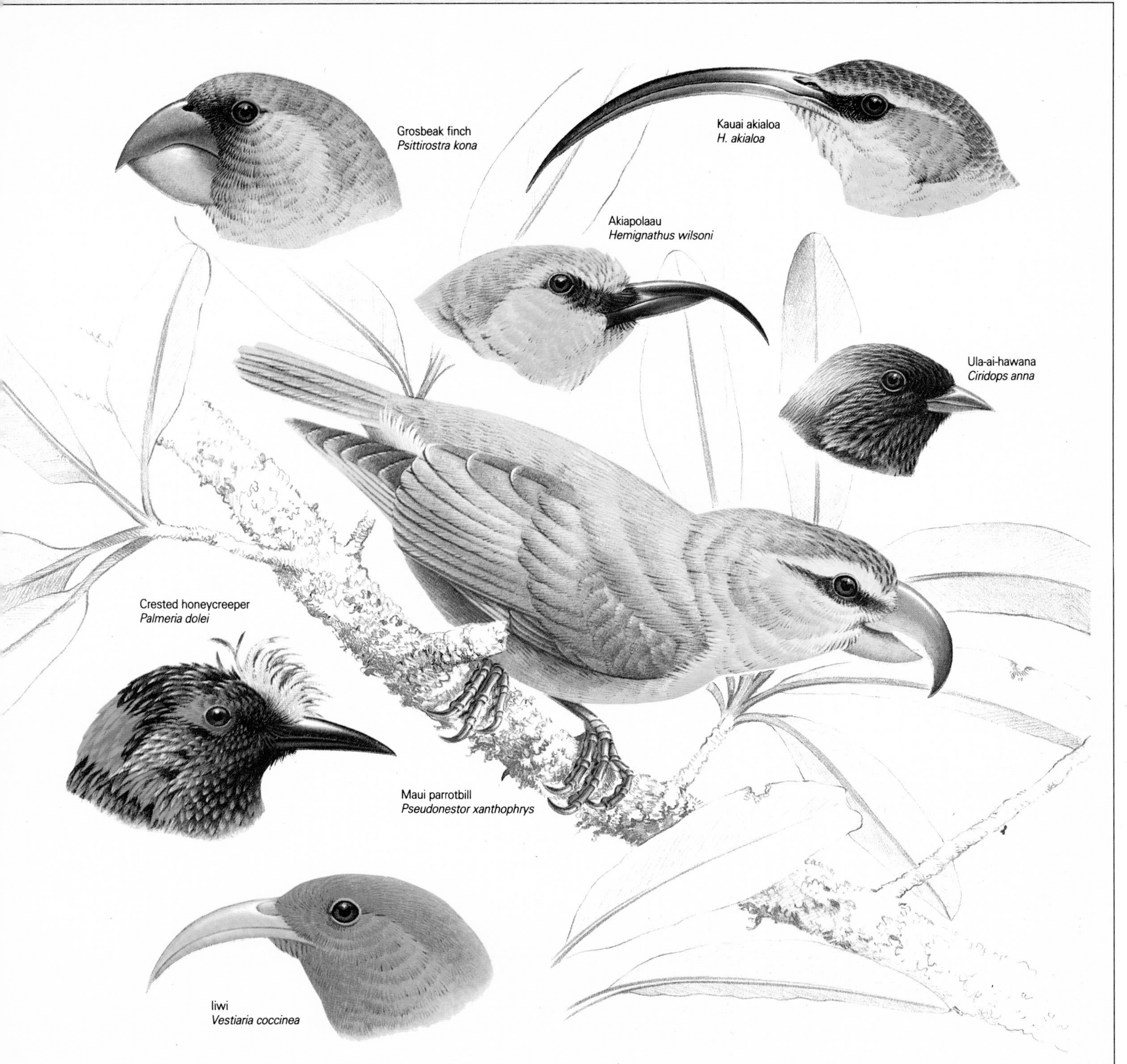

than 25 percent of the original forest cover survived, and the dwindling populations of honeycreepers and honeyeaters had been confined to areas at altitudes of over 600 m (2,000 ft), at which height mosquitoes cannot survive. These alien insects first colonized the island of Maui in 1826 and spread to other islands. Some are carriers of a form of malaria that affects birds, and the disease is known to have become established in the islands early this century. Most of the endemic birds had no resistance to it, setting off a wave of extinctions about 50 years ago.

Diseases such as bumblefoot (avian pox) and parasites such as worms also had significant effects on bird populations. A few species were probably exterminated by human hunters, particularly those species on the brink of extinction that became sought after by collectors. Cats also took their toll, and finally there was competition for food sources from non-native birds such as Pekin robins, now abundant and widespread, which had greater resistance to disease.

The process of extinction continues today. The akioloa became extinct on Kauai in about 1965 and the alauwahio died out on Molokai in 1970. The Kauai o-o was reduced to a single population by the late 1970s, though a single population of Bishop's o-o, thought to have become extinct in 1904, was rediscovered in 1981. Even species such as the Maui parrotbill and the ou, which may number a few hundred, are thought to be declining. It is difficult not to be pessimistic about the ultimate fate of the few survivors of Hawaii's unique bird life, but even harder to suggest any effective solution to their decline.

WETLAND WILDLIFE

The wetlands of the United States support a wide range of wildlife. The waterlogged summer tundra of Alaska and the water-filled depressions – "potholes" – of the northern prairies provide the breeding grounds for most of the country's ducks. Farther south the lakes, pools and channels of the Mississippi floodplain are inhabited by otters, mink, muskrats and numerous populations of fish-eating and insect-eating birds. Alligators, snakes, turtles, raccoons and waterfowl thrive in the warm, humid swamps of the Mississippi delta. To the southeast the Everglades and the mangrove-fringed Florida coast are a subtropical animal paradise; they are the only remaining North American stronghold of the American crocodile, though the American alligator is much more widespread.

Landscape architects

When the first Europeans reached North America, beavers were found in nearly every lake and river, but the early settlers ruthlessly hunted them for their skins, which were used to make the beaver hats then fashionable in Europe. Now protected by law, they are returning to much of their former range. The beaver is a natural wetland engineer and one of the very few animals capable of modifying the landscape on a large scale to suit its needs. By felling trees and using these to build dams it creates a swampy environment to promote the growth of the vegetation on which it feeds.

The beaver is superbly adapted to its environment: its dense waterproof fur insulates it and keeps it dry; its large feet are webbed; it has a large, flat paddlelike tail; and its well-developed incisor teeth can fell small trees easily. It can digest bark, in addition to more tender plant material, and in areas where the ponds freeze over in winter it gathers together a food supply of twigs and branches and stockpiles them underwater.

The United States has a wide range of freshwater turtles (known in the Old World as terrapins). Largest is the Alligator snapping turtle, which can weigh over 90 kg (198 lb). It has a red, wormlike tongue with which it lures its prey. It waits motionless in the water with its mouth open until a fish or other animal falls for the bait, whereupon its jaws snap

The Red-eared turtle (*above*) is a carnivore when young. It switches to an omnivorous diet as it matures, and may even become largely vegetarian in old age. It rarely leaves the water, except to nest or to migrate to another pond.

Predator and protector (*right*) The American alligator unwittingly protects some of the other animals that share its habitat. Smaller predators are deterred by its presence, leaving birds to nest and turtle eggs to hatch in relative safety.

A deceitful display (*below*) This Northern red salamander, though not poisonous itself, is mimicking the highly poisonous Red-spotted newt whose skin secretions are lethal to predators. Both amphibians live in the same habitat.

A KEYSTONE SPECIES

Florida's warm subtropical mangrove swamps support many thousands of birds, fish, turtles, snakes and deer. These are prey for many species of predator, including the bobcat and the Florida panther. Preying on them all is the American alligator.

America's largest reptile, the alligator grows to a length of over 5.8 m (19 ft). Its range extends through the swamps north to North Carolina and Texas, and it plays an essential role in maintaining the wetland habitat. Winding through the pools, its trails keep tracts of open water free of invading vegetation, clearing routeways that are exploited by the other animals of the swamps.

They affect the habitat in other ways too. In times of drought they dig deep holes ("gator holes") that fill with water or mud. These are vital refuges for other aquatic species. Fertilized by the alligator's droppings, the holes become rich marshes in miniature, filled with lilies and other aquatic plants. The bushes that spring up around them provide shelter for nesting birds. Herons and egrets nest near the sunbathing spots favored by the alligators – the presence of these larger predators acts as an effective deterrent against unwanted visitors.

An animal such as the alligator is often referred to as a keystone species. Robbed of its presence, there would be significant changes in the habitats, with important consequences for the local wildlife at every level.

shut. Algae grow on its knobbly shell so that it looks like a submerged rock, helping to disguise it from its prey.

Turtles are well adapted for an underwater life. Their shells are flattened and streamlined for swimming, and their feet are partly webbed. Sea turtles' limbs are flattened for use as paddles. Some freshwater turtles are flat enough to submerge themselves in the mud. But in swamps such as the Everglades, where alligators and other powerful predators abound, species such as the Florida redbelly turtle have abandoned the advantage of streamlining and evolved tall domed and buttressed shells for extra protection.

Fish and their predators

There are a number of living fossils among the fish that live in these swamps. The slow-swimming bowfin and the Alligator gar – which can reach 3 m (10 ft) in length and may weigh 136 kg (300 lb) – belong to an ancient and almost extinct family. The gars cruise along sedately, resembling drifting logs. These fish can live in stagnant water, using their swimbladders as lungs. The bowfin can survive for up to a day out of water.

Sturgeon, descendants of fish that lived 135 million years ago, are also found in the United States. The White sturgeon is North America's largest freshwater fish, attaining a length of up to 6 m (20 ft). It is a bottom feeder, probing the mud with the sensitive feelers on its chin. Perhaps the most bizarre of these ancient fish is the paddlefish, which has a huge shovel-shaped snout. It feeds by night, swimming near the surface with its huge mouth open to scoop up plankton and other small invertebrates.

Large numbers of waterfowl and wading birds can live in the same wetland areas because their sources of food and feeding techniques differ. Shallower waters support large wading birds such as the herons, which stalk and ambush their prey. The Snowy egret stirs up the water with one foot while in flight to flush out its prey, while the Reddish egret catches fish with its wings spread to cast a shadow over the water, attracting the prey and removing reflections.

It does not pay to be too much of a specialist, however. The Snail kite of the Everglades has a finely curved bill ideally suited to feeding on the Apple snail, its only prey. The snail was once very common in these swamps, but the drainage of wetlands has left the kite desperately short of food; it does not seem to have adapted to any other diet.

ANIMALS OF THE SEAS

The marine life found off the coasts of the United States is as diverse as any in the world. Polar bears and walruses hunt among the Alaskan ice floes, and seals breed on the sea ice. The very dense kelp beds of the Alaskan and southwest coast, rich in fish and shellfish, are hunting grounds for the Sea otter. Sea lions, elephant seals and rarities such as the Hawaiian monk seal and the Guadalupe fur seal inhabit the beaches, and farther out in the Pacific are various species of great whale: Finback, Minke, Bowhead, Humpback, Gray, Killer and Right whales and the giant Blue whale. The mangrove swamps and coral reefs of Florida and the Gulf coast support very different animals, including the strange manatee and several species of sea turtle. Seabirds abound all along the coastlines.

Tidal rhythms
In spring and summer, during the highest spring tides, people are drawn from far and wide to witness a spectacle that takes place on the Californian coast. This is when grunions, small, silvery fish, come ashore to spawn. On the second night after full moon, just after the tides reach their highest point, the moonlit beaches are covered with a glistening mass of writhing fish. Using the waves to help them, they make their way to the highest part of the beaches; there the females burrow in the wet sand, and shed their eggs, which are fertilized by the males. Hidden under the sand, the eggs develop out of sight of predators and out of reach of the waves, until the next spring tide washes them out. Then the eggs hatch simultaneously, and the young are swept into the surf. A possible reason for this unusual behavior among fish is that the eggs are safeguarded from a multitude of ocean-swimming predators.

A similar phenomenon can be seen on the Atlantic coast: the spawning of the Horseshoe crab (sometimes known as the King crab). This behavior is of ancient origin, for these living fossils date back 300 million years; they are not crabs at all – their nearest relatives are probably spiders. The signals these animals use to time their shoreward migrations are not known for certain. They probably detect the phases of the moon, perhaps using them to synchronize an internal "tidal clock", a mechanism that is quite common among marine invertebrates.

Although they spend long periods at sea, most marine mammals (except the whales and dolphins) are compelled to come ashore to give birth; the fur of the

THE SKILLFUL SEA OTTER

The Sea otter, whose range extends up the west coast from California, is renowned for its dexterity. It picks up a flat stone from the sea bed, balancing it on its chest as it lies back in the water, and uses it as an anvil on which to crack open a sea urchin or an abalone. This appealing animal was hunted close to extinction for its pelt, but it is now protected.

Recovered to some 1,600 individuals, the Californian population of Sea otter now faces new threats from coastal construction and pollution, and from fishermen, many of whom regard it as a pest, competing with them for dwindling fish stocks. However, there is growing evidence that the Sea otter helps to keep in check the sea urchins that would otherwise devastate the kelp beds that fish populations depend on and use as a nursery for their young. Giant kelp farming is a valuable industry, and controlling the sea urchins by other means can be expensive.

newborn mammal is not waterproof, and it does not yet have sufficient blubber to retain heat in the water. Most true seals breed on the sea ice, but the eared seals – the sea lions and fur seals – and elephant seals – breed on rocky shores. They are naturally gregarious, living in vast breeding congregations, though human disturbance has reduced the number of secure shores where the seals can safely establish themselves. They disperse over wide areas at sea, but always return to the same colonies to breed. The huge bulls fight each other in fierce battles for the best territory on the shore, which gives them access to the most females. Within a few days of giving birth the females are ready to mate again.

Migrations at sea

The breeding places are not necessarily in areas rich in food, and many seals migrate long distances between feeding and breeding grounds. The Northern fur seal that breeds off the Alaskan coast has one of the longest seal migrations. The males are much larger than the females, and their reduced surface area to volume ratio means that they are better able to conserve heat. The males remain in the far north to feed, but the females migrate up to 2,800 km (1,700 mi) to warmer southern waters to feed in winter.

The great whales, which spend all their lives in water, make even longer journeys, traveling to the far north to take advantage of the great swarms of plankton and other invertebrates that flourish there in summer, then returning to subtropical or tropical waters in winter. The Humpback whale of the North Pacific migrates up to 16,000 km (10,000 mi) to the Indian Ocean or Mexican waters in winter, and the Gray whale, which breeds in the coastal lagoons of Baja California in Mexico, migrates to the Arctic Ocean to take advantage of the plentiful plankton.

Whales make similar journeys in the Atlantic Ocean. Coastal fish and invertebrates make lesser migrations. Every fall off the coast of Florida, up to 100,000 Spiny lobsters leave for deeper water, moving in single-file columns some 60 lobsters long. Keeping touch with their antennae, they travel up to 50 km (30 mi), perhaps stimulated by the sharp drop in temperature that accompanies the first fall storm. Although food is less abundant in the deeper water, they escape the harsher conditions of the coastal winter.

Alaskan fur seal colony (*above*) Pups play together in the nursery, while their parents bask on the beach or fish out at sea. The gathering of huge colonies of fur seals on the breeding beaches made them highly vulnerable to sealers, but with careful management the stocks are now recovering.

A leisurely meal (*left*) Apart from primates the Sea otter is the only mammal reported to use a tool while foraging.This California sea otter is balancing a large stone on its belly, which it uses to crack open a clam. Half of its diet is sea urchins, but it is also partial to abalones and fish. The otter has wrapped itself in kelp fronds to prevent it from being washed out to sea.

Brown pelicans (*right*) breed in summer all around the southern shores of the United States. Both parents share the task of bringing up the young; the pelican may have to fly 240 km (150 mi) or more to fish, so a single parent would have to abandon the chicks for long periods. The chicks grow rapidly and the parents need to supply large quantities of food.

THE HUMAN IMPACT

About 11,000 years ago, as the last of the ice sheets retreated, many large mammals became extinct in North America. This huge wave of extinction took place during the same period as the *atl-atl*, or spear thrower, became widely used by the early hunters of the region. Within perhaps just 2,000 years camels, horses, Giant bison, Giant beaver, ground sloths, peccaries, Four-horned antelope, mastodons and mammoths disappeared over much of North America. This process had drawn to a close by about 8,000 years ago, and the species that remained were basically the same as those that were found by the first European settlers.

Exploiting the wildlife

The colonizers who opened up the New World to economic activity exploited its rich natural resources. Many animal species were hunted for their skins and furs, which were exported in their millions to Europe. Beavers were the chief quarry of the trappers, but the pelts of otters, martens, fishers, wolves, lynx, bobcats, weasels and deer were all also in demand. Some species, such as the Sea mink, were hunted to extinction.

The Trumpeter swan, whose feathers were used to trim garments and line ladies' muffs, also came perilously close to extinction, but it was saved in the Yellowstone and Red Rock Lakes refuge in Montana in the northwest. Commercial hunters who slaughtered ducks and geese in their thousands were probably responsible for the extermination of the Labrador duck, which had gone by 1875, and of the Eskimo curlew, ruthlessly hunted as the birds migrated south in tens of thousands from their Arctic breeding grounds. Other species that have been completely or nearly wiped out as a result of human persecution include the Great auk, Passenger pigeon, Carolina parakeet, Ivory-billed woodpecker and California condor. The list of species that have become extinct in historical times on the Hawaiian Islands is even longer.

Virtually all the rivers of the Atlantic and Pacific coasts once teemed with fish. Now, however, this abundant resource has been destroyed. The combined effects of overfishing by European settlers, and pollution from their burgeoning industrial developments, wrought havoc among fish populations. Salmon and sturgeon were among the species that were most severely depleted.

By the end of the 19th century attitudes were changing. Wilderness and wildlife were coming to be recognized as having intrinsic value, and the modern conservation movement was born. In the United States hunting for sport is today a multibillion-dollar industry and a major force in preserving large areas of the countryside from development.

New arrivals

Many non-native (exotic) animals were introduced to the United States by early European settlers. Some were escaped domestic animals; herds of semiwild mustangs descended from horses belonging to the Spanish became established in the prairies, and wild donkeys (burros) are found in more arid areas. Rats and mice crossed from Europe aboard ships, but other species such as domesticated pigeons were deliberately introduced – often with devastating effects. Homesick Europeans brought over their native goldfinches, skylarks, House sparrows and starlings. Wild boar were introduced into the Great Smoky Mountains from

Urban aliens The Common raccoon scavenges in garbage dumps and campsites throughout most of the United States. The House mouse and the Common rat have followed human travelers around the world, concealed in cargo and slipping ashore from ships. The ever-increasing amount of waste from cities and towns provides food for growing populations of the Herring gull and the Black-headed gull.

THE DEMISE OF THE CALIFORNIA CONDOR

Grasslands and semiarid scrublands are ideal places for airborne scavengers, particularly vultures. The lack of cover makes it easy to spot a carcass from a long distance. The California condor – among the largest of flying birds with a wing span of up to 3 m (10 ft) – once ranged the length of the west coast and across to Florida, but it is now almost extinct and confined to California.

The main causes of its demise are thought to be lead poisoning from the presence of hunters' shot in carcasses, and the accumulation of toxins from feeding on the carcasses of coyotes poisoned by farmers. Pesticides, which lead to the formation of thin eggshells, have also had a dramatic effect on breeding success. Condors rear only one young every other year, and do not breed until they are six years old. This slow reproductive rate is compensated for by a long lifespan. Despite a nationwide conservation campaign, the last female condor in the wild died of lead poisoning in 1986.

The California condor's struggle for survival was dealt a further blow by a prolonged controversy over whether or not to bring the last three remaining birds – all males – into captivity. For some years eggs had been taken from the wild and hand-reared: the female would lay another egg, so more young could be reared. In April 1988 the first fully captive-bred condor hatched, so the species may yet be saved.

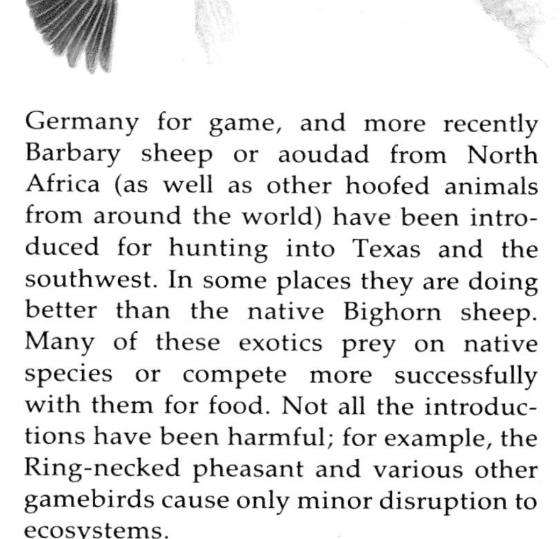

City birds (*right*) Feral pigeons, descended from the wild Rock pigeon, despoil buildings and sidewalks with their droppings. The House sparrow and Common starling were introduced to North America by European settlers. The House sparrow is a grain feeder, while the starling feeds on invertebrates, finding its food in suburban gardens, but roosting communally in huge numbers in city centers.

Germany for game, and more recently Barbary sheep or aoudad from North Africa (as well as other hoofed animals from around the world) have been introduced for hunting into Texas and the southwest. In some places they are doing better than the native Bighorn sheep. Many of these exotics prey on native species or compete more successfully with them for food. Not all the introductions have been harmful; for example, the Ring-necked pheasant and various other gamebirds cause only minor disruption to ecosystems.

In Florida, perhaps more than anywhere else, exotic species have caused the greatest damage to native wildlife. By the mid-1980s there were 23 species of exotic reptiles breeding here, including the Spectacled caiman from South America. Originally brought in as pets, they either escaped or were deliberately released into the wild. The flourishing trade in tropical fish and exotic birds means that the problem continues.

In the last 25 years many states have devised imaginative programs to reintroduce native species that have suffered as a result of these new arrivals. Peregrines, Prairie falcons and Bald eagles have all been bred in captivity and successfully released into the wild. Similar programs have reestablished wild turkeys in much of the eastern United States, and beavers are once again widespread throughout most of their former range. But the animals of the United States are still under threat from less benign human interference: ever-encroaching urbanization and the drainage and pollution of wetland habitats.

Mustangs in the wild (*left*) live in bands of between 3 and 20 animals. Horses are thought to have evolved in the Americas. After colonizing the Old World, they became extinct in the New World and were later introduced by Europeans.

Giant bears of North America

Grizzly bears once roamed almost all of North America from the Mississippi river westward, and the length of the continent from Mexico to Alaska. Although there were many other sightings of the animals by explorers, who brought back tales of their ferocity that gave them a reputation that survives to this day, it was not until the early decades of the 19th century that the first scientific studies of the species were made. This coincided with the intensification of the hunter's onslaught against the bear, and the development of the high-powered rifle brought about their rapid demise.

The Grizzly bear is still the official state animal of California, but has long vanished from there. Today grizzlies survive in only a few remote places in the north of the continent. A tiny remnant of fewer than 900 bears lives in Montana; Western Canada has more than 10,000; but the last stronghold of the species is Alaska, where it is thought there may be as many as 40,000 individuals.

The grizzly is generally considered to belong to the same species as the Brown bear of Eurasia. In the past there were believed to be many separate species of North American grizzly, and there is great variation between local populations. The largest bears, which can reach up to 545 kg (1,200 lb) in weight, making them the largest living land carnivores, are found on the Alaska Peninsula and near Kodiak Island. Male grizzlies are normally about 400 kg (900 lb) in weight.

Grizzly bears are solitary animals. Except for mating pairs or mothers with cubs, they usually avoid one another, and may fight if their paths cross. They hibernate during the winter months in dens, which may be a cave, a hollow tree, or a

A resting grizzly (*right*) may look placid and harmless, but unwanted intrusions by humans can provoke disastrous conflicts. Even in national parks, some bears have had to be shot after attacking tourists who have attempted to feed them.

Salmon fishing (*below*) The fall run of salmon as they head toward their spawning grounds in Alaskan rivers provides the bears with welcome nourishment before their long winter sleep. Bears will frequently fight over the best places to fish.

LIFE IN THE SLOW LANE

Sloths are the commonest large mammals in the Central American rainforests, though they are seldom seen. Their lazy existence is a fascinating adaptation to a life spent feeding on the forest's least nutritious foliage – a diet for which they encounter very little competition. Like ruminant mammals such as cattle or deer, sloths have continuously growing teeth, a slow digestive system and a many-chambered stomach that houses bacteria to help them digest plant cellulose.

Sloths move extremely sluggishly, and sleep for up to 19 hours a day. Their slow habits conserve energy, thus minimizing food requirements. Unlike most mammals, the sloth has a body temperature that fluctuates markedly, dropping at night and in wet weather, so reducing the need to generate heat. Its thick fur, together with a complex arrangement of blood vessels rather similar to those of marine mammals, also helps to conserve heat.

The sloth's long hair is a camouflage mixture of black, yellow and white. Algae live in grooves in the hairs, turning yellow in periods of drought and green in damp weather, further enhancing the camouflage.

The animals spend most of their time hanging upside down. This posture misleads predators looking out for the typical upright outline of a four-footed animal on a branch. The sloth descends to the ground only to defecate, about every six days. Its powerful grip persists even in death; the long, curved claws remain firmly clasped around the branch until its body rots. This deters many local people from hunting it.

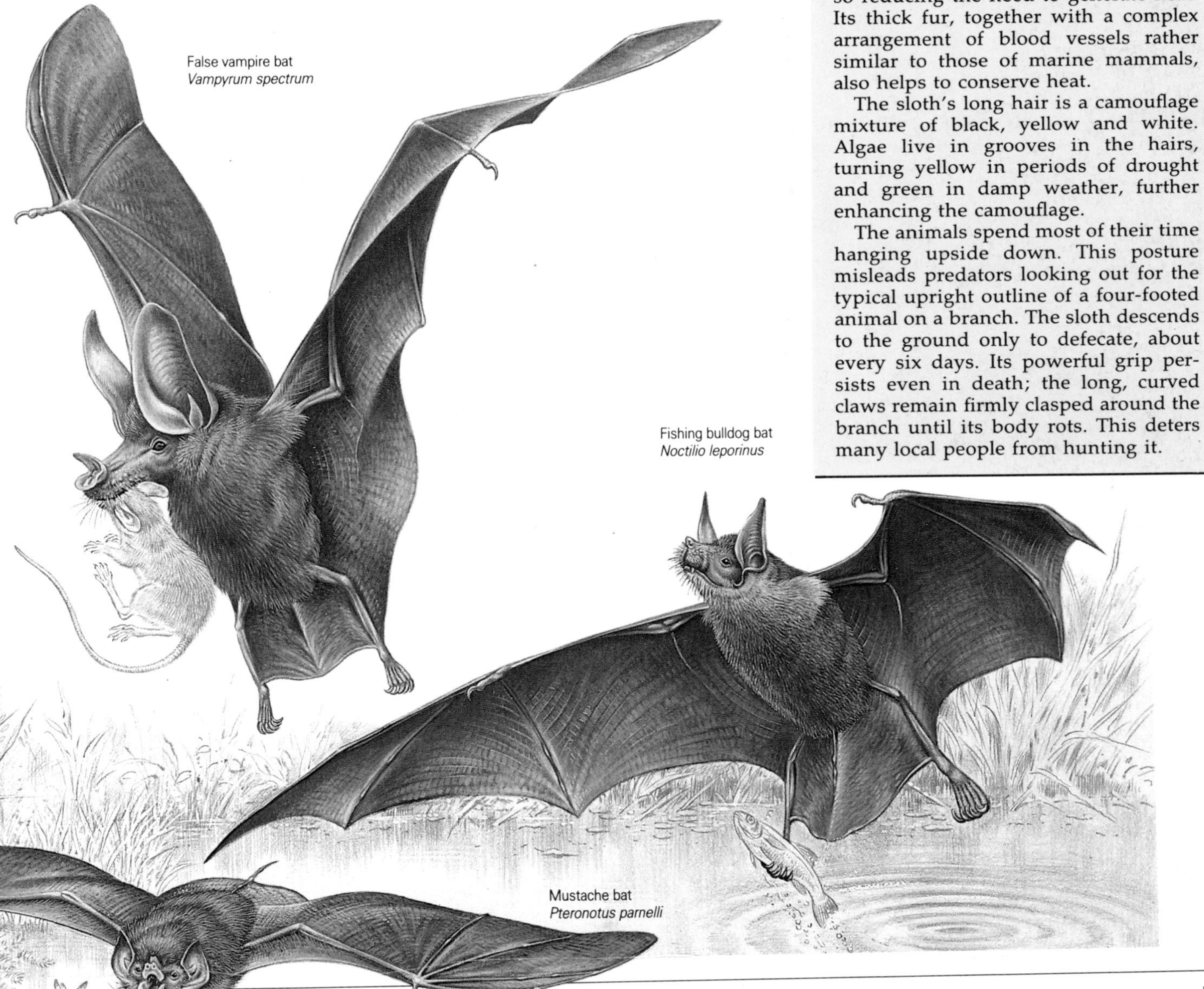

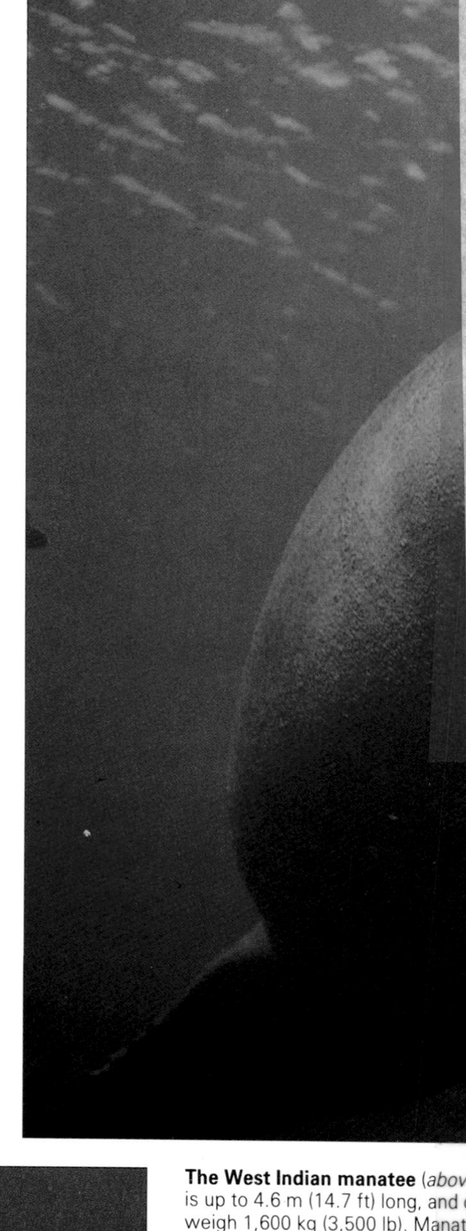

HABITAT LOSS AND CONSERVATION

Like many tropical regions, Central America has a rapidly increasing human population. The need to plant crops and rear livestock inevitably destroys natural habitats. However, protected areas have been established, and more are planned. Some are specifically designed to protect particular species, such as the jaguar in Belize and the quetzal (a bird with iridescent plumage) in Guatemala.

Most of the Caribbean islands are fringed by mangrove swamps and coral reefs rich in marine life. Belize on the mainland has the second largest coral reef in the world. Reefs are mined for building materials, and their fish and shellfish hunted for sport and food.

Although Caribbean governments have brought in conservation legislation and created marine nature reserves, the rapid growth of tourism, especially of diving and watersports, has brought new threats to the reefs. Shell- and coral-collecting for souvenirs damages fragile corals that may have taken hundreds of years to grow. Manatees (herbivorous sea mammals) are injured by boat propellers as they graze in the shallow sea grass beds. Tourism has created fewer problems on the Pacific coast of Baja California.

Conservation efforts have been made to protect the many sea turtles – Loggerhead, Hawksbill, Green, Ridley and Leatherback – that visit the beaches of the Caribbean. Local people help to guard beaches against egg collectors; the eggs are removed to safe incubators, and the hatchlings later returned to the sea. This achieves a much higher survival rate than for those young turtles that are allowed to hatch in the wild.

Many Central American species are threatened by the trade in animals and animal skins. Although it is banned by most countries in the region, the illegal export of the pelts of ocelot, margay and other small cats continues. The pet trade damages not only wild birds such as parrots, but also the Indigo snake and the Red-kneed tarantula. Particularly at risk are the colorful parrots of the genus *Amazona*, some species of which are confined to individual Caribbean islands. Fourteen species are already extinct.

Farming the forest

The loss of their forest habitats threatens hundreds of species. An attempt is being made to save the iguanas and their habitat by developing a form of ranching. The meat of iguanas – fast-growing tree lizards up to 2 m (6.5 ft) long – is valued for eating. Today only about 5 percent of the original iguana population remains in the rainforests. A single hectare (2.5 acres) of cleared land provides poor grazing for

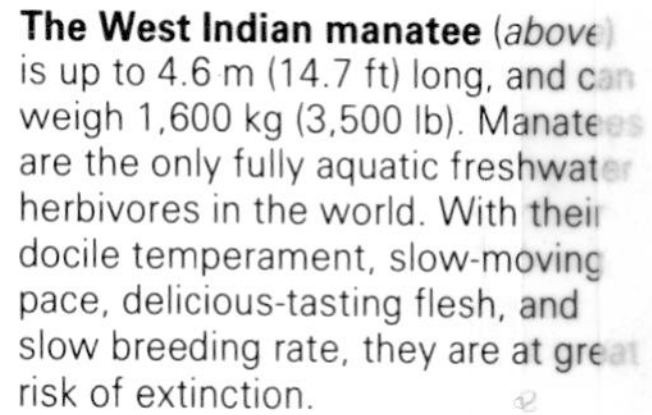

The West Indian manatee (*above*) is up to 4.6 m (14.7 ft) long, and can weigh 1,600 kg (3,500 lb). Manatees are the only fully aquatic freshwater herbivores in the world. With their docile temperament, slow-moving pace, delicious-tasting flesh, and slow breeding rate, they are at great risk of extinction.

A school of French grunts (*left*) cruises over a West Indian coral reef. Grunts are so-named because of their ability to make grunts and drumlike noises by vibrating powerful muscles on either side of the swimbladder. These noises are loud enough to be heard out of the water, especially in the breeding season, when huge groups of fish drum furiously together.

A solitary hunter (*right*) A margay stalks the forest floor in Belize. Together with the larger, even more sought-after ocelot, the margay has been hunted almost to extinction for its beautiful pelt. About the size of a domestic cat, the margay is equally at home on the ground and climbing among the trees.

BERMUDA'S EMBATTLED CAHOW

The cahow, Bermuda's endemic petrel, was only recently rescued from the brink of extinction. It was first discovered by sailors in 1603; early European settlers took a heavy toll of the birds, hunting them for food, and introducing pigs and cats that, together with the rats that came ashore from their ships, preyed on the hatchlings. The cahows survived only on small outlying rocky islands. In 1609 a plague of rats caused a famine in Bermuda, and the hunt for the birds spread to these islands. For more than 300 years the birds were believed to be extinct.

However, in 1951 about 18 breeding pairs were found. The cahows faced competition for nest sites from tropicbirds, which killed their young and took over their burrows during the day while the cahows were at sea. Conservationists poisoned the rats and erected baffles at the nest entrances to keep out the larger tropicbirds. The breeding rate of the cahows improved – for a short time. The number of surviving young then declined markedly; the adults were taking in DDT and other pesticides while feeding in the North Atlantic. The North American pesticide ban of the early 1970s came just in time, and the cahows are now recovering. Nevertheless, rats and feral cats still prevent them from thriving in their former main island haunts.

cattle, yielding only 12.5 kg (27.5 lb) of beef. As forest it can yield over 225 kg (495 lb) of iguana meat. Encouraging farmers to keep patches of forest on their land to rear iguanas benefits many other species.

Some animals have gained from the presence of people. The Marine toad has invaded sugarcane plantations to feast on the insect pests there. The Cattle egret, originally from Africa, feeds well on the insects disturbed by livestock.

Chains of distress

Disturbance to one part of an ecosystem can set in train a series of consequences that may affect humans as well as wildlife. For example, Guatemala's beautiful Lake Atitlan is the only home of the Giant pie-billed grebe. This flightless waterbird has a sturdy head and bill, and legs set so far back on its body that it has great difficulty in walking. It nests in the reed beds that fringe the lake's shores, and probably evolved to take advantage of the crabs that were once abundant near the shore. Today, however, its main food source is small fish.

In 1958 Large-mouth bass and three other species of exotic fish were introduced to the lake for sport fishing. The bass flourished, outcompeting the grebes for fish and crabs, and eating the young birds; the population rapidly declined. Some 50,000 Indian people live around the shore of the lake. They depend on fishing, and their livelihood suffered as the bass took over. The crabs retreated into deeper water, and the local fishing collapsed. As the bass multiplied their numbers also stabilized, and the grebes recovered. But by then city people were building vacation houses on the lake shores, cutting the reeds to clear beaches and build boathouses. The wash from their motorboats and waterskiing activities damaged the nesting sites and greatly disturbed the birds, and their sewage began to pollute the lake.

Conservationists blocked a scheme to dam the lake for hydroelectric power. But they could not protect the lake from the effects of an earthquake in 1976 that led to the water level falling by more than 3 m (10 ft). More reed beds were lost, and the grebe population has fallen to fewer than 50 birds. Renewed conservation is put under threat by the region's political instability. Not only the grebe but a whole ecosystem and the livelihood of 50,000 indigenous people are at risk.

Ants – the teeming armies

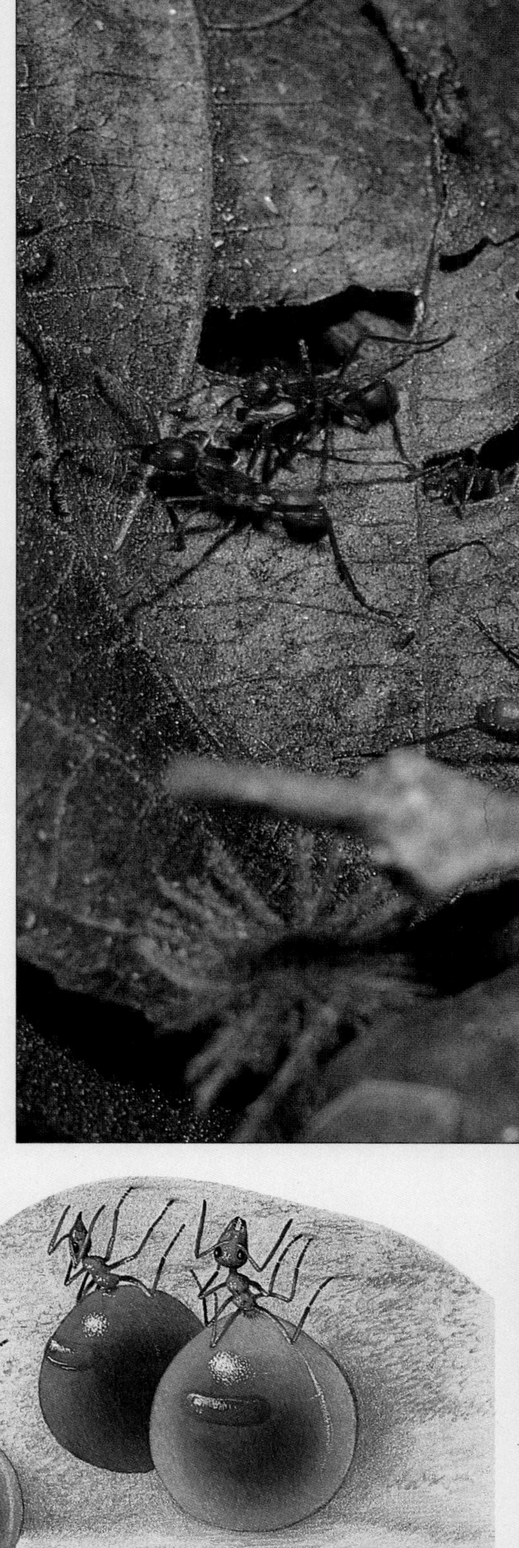

There are several thousand species of ant worldwide, and perhaps 10 million billion individual ants alive at any one time. Throughout Central America, as in all other tropical regions of the world, ants are found everywhere and in vast numbers. They are social insects, living in large colonies. One or more very large queens form the heart of the colony, spending almost their entire lives laying eggs. Queens are produced as a result of special feeding of the larvae from fertilized eggs. The rest of the colony is made up chiefly of nonreproducing worker females with specific roles.

Worker ants have clearly defined tasks – brood rearing, tending the queen, foraging – usually allocated by age. In some species there are distinct castes of workers; soldier army ants are much larger than other workers, and have massive heads with powerful jaws. Leafcutter ants have extra small workers; they ride on the leaves being carried by larger workers and fend off parasitic insects.

Colonies vary in size from a few thousand to several million ants. The nests may take the form of networks of underground chambers, vast mud-plastered balls suspended from branches by tiny hollow thorns. The behavior of the colony is controlled by chemical signals (pheromones) given off by the queen. Each ant

Honeydew lovers (*above*) Aphids and other sap-sucking insects ooze a sugary solution, honeydew, that many ants like to eat. The ants stimulate the insects to produce the honeydew by stroking them with their antennae. In return, the sap-suckers are defended against predators and may be herded down to the ant nest for protection at night.

Army ant soldiers (*right*) guard the marching column on its daily search for prey. When the ants need to cross a gap, they form a living bridge, linking claws to form a chain, and allowing other ants to walk over them. They breed in a 35-day cycle; while the larvae are growing they move to a new bivouac site every night, but while the pupae are maturing they remain in the same place until the new workers emerge.

Living larders (*below*) Honeypot ants have a caste, the "gasters", that are fed with nectar and honeydew, which they store in hugely distended crops. Never leaving the nest, the gasters serve as living larders for the colony.

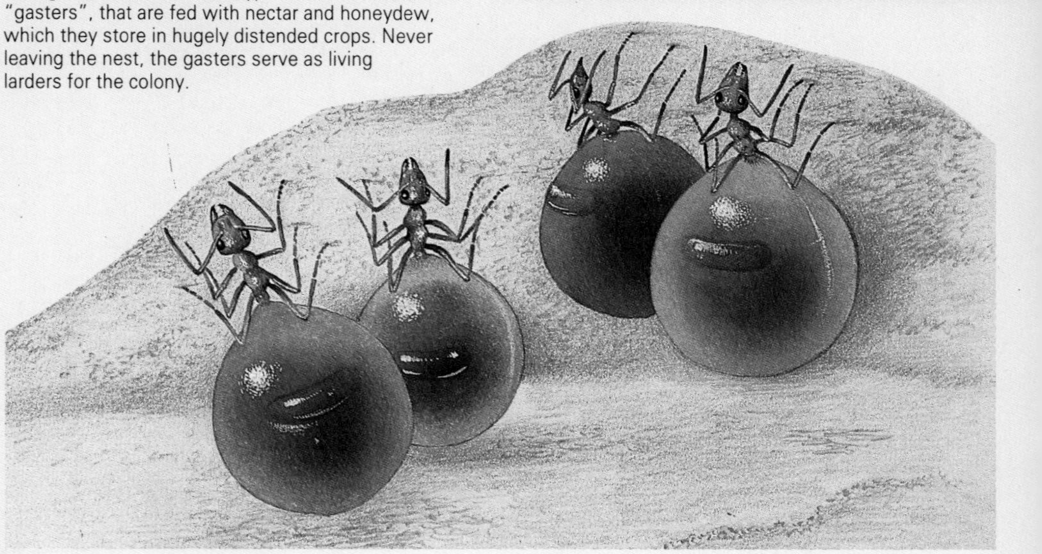

acquires the distinctive smell of its own colony, and may attack ants from other colonies. Ants also communicate by touch, using their antennae, and workers often lay chemical trails to guide others to an abundant food source. Ants have a second stomach or crop in which they store food that can be regurgitated to feed other ants of the brood.

Farmers, gardeners and soldiers
Many ants are herbivores. Some of the commonest ants in the tropical forest are the leafcutters, which bite off pieces of leaf and take them back to their underground nest. Chewed and coated with digestive juices, they serve as food for a fungus found only in ant nests. It produces spherical fruiting bodies, which the ants eat and feed to their larvae.

The leafcutters have evolved foraging strategies that ensure they do not deplete their food supply. They vary their route frequently, and never forage very close to the colony. Experiments have shown that leafcutters favor a varied diet. When offered a choice of food, they prefer one particular kind of leaf for a time, but will eventually tire of it and switch to another species. This gives the first species time to recover. They are also careful to space out their nests, never siting them too close to another colony of the same species.

Other ants also keep food in their nests. Ants of desert areas have a rather unpredictable source of food. Harvester ants collect seeds and grass cuttings to store in the nest for times of scarcity.

Army ants do not build permanent nests but bivouac on the trail. With the queen and brood in the center, army ant columns may be up to 20 m (66 ft) wide and contain some 200,000 individuals. They fan out over the forest floor, devouring any creature not quick enough to make an escape.

Some ant species have formed intimate associations with plants. Many forest plants have hollow stems or leaf nodes that house colonies of ants. The plants produce food bodies and nectaries for them on stems or leaves; in return the ants defend the plants against hungry insects and larger herbivores.

Attack and counterattack

Wherever so many predators and prey live in close proximity, some bizarre defenses evolve. In Central American rainforests many plants accumulate poisons in their leaves to deter caterpillars and other plant-eating animals; some animals manage to accumulate these chemicals in their bodies and use them for their own defense. Flower bugs and *Heliconius* butterflies ingest the poisons from certain passionflower vines.

Heliconius butterflies and their caterpillars may have overcome the poisons of the passionflower, but that is not the end of the story. The female butterflies are very choosy about which passionflower leaves they lay their eggs on – they will not lay on any leaf already occupied by an egg. The passionflower develops tiny outgrowths on its leaves that mimic the eggs of the butterflies, so reducing the predation.

Poisonous insects are usually clearly identified with bright warning patterns of black and red, orange or yellow. After one unpleasant meal, a predator learns to avoid these color combinations. Other insects have taken advantage of this, evolving a similar appearance, even though they are quite palatable. As long as there are more poisonous "models" than "mimics", so that most encounters with these forms are unpleasant, the mimics gain protection. Similar strategies have evolved among harmless and poisonous snakes.

Warning colors The bright colors of these Mexican grasshoppers signal that they have poisonous flesh. When in flight their bright red underwings reinforce the warning.

SPECIALISTS IN SURVIVAL

AN ANCIENT HISTORY · WHERE DIVERSITY REIGNS · COOPERATION AND EXPLOITATION

South America contains some 3,000 species of birds – one-third of the world's total – 80 percent of which are unique to the region. But birds are only part of South America's animal riches. Its wide range of habitats – coastal desert, mountains, grasslands, rainforests, rocky shores and coral reefs – means that a unique collection of land and aquatic animals evolved to take advantage of all the opportunities for survival. In moist tropical rainforests in particular, insects, amphibians and fish all developed unique specialized lifestyles. This diversity is partly due to South America's unusual geological history: the fragmentation of the continents over 100 million years ago meant that its animals evolved in isolation for millions of years, until the creation of the land bridge with North America let other species in.

Bird paradise *(left)* The South American tropics are renowned for their wealth of bird life. Scarlet ibises roost in tree clumps scattered across the llanos grassland of the Orinoco river basin, the bird life of which is remarkably similar to that of the African savannas: rheas, seriemas and tinamous are South America's equivalent of Africa's ostriches, Secretary birds, francolins and guinea fowl. As in Africa, the grasslands abound with pipits, kites, hawks, falcons and vultures.

A highly adaptable cat *(right)*, the jaguar is a good climber and swimmer, at home in all terrains. It has been hunted almost to extinction in the more open habitats because of the value of its exquisite pelt; today it survives chiefly in deep inaccessible forest. Jaguars are mainly active at night and take a wide variety of prey. They are normally solitary animals, coming together only to breed.

COUNTRIES IN THE REGION

Argentina, Bolivia, Brazil, Chile, Colombia, Ecuador, Guyana, Paraguay, Peru, Surinam, Uruguay, Venezuela

ENDEMISM AND DIVERSITY

Diversity Very high – probably highest in the world.
Endemism High to very high (a number of species shared with Central America)

SPECIES

	Total	Threatened	Extinct†
Mammals	800	70	4
Birds	3,000*	296	2
Others	unknown	83	0

† *species extinct since 1600 including Falklands Islands wolf* (Dusicyon australis), *Glaucous macaw* (Anodorhynchus glaucus), *Colombian grebe* (Podiceps andinus)
* *breeding and regular non-breeding species*

NOTABLE THREATENED ENDEMIC SPECIES

Mammals Golden lion tamarin *(Leontopithecus rosalia)*, Emperor tamarin *(Saguinus imperator)*, Woolly spider monkey *(Brachyteles arachnoides)*, Maned wolf *(Chrysocyon brachyurus)*, Giant otter *(Pteronura brasiliensis)*, Mountain tapir *(Tapirus pinchaque)*, Marsh deer *(Blastocerus dichotomus)*
Birds Junin grebe *(Podiceps taczanowskii)*, White-winged guan *(Penelope albipennis)*, Little blue macaw *(Cyanopsitta spixii)*, Esmereldas woodstar *(Acestrura berlepschi)*
Others South American river turtle *(Podocnemis expansa)*, Black caiman *(Melanosuchus niger)*, Ginger pearlfish *(Cynolebias marmoratus)*, Galapagos land snails *(Bulimulus)*

NOTABLE THREATENED NON-ENDEMIC SPECIES

Mammals Caribbean manatee *(Trichechus manatus)*, Giant anteater *(Myrmecophaga tridactyla)*, jaguar *(Panthera onca)*
Birds Dark-rumped petrel *(Pterodroma phaeopygia)*, Orange-breasted falcon *(Falco deiroleucus)*, Harpy eagle *(Harpia harpyja)*
Others Leatherback turtle *(Dermochelys coriacea)*, Hawksbill turtle *(Eretmochelys imbricata)*

DOMESTICATED ANIMALS (originating in region)

llama *(Lama 'glama')*, alpaca *(Lama 'pacos')*, Guinea-pig *(Cavia porcellus)*, Muscovy duck *(Cairina moschata)*

AN ANCIENT HISTORY

South America is the only landmass in which marsupials (pouched mammals) and placental mammals, which evolved later, coexist in large numbers – the result of its long period of separation from other continents after the breakup of the supercontinent of Gondwanaland. Both forms of mammal were already present in the region, and subsequently evolved into a large number of unique species. Sloths, armadillos and anteaters, which died out elsewhere, survive here, and llamas, guanacos and vicuñas, related to camels, graze the grasslands instead of the antelopes or bison that are found in similar habitats in other parts of the world. When the land bridge with North America formed about 3.5 million years ago, a wide range of placental mammals migrated southward to take advantage of the new territory.

Among its marsupials South America supports a variety of opossums, which have continued to evolve and are still extending their range – the Common opossum has spread through Central and North America to reach Canada. South America also contains the only aquatic web-footed opossum, the yapok. Other opossums, with their prehensile tails, are well adapted to forest life, while the marmosas (small ratlike marsupials) forage on the forest floor.

As the continent of Africa began to break away from Gondwanaland, rodents and monkeys probably island-hopped to South America. Here, the monkeys have since evolved into two distinct groups, the tamarins and marmosets (small-sized fruit-eaters) and the cebids. This latter group includes the noisy howler and slender spider monkeys.

South America's main groups of fish – the catfish, characins and cichlids – as well as the ancient lungfish, had already evolved before the continents separated. The creation of the Andes mountain range blocked the path of the Amazon river to the Pacific. Until it cut a new path to the Atlantic, it formed a vast inland sea. Many former marine species, such as dolphins and stingrays, evolved into freshwater species.

An array of species

South America's wide range of habitats, from mountains to savannas, tropical rainforest to desert, has allowed the evolution and coexistence of numerous specialized animals, especially in the tropics. The rivers today contain a diversity of species – voracious piranhas, electric fish, strange armored fish and the giant arapaima that can leap out of the water to capture insects and even small birds on the forest branches above. These share the water with crocodiles, caimans and anacondas – the largest snakes in the world. Capybaras, the largest of all rodents, feed on the lush vegetation that grows at the water's edge.

Many amphibians inhabit the region's wetlands and moist tropical forests. They

include the poisonous dendrobatid frogs that carry their eggs and tadpoles on their backs, the beautifully camouflaged tree frogs with suction pads on their toes and the marsupial frogs, which carry their eggs and tadpoles in pouches. Insect life is similarly various – there are countless species of butterflies, moths, ants, bees, wasps, beetles and other insects – many, perhaps most, still unknown to science.

More than twenty families of birds are endemic to South America. Among them are the predominantly ground-living tinamous and the trumpeters, hoatzins (related to cuckoos) and the two species of cock-of-the rock, which build a bracket-shaped nest fixed to sheer rock faces. The forests provide insects for New World flycatchers, antbirds and ovenbirds, flowers for hummingbirds and fruits for, among others, toucans, brightly colored tanagers and even more brilliant trogons.

Tapirs and a wide range of rodents browse the forests. Sloths, monkeys and tree-climbing rats and mice feed in the tree canopy, along with a multitude of birds and insects. Both armadillos and anteaters have become specialized for feeding on those most abundant of insects, the ants and termites. The tamandua, a forest-dwelling anteater, has a prehensile tail that enables it to raid ant and termite nests hanging from high branches. While jaguars, jaguarundis and ocelots (small wildcats) are agile climbers, they prefer to stalk the forest floor for their prey, where bands of peccaries (piglike animals) root for tubers.

Even the inhabitants of the South American grasslands are quite unlike those of other continents. Guanacos and vicuñas wander across the high plains. Another fast-moving grassland grazer is the rhea, a large flightless bird related to ostriches and emus. There are only a few species of deer, including the Andean deer or huemul, which can survive harsh extremes of climate at great height.

Monkeys of the family Cebidae (capuchin-like monkeys): a characteristic of many New World monkeys is a prehensile tail that acts as a fifth limb when climbing; the Guianan or White-faced saki; the Red uakar – uakaris are the only short-tailed primates in the New World; the Dusky titi and the Squirrel monkey – both move chiefly by leaping, and do not have prehensile tails; the Brown capuchin; the Female Black howler monkey; the Back-handed spider monkey – it uses its tail for picking up food; the Night monkey, whose nocturnal habit is rare among monkeys; the Humboldt's or Smokey woolly monkey.

WHERE DIVERSITY REIGNS

In the tropical rainforests evolution occurs rapidly. The animals living in this environment have specialized to fill every microhabitat and to exploit every possible source of food. Yet some ancient species have survived from before the region's early period of isolation. One of these is the velvet worm, *Peripatus*, which has short stumpy legs and a segmented body, similar to a fleshy millipede. It appears to be an intermediate form between annelid worms (such as earthworms and leeches) and arthropods (insects, crustaceans, and so on); it may have shared an ancestor with both.

Some insects in South America can reach unusually large proportions. The Hercules or Rhinoceros beetle, for example, has long horns and grows to 150 mm (6 in) in length. Another recordholder is the Giant ghost moth, which has the

Brown capuchin
Cebus apella

Black howler monkey
Alouatta caraya

greatest wing-span of any butterfly or moth in the world: 28 cm (11 in).

In the tree canopy

An astonishing array of animals with prehensile tails have evolved in the South American rainforests. Not only monkeys, but also porcupines, mice, rats, opossums, a raccoon – the fruit-eating kinkajou – and even some tree snakes have this useful "extra" limb. One reason for this may be the tendency for large areas of the South American forests to become flooded at certain times of year, compelling animals to travel from branch to branch rather than along the ground.

By feeding on different kinds of food and at different levels in the forest, a wide range of monkeys can coexist here. They belong to two families, the Callitrichidae (the tamarins and marmosets) and the Cebidae (capuchins and other monkeys), which are found only in South and Central America and are distinguished from their African and Asian counterparts by having widely spaced forward-directed nostrils. Some species, such as the spider monkeys, are also unique in having prehensile tails. The small, fruit-eating marmosets and tamarins are monkeys with silky coats and sharp, curved claws for climbing trees. Tamarins and marmosets hate water and have thus evolved into many different species separated by the rivers that fragment the rainforest. The family includes some of the most endangered primates, such as the Emperor tamarin, the Golden lion tamarin and the Cotton-top tamarin. The Pygmy marmoset, which weighs less than 200 g (7 oz), is the smallest monkey in the world.

The capuchins and their relatives eat a range of food, from fruits and leaves to birds' eggs, insects and other animals. They include the only nocturnal monkeys in the world, the douracoulis, and the howler monkeys. They gain their name from the call the males make – among the loudest produced by any mammal.

The South American rainforests are home to over 300 species of hummingbird. These brilliant iridescent birds have evolved into a variety of species ranging in length from 5.8–21.7 cm (2.3–8.5 in), with a wide range of different bill shapes. They are supreme fliers, able even to fly backward or to hover motionless while feeding from the sugary nectar of countless different flowers, which supply them with a store of energy. Hovering is the most energy-demanding activity in the animal kingdom. Hummingbirds will typically consume half their weight in sugar a day; they also eat flies and spiders to gain protein and other nutrients. The humming sound is due to the amazing speed at which they vibrate their wings, up to 200 times per second.

Specialist lifestyles

One of South America's strangest birds is the oilbird – a cinnamon-brown, crow-sized bird that roosts in colonies in mountain caves throughout much of the region. It finds its way out of the cave in total darkness, using a sonar system similar to that of bats, an unusual feat for a bird. At night oilbirds leave their roosts in thousands to forage for the oily fruits of palms and laurels, sometimes traveling more than 80 km (50 mi) in one night. The nestlings, stuffed with these fruits, develop masses of fat and weigh up to twice as much as the adults. For centuries Amerindians have used the oil from the chicks for cooking, hence the name.

South America is also home to an impressive variety of amphibians. Some frogs carry their tadpoles in a pouch on

THE HOATZIN

Among the notable endemic birds of South America is the hoatzin. It has so many unique features that only recently has it been recognized as closely allied to the cuckoos. It is a slender bird, similar to a pheasant, with a distinctive crest of stiff narrow feathers and a row of bristles that form mammal-like eyelashes. It lives in riverine swamps and forests throughout the Amazon and Orinoco river basins.

The hoatzin's nest is a platform of sticks above the water, where it lays two or three eggs. Both parents take care of the offspring, aided by the young from previous years. One remarkable peculiarity of the chicks has long intrigued ornithologists: they have two well-developed functional claws on each wing that enable them to leave the nest and clamber along branches using all four limbs in reptilian fashion. The claws are an advanced feature that has evolved in response to their way of life. When threatened by a potential predator, the chicks readily plunge into the water – they are not only able swimmers, but also accomplished divers. When danger is past, they climb back up the tree to the nest.

The hoatzin feeds mainly on green leaves, most unusually for a flying bird. It digests the fibrous plant matter in much the same way as cows do; this probably gives the bird its characteristic odor of fresh cow dung. The hoatzin selects the most nutritious leaves and retains the food for as long as 43 hours for fermentation to take place. Geese, by contrast, pass leaves through in one and a half hours, but have to eat prodigious quantities. The drawback of the hoatzin's system is that the breastbone is undersized to make room for the digestive organs, and hoatzins are consequently poor fliers.

The tiny Golden lion tamarin is found only in a small area of forest in eastern Brazil. It is in serious danger of becoming extinct as its nature reserve home is being invaded by squatters. The only other population is a colony of some 25 individuals at the Rio de Janeiro Primate Center.

their backs until they metamorphose and can be released; this avoids the intense competition and predation in the ponds. Other frogs live in the pools of water contained in the rosettes of bromeliads, plants that grow high on the branches of forest trees. There is also a species aptly named the Paradoxical frog, whose tadpole can be up to 28 cm (11 in) long, more than three times the size of the adult. The brilliant colors of the poison-arrow frogs warn that they are highly toxic, threatening any would-be predator.

The absence of hoofed mammals during much of South America's history left many vacant ecological niches. These have mostly been filled by the large rodents, such as cavies, found only in Central and South America. The mara of the Patagonian scrub desert is a monogamous animal with a harelike head and legs that resemble those of deer. It weighs about 4 kg (9 lb). Maras also perform the "stot", a high jump typical of some deer species. While stotting, they expose a white rump patch, which can also be seen on deer.

The capybara is the largest living rodent weighing 50 kg (110 lb) and measuring 1.2 m (4 ft) long. It grazes the water meadows, but is always ready to retreat into the water at the first sign of danger. The coypu is even more aquatic; the female has nipples on her flanks to suckle her young while swimming.

Fleet-footed pacas live on the forest floor, along with Guinea pigs and agoutis, which are armed with large claws for digging. The tuco-tuco of the plains also has huge claws for burrowing, like its North American equivalent, the Pocket gopher. High mountain pastures are home to the chinchillas, with their highly prized long, soft fur.

COOPERATION AND EXPLOITATION

Long before the arrival of Europeans in South America in the early 16th century, the indigenous Amerindians had domesticated various animals. Archaeological evidence indicates that the Guinea pig has been raised in farms in Peru since 2500 BC. Today, up to 20,000 tonnes of meat per year are produced from a population of 20 million of these animals. The success of Guinea-pig farming lies in their high reproductive rate, typical of rodents: they produce up to four litters of three or four young every year.

Alpacas and llamas were first domesticated in the high Andean plains around Lake Titicaca on the Bolivia–Peru border as long ago as 3000 BC. Alpacas produce high-quality wool, while llamas are used mainly as pack animals. There is evidence to suggest that Amerindians had domesticated a stock of dogs before the arrival of the European dog.

Parrots, parakeets and toucans have long been favored as pets in South America. The Amerindians also tamed trumpeter birds, which they caught in the forest. As that name suggests, they have loud, strident voices and were often kept around dwellings to act as "watchdogs".

Wild animals are an important source of food for the indigenous human population. The honey produced by native bees and the roasted larvae or adults of many wasps, ants and termites provide highly esteemed and nutritious meals – even large caterpillars and tarantulas are sometimes eaten. Turtle eggs and hatchlings are taken by the thousand from nesting beaches – a practice that has seriously reduced populations of some species in recent years.

Trade in wildlife

Since the Spanish and Portuguese conquest of South America, the region has had a long history of animal use and abuse. Egrets and herons were almost exterminated in the northern floodplains when hats decorated with long feathers were the height of fashion among American and European ladies. Hummingbirds suffered a similar fate; millions of skins were shipped to European markets to be made into ornamental pins and other decorations. However, when the fashion changed, the exporters went out of business, and the bird populations recovered their numbers.

Reptile skins were a later fashion, resulting in the virtual extermination of the crocodiles of the Orinoco. They are slow breeders and have been unable to recover. Now the small caimans are being exploited, though in a more controlled way as a result of the efforts of conservationists. Crocodiles are now bred in captivity for reintroduction into their natural habitat. The large tegu lizards are also bred for their skin, which may conserve the species.

Current trade in South American birds and mammals – notably parrots, macaws, hummingbirds and monkeys – involves large numbers of live animals intended for the pet markets of North America and Europe. Most of this trade is now illegal, but such trafficking has brought several species of South American parrot to the brink of extinction.

Problems of coexistence

A number of South American animals carry diseases that plague humans. Malaria is transmitted by mosquitoes; Chagas' disease (similar to sleeping sickness) is produced by a protozoan carried by kissing bugs; and bilharzia (river blindness) is caused by a parasitic flatworm found in rivers and streams. True vampire bats – those that feed on blood – are a mainly South American group, feeding at night on warm-blooded birds and mammals, including horses, cattle and occasionally humans. They transmit rabies through their bites, but this has been greatly reduced by the use of medicines and through the largescale slaughter of vampires and many other species of bat.

A growing – and now the gravest – threat to the wildlife of tropical South America is the loss of wild habitat. The rainforests are being felled at an alarming rate, and the rivers and streams that flow through them are being polluted. On the extensive grasslands, native grasses have been replaced by alien species to feed livestock that have displaced the native grazing mammals. For how much longer will the continent remain a vast treasure house of rare and fascinating species?

FISH OF THE FLOODPLAIN

In much of the tropics widespread flooding occurs in the wet season and severe drought in the dry season. The fish living in these regions have had to adapt to the contrasting conditions. Many species move into the flooded forests and reproduce prolifically during the period of peak flooding. In the dry season they concentrate in the retreating shallows, while fish-eating birds – for example, terns, gulls and skimmers – nest on the newly exposed sandbanks, taking advantage of the plentiful food.

The Amazon has at least 20 species of fish that breathe atmospheric oxygen when pools dry up. One of these is the Swamp eel. Some, including the Giant electric eel, have mouths that are well-supplied with blood vessels, enabling them to absorb oxygen. Other fish, including the lungfish, which evolved 150 million years ago, use an air bladder for breathing; in the dry season the lungfish will bury itself in the mud.

Many species have evolved ways of finding food in the murky floodwaters. The electric fish uses weak pulses of electricity, while the Giant electric eel uses its electricity to stun its prey. It can produce up to 500 volts of low current. River dolphins use echolocation for hunting, and catfish feel for and taste their prey with sensitive barbels on their chins. Some fish, which feed on fruit and seeds that fall into the water, may act as important dispersers of plants to new locations.

The world's largest rodents These capybaras are sharing their river bank feeding grounds with a group of Spectacled caiman. Capybaras feed in the cool of evening, and sleep or bathe during the hottest hours of the day.

Wildlife of the Galapagos

Some 960 km (600 mi) west of the coast of Ecuador lie the Galapagos Islands, famed for the uniqueness of their plant and animal life. The Galapagos began to rise as volcanoes from the ocean floor more than 3 million years ago and have never been connected to South America. Their animals are derived from ancestors that colonized from the mainland.

The islands' land birds are dominated by the Galapagos finches, which were important in molding Charles Darwin's ideas on evolution by natural selection. The finches constitute a group of 13 species found nowhere else. They are believed to have evolved in the Galapagos from an original South American ancestor. Lacking competition from other birds, they were able to colonize many different ecological niches.

The finches' bills evolved into a wide range of shapes and sizes suitable for dealing with various types of food; different species assumed roles of seed-eaters, woodpeckers, warblers and even (in the case of the ground finches) blood-suckers. Several Galapagos finch species are remarkable among birds in their use of tools: the Cactus finch holds a long cactus spine in its beak and probes with it into the crevices of plant stems for the insects hidden there, and the Mangrove and Woodpecker finches use thorns or slender twigs.

Galapagos reptiles have also undergone extensive evolution. The different races of the giant tortoise are so distinctive that sailors who exploited them for food could easily identify the island of origin from the tortoise's carapace or shell. The land iguanas of the islands are large lizards that specialize in eating prickly pear cacti. The Galapagos Islands also boast the Marine iguanas, the world's only marine lizards; these have blunt noses and serrated teeth for feeding on seaweed, as well as sharp claws for clinging to wave-battered rocks. They control their salt level by expelling strong saline solution in spurts from their nostrils. However, as Darwin observed, when threatened they head for land, not water; until the arrival of humans and their animals, these lizards never had any predators on land.

The cool, nutrient-rich waters of the Peru (Humboldt) Current that sweep northward past the Galapagos support a rich underwater life. Galapagos sea lions and fur seals breed along the coast, and a great variety of whales and dolphins feed offshore. Almost three-quarters of a million seabirds nest on the islands, including nearly the entire world population of the Waved albatross and some boobies. The flightless Galapagos cormorant is an excellent swimmer and diver; lacking predators on land, it has no need of full-sized wings. However, it still holds out its vestigial wings to dry after a swim, betraying its ancestry.

Dragon of the sea (*right*) The Marine iguana is the world's only seagoing lizard. Its blunt nose and serrated teeth enable it to scrape algae from the rocks to feed on, and with its sharp, curved claws it can cling to rocks in the surf. It copes with sea water by having salt-secreting nostril glands that expel saline fluid.

A living address book (*below*) The shells or carapaces of the Giant land tortoises of the Galapagos Islands have evolved into different shapes (for different subspecies) on particular islands.

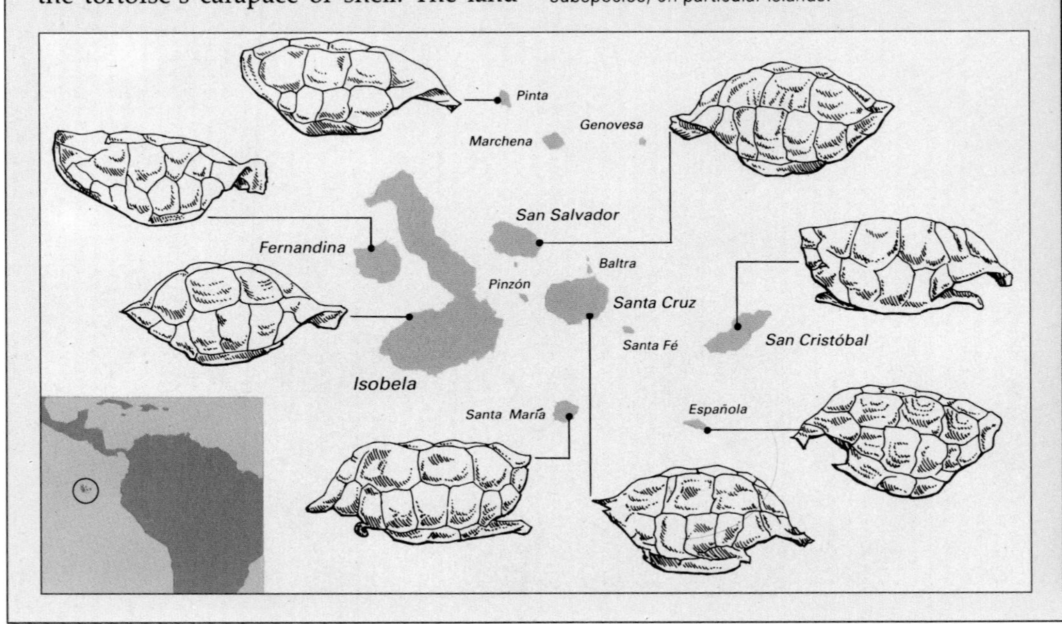

Sally lightfoot crabs share the wave-dashed rocks with the Marine iguanas. Scuttling around the intertidal zone, they scavenge for scraps of food and the washed-up corpses of small marine animals, their stalked eyes alert for airborne danger in the form of hungry gulls.

Human disruption

Until people came to these islands, the animals of the Galapagos had few predators. The first threat came from passing sailors, who slaughtered the islands' tortoises for food. Then came whalers and sealers hunting for furs. As the islands were settled, introduced animals caused havoc to the native wildlife. Rats and pigs dug up the eggs of tortoises, sea turtles and iguanas; cats and dogs preyed on their young; and goats ravaged the vegetation. A massive program to eradicate these intruders has been partially effective, and some of the more threatened species have been successfully bred in captivity and reintroduced to their former haunts. Today 97 percent of the islands' land area is a national park, and the animals have been afforded legal protection.

The islands' human population is growing at 12 to 15 percent a year as settlers from the mainland arrive, seeking to profit from the tourist boom. This, together with some 60,000 annual visitors (despite the official limit of 25,000), causes pollution and puts pressure on natural habitats as the settlements expand, and shellfish and other marine animals are in greater demand. With such large numbers of people visiting the Galapagos Islands, alien species are often introduced unintentionally: in the past decade alone, at least 50 species of plants, several dozens of insect species, a lizard and a bird have all arrived.

Kings of the south

The cold waters from the Antarctic that stream up each side of southern South America, as the Peru Current on the west and the Falkland Current on the east, are rich feeding grounds for several species of penguins, as well as many other kinds of seabirds. Penguins are superbly adapted to exploit this wealth of food. Awkward on land, they are expert swimmers, and can spend weeks at sea. The torpedo-shaped body is streamlined with a layer of fat and a sleek coat of feathers, both of which keep the penguin warm during its long immersion in chill water. Penguins swim with powerful beats of stiff, bladelike wings or flippers, and are steered by the feet and tail acting as multiple rudders. Like whales and seals, they are physiologically adapted for deep diving. The King penguin can dive to over 240 m (800 ft) as it searches for fish and squid.

Penguins are very sociable during the breeding season; King penguins gather to rear their young in colonies on the Falkland Islands, South Georgia and other subantarctic islands. They pair to breed. The single egg is laid in November or December, and hatches in two months. The young penguin is then fed by both parents through the winter and launched into independence a year after the egg was laid. The parents do not lay again until the following summer, so the breeding cycle of the King penguin is unusual because pairs can rear young only once every three years.

The regal plumage of these King penguins on South Georgia becomes mudstained as they shelter their young from the wet weather. Some have quite well-grown young, while others are still incubating eggs balanced on their large feet.

A NORTHERN REFUGE

SEASHORE, FARMLAND AND FOREST · THE LONG WINTER · PROTECTED OR PERSECUTED?

Thinly populated, with rugged mountains and dense forests, the Nordic countries are a last refuge for rare animals such as the Brown bear and the lynx, which have almost disappeared elsewhere in Europe. The region also supports large numbers of elk (moose) and reindeer, and other animals that had a more southerly distribution before the climate warmed. The Svalbard archipelago, which includes Spitsbergen, is a major denning area for Polar bears. The nutrient-rich waters of the Norwegian Sea support huge breeding colonies of seabirds along the cliffs of Norway, Svalbard, the Faeroe Islands and Iceland. The numbers of some animals are so small, however, that they disappear from the region from time to time – the entire population of wolverines in Finland occasionally crosses into the Soviet Union.

COUNTRIES IN THE REGION

Denmark, Finland, Iceland, Norway, Sweden

ENDEMISM AND DIVERSITY

Diversity Very low to low
Endemism Very low

SPECIES

	Total	Threatened	Extinct†
Mammals	69	10	0
Birds	300*	4	1
Others	unknown	22	0

† *species extinct since 1600 - Great auk* (Alca impennis)
* *breeding and regular non-breeding species*

NOTABLE THREATENED ENDEMIC SPECIES

Mammals none
Birds none
Others none

NOTABLE THREATENED NON-ENDEMIC SPECIES

Mammals Gray wolf *(Canis lupus)*, wolverine *(Gulo gulo)*, Polar bear *(Ursus maritimus)*, Harbor porpoise *(Phocoena phocoena)*, Northern bottlenose whale *(Hyperoodon ampullatus)*, Fin whale *(Balaenoptera physalus)*, Blue whale *(Balaenoptera musculus)*, Bowhead whale *(Balaena mysticetus)*, Humpback whale *(Megaptera novangliae)*, narwhal *(Monodon monoceros)*
Birds Lesser white-fronted goose *(Anser erythropus)*, Red kite *(Milvus migrans)*, White-tailed sea eagle *(Haliaeetus albicilla)*, corncrake *(Crex crex)*
Others Hermit beetle *(Osmoderma eremita)*, Tree snail *(Balea perversa)*, Large blue butterfly *(Maculinea arion)*, Noble crayfish *(Astacus astacus)*

DOMESTICATED ANIMALS (originating in region)

reindeer *(Rangifer tarandus)*

SEASHORE, FARMLAND AND FOREST

A wide variety of marine species thrive along Scandinavia's extensive coastline. The waters of the Atlantic continental shelf, influenced by both cold Arctic currents and the warm Gulf Stream, are extremely rich in plankton, fish and shellfish. The number of different large marine species found at Rustaadirka, a marine national park in Norway, rivals that of the Caribbean and includes the Conger eel, the Basking shark and the Killer whale.

Fjord and seashore

The fjords of Norway's coast, shaped by glaciers and beaten by ocean gales, shelter Common and Gray seals, dolphins and other whales. The cliffs support large colonies of seabirds. On the Lofoten islands in the Norwegian Sea, puffins, guillemots and razorbills live alongside fulmars and kittiwakes. Black guillemots and Storm petrels also breed here, while Arctic skuas nest on the hillsides, preying on the terns and the shorebirds. White-tailed eagles patrol the shore hunting fish, other birds and mammals. A little way inland male and female Red-necked phalaropes "spin" on calm pools as they look for food. Iceland's Lake Mývatn is one of the most important breeding places for wild ducks in northern Europe, with as many as 16 species.

The Baltic seashore is almost completely tideless and is sheltered from the battering ocean storms, though it can be very cold. About 10,000 Ringed seals live in the Gulf of Bothnia and the Gulf of Finland, to the north of the Baltic Sea. However, the population is in decline: the Baltic has become so polluted that many females are unable to breed. The Ringed seal, the smallest of the seals found in Arctic waters, makes breathing holes in the ice as the sea freezes over during the winter. Denmark's lagoons and beaches afford shelter for Common seals, which are similarly threatened by pollution from the North Sea.

Animals of north and south

In the far north Polar bears from the Svalbard islands, including Spitsbergen, some 800 km (500 mi) north of Norway, may sometimes drift into Scandinavian waters, traveling on ice floes. In these

waters, the bird life is distinctly Arctic rather than northern European. The White-billed diver and the Great northern diver fly in from Arctic Asia and Iceland or Canada respectively to overwinter at Varangerfjord in the extreme northeast of Norway. Both King eider and Steller's eider can often be seen near the shore. Common eider are so numerous that local people still collect eiderdown from the birds' nests during the spring, as do Icelanders. The bird life of the tundra and mountains of Iceland, Norway and the northern parts of Sweden and Finland includes breeding shorebirds and geese. During the Arctic summer the gyrfalcon and the bluethroat – the "nightingale of the north" – can also be seen.

Far to the south, in the gentler climate of Denmark and southern Sweden, typically European animals such as the hedgehog, mole and badger live on the northern edge of their natural range. The rabbit is also able to survive here. Rolling farmland is interspersed with patches of mixed woodland, to the advantage of 12 species of bat; but these mammals only survive in the Nordic countries in places where the climate is favorable. Bats such as Bechstein's bat, Natterer's bat and the Noctule bat are particularly sensitive to the cold. Farther north they are replaced

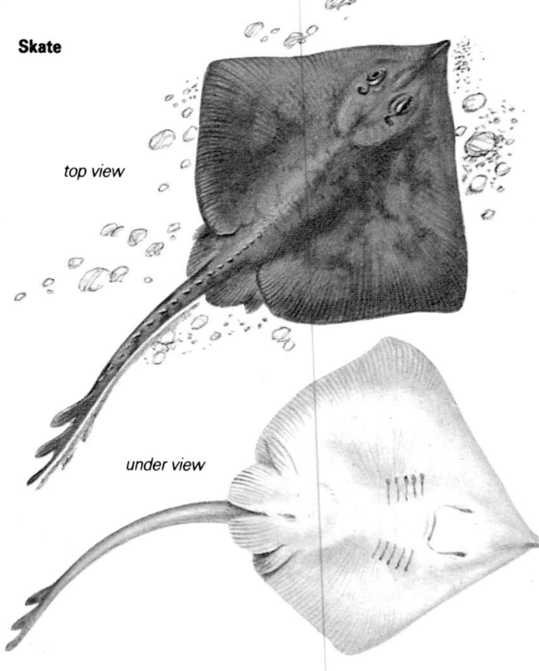

Inhabitant of the sea bed (*above*) The Common skate viewed from above and below. Skates are bottom feeders, preying on mollusks, crustaceans and fish, and they also scavenge carcasses. Both the mouth and gill slits are on its undersurface.

The cry of the north (*above*) The evocative call of the Red-throated diver or loon is characteristic of northern waters. In summer the bird moves from the sea to inland lakes and tundra pools to breed. Its dense plumage aids both insulation and buoyancy.

Bird of the cliff tops (*right*) In the Nordic region the puffin nests in crevices in the rocks. Its colorful bill serves as a weapon and an advertisement to prospective mates; it changes to a duller brown outside the breeding season.

by the Northern bat, a cold-tolerant species that is found inside the Arctic Circle.

The dark Nordic forests of spruce and pine support the European red squirrel and, in northern Finland, the Flying squirrel. Woodpeckers and redstarts feed and nest among the trees, while capercaillies and Hazel grouse patrol the forest floor. Many owls hunt here: the Little, Tawny and Barn owls are restricted to the south, but the Eagle, Hawk, Pygmy, Long-eared and Tengmalm's owls thrive in colder conditions. On bare mountaintops and on the Arctic tundra the magnificent Snowy owl reigns supreme.

The wolverine and the Gray wolf have suffered badly from overhunting, but the Brown bear can still be found in remote forested areas of Finland, Norway and Sweden, where it is protected. Conservation action has also helped a small population of lynx to stabilize.

Musk ox
Ovibos moschatus

THE LONG WINTER

Few animals can survive the harsh climate of the far north, with its long, bitterly cold winters of almost continuous night, without special adaptations. They need extra calories during the winter in order to keep warm, at a time when there is very little food to be found. Most either hibernate, building up a store of fat when food is freely available, or they migrate to warmer climates.

Migrant birds

An effective adaptation is the ability to leave before conditions become too harsh. Having used the long Arctic summer days to feed and raise their young, most birds migrate away from the colder weather. Some species, such as the Great northern diver, move just a little farther south to find unfrozen sea inlets. Others make long journeys, perhaps as far as southern Africa, to wintering grounds that suit their needs. The long distance record is probably held by the Arctic tern, which covers more than 17,000 km (11,000 mi) to reach its alternative residence in the Antarctic; in this way it passes its life enjoying a continual summer, scarcely experiencing long dark nights.

Migration enables birds to exploit the seasonal resources to the full. The Arctic provides abundant food in the summer, as millions of insects breed in meltwater pools. At the same time, the departure of migrants from the warmer south ensures that resident birds are left with a plentiful supply of food to raise their young.

Birds have migrated for many millions of years, their routes and destinations influenced by the drifting of continents and the effect of ice ages on the landscape. Their knowledge is in part genetic, formed by the trial and error of billions of journeys. The migration patterns of different species vary considerably. Willow and Arctic warblers, for example, both undertake long migrations from the region – the Willow warbler migrates to Africa, while the Arctic warbler overwinters in Southeast Asia.

So great is the passage of bird migrants that Falsterbo, on the southern coast of Sweden, has become an important place for birdwatchers. Each year they come to see millions of birds – geese and ducks, birds of prey, storks and cranes, as well as countless small songbirds – streaming in across the narrow waters of Öresund. Although it might seem that migration is the perfect answer to the severe rigors of winter, it is in fact a very dangerous business. Some birds depart carrying up to 50 percent of their body weight in fat – the fuel for their journey. However, if they are diverted by bad weather, or perhaps run out of energy over water, hundreds or even thousands of individual birds can perish.

Winter survival

During the last ice age the musk ox was driven south by the dramatic change in climate and there is now only a small reintroduced population in Norway. Elk and Musk ox build up reserves of fat for the winter, with a thick coat of highly insulating fur covering the extra flesh. The Musk ox has a particularly long coat that grows luxuriantly to more than 60 cm (2 ft) on the neck. It also has extremely fine underfur.

Mobility can mean the difference between life and death in winter. Ptarmigan grow extra feathers on their feet, which act like snowshoes. The reindeer is similarly adapted to the snow; its hooves are broad with the toes splayed far apart, enabling it to cross snow-covered ground. This adaptation is also useful when the tundra thaws in summer; the water forms marshes and swampy areas, and the reindeer's broad feet prevent it sinking. Wolves and wolverines can run across the snow's crust, driving heavy prey such as elk into snowdrifts, where they quickly become exhausted.

The Northern birch mouse survives in the Arctic by hibernating from October through May or even June. Even in summer it spends cold days so deeply asleep in its nest that, if uncovered and picked up by a warm human hand, it will take some time to wake up. Using the techniques of hibernation to conserve energy for those rare Arctic days of sunshine, the Birch mouse reduces to the minimum its need to forage for food.

Snow has extremely effective insulating properties. Voles and lemmings can survive and even raise young in nests and tunnels under the snow, protected from the cold by a layer of air between the snow and the ground, and out of sight of predators. At night it may be −30°C (−22°F) at the surface, but under a snow blanket the temperature remains constant at around freezing point, and there is a plentiful supply of fresh roots and buds for small mammals to eat.

Arctic animals tend to have shorter ears, noses and tails than their relatives in warmer climates. With the animal's overall surface area thus reduced, less heat is lost from the body. A white coat provides camouflage and highly efficient insulation, as colorless hair is hollow; every fall Willow ptarmigan, Arctic foxes, stoats and weasels molt their brownish coats for white ones.

Animals of taiga and tundra The Musk ox has large hooves with sharp rims for digging in the snow. It is able to survive on very sparse vegetation. The Snowy owl travels vast distances in search of its staple prey: lemmings. The wolverine uses its keen sense of smell to find hibernating rodents in their burrows. The elk (moose) has large split hooves that prevent it from sinking in marshy ground in summer and deep snow in winter.

THE LEMMING AND THE SNOWY OWL

Arctic lemming and vole populations fluctuate on roughly three- to four-year cycles, and are synchronized to the activity of the plants on which they feed. In years when the tundra plants fail to flower, they store carbohydrates in their roots. In some winters the roots are extremely nutritious and the lemmings are able to raise large litters under the snow. By the spring the lemming population can be so large that the animals leave in mass migrations to seek food elsewhere – hence the myth of lemming "suicide" runs.

Lemmings and voles form an important food source for many Arctic predators. In years with high rodent numbers, most predators raise extra-large families that will eventually help to reduce the rodent population. However, if the lemming population is low the predators may not breed at all: a female Snowy owl needs "presents" of lemmings from her mate to persuade her even to nest.

Snowy owls make long migrations in the spring to places where lemmings are abundant. The rodent cycles vary from place to place. One year the burrows might be relatively empty in Norway, but crowded in Sweden; but wherever the rodents occur, Snowy owls will also be found.

Lemming behavior

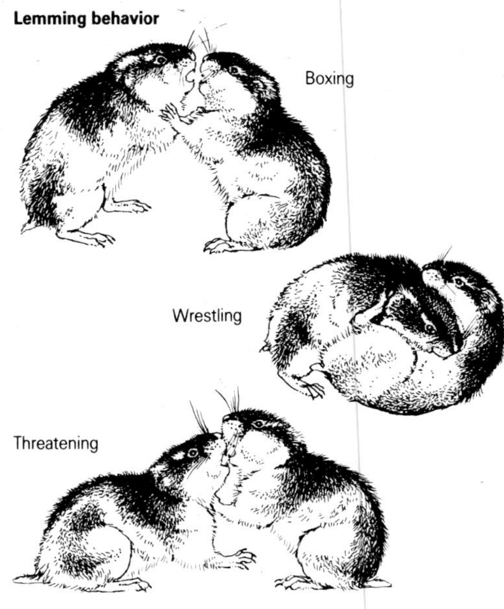

Population pressure When numbers of Norway lemmings increase, so does their aggressive behavior. Here two males wrestle, box and threaten. When the population density becomes too great, lemmings emigrate from their home areas in huge numbers in search of better grazing and more living space.

PROTECTED OR PERSECUTED?

Hillwalking and birdwatching are both popular pursuits in the Nordic countries. Every year in April several hundred birdwatchers gather at Hornborgasjön in Sweden to witness the arrival of the Common crane. These are the first birds to return from the warm south after the winter, and their deep, flutelike calls signal an end to the dark days and bitterly cold nights. The arrival of the cranes is announced on television – an occasion sometimes marked by the giving of small presents to celebrate the days of sunshine to come.

Lapland reindeer

An intimate relationship exists between the people of the north and the reindeer. Nomadic Lapp hunters once trailed the migrating herds of wild European reindeer on which they depended for food. They learned pastoral skills, evolving a culture based on the reindeer, its skin and meat. Those traditions are still practiced in a few conservation areas; but most of Scandinavia's reindeer are now owned by wealthy Lapp farmers, who drive the herds using snowmobiles or helicopters. A few hundred wild reindeer survive in Finland, where attempts are being made to protect some areas of their mixed woodland habitat from being taken over by commercial forestry.

Thousands of reindeer recently had to be slaughtered when radioactive contamination from the 1986 nuclear accident at Chernobyl in the Soviet Union was absorbed by the tundra plants, and hence the reindeer, rendering their meat unfit for consumption. Such contamination may well persist in slow-growing lichens and tundra plants for decades.

Fear of wolves

Wolves have been feared and hated by humans for thousands of years. Persecution and habitat destruction have brought them to the brink of extinction in Norway and Sweden, where they are now dependent on human protection for survival. Nobody in Sweden has been attacked by a wolf since 1821, but this does little to reassure farmers and weekenders in the wilderness, who have been brought up on stories of ravening wolves.

In 1983 a pair of wolves met to breed in Värmland, several hundred kilometers south of the Arctic Circle. The attraction to the pair lay in the extensive forests,

PILOT WHALES

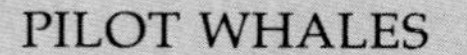

Pilot whales have provided the people of the Faeroe Islands – which lie midway between Norway and Iceland – with meat and oil since the islands were first inhabited by the Vikings more than a thousand years ago. Records dating back to 1584 show that a total of 117,456 whales were killed over the next 300 years. Despite worldwide efforts to conserve whale stocks, the hunt continues to this day. When schools of whales are sighted offshore, the islanders take to their boats for the *grindabod*. By shouting, whistling and throwing stones on leather thongs into the water, they confuse the whales with noise (whales have excellent hearing and their calls can be heard over many miles through water). Then they drive them into shallow bays where they are killed with knives and spears.

The people of the Faeroe Islands have little need for the whalemeat. Traditionally it was shared out among the village commune; today it is still second in importance in their diet to lamb and mutton. Many schools of Pilot whales are driven ashore each year, and about 1,500 individuals are killed.

Sea of blood The slaughter of Pilot whales in the Faeroe Islands is a tradition that dates back many hundreds of years.

Nomads of the north (*above*) Reindeer spend the summer on the mountain tundra, feeding mainly on lichen and grass. In fall they migrate long distances southward to the shelter of their native forests. Domesticated reindeer and their herders do likewise. The genetic integrity of the wild forest reindeer is now considered to be at risk from interbreeding with domesticated reindeer.

Heralds of spring (*right*) Common cranes on a field near Hornborgasjön, Sweden, on their way north in spring, take time off to feed and rest. The cranes have flown 5,000 km (3,000 mi) up through Europe to breed among the peaceful lakes and woods of Scandinavia. The potatoes in the surrounding fields are grown specially, and subsidised by the state. This easily accessible food source diverts the cranes from eating seed corn.

which contain the highest density of elk found in Sweden. Regrettably, these and other young wolves have since been killed by humans – one was hit by an automobile in Stockholm, while several others were shot in Norway and Sweden. A blaze of publicity caused passions to run high; a society for the destruction of the wolf was formed as wolf hysteria gripped the countries. Since then 15 more wolves have been publicly hunted down, and no punishment has been meted out to the killers. The total wolf population of Norway and Sweden has now been reduced to fewer than 10 animals. Both countries have ratified the Bern Convention on Endangered Species, but it would seem that this cannot, in practice, offer adequate protection to the greatly maligned wolf.

Among the lakes

Many of the numerous lakes and rivers of the Nordic countries are now devoid of fish and other aquatic life. This natural disaster has been caused by clouds containing acid rain, produced mainly by the pollutants emitted by industry and motor vehicles in Britain and on the European mainland, which are blown over the Nordic countries. The rain that falls reacts with chemicals in the soil to acidify water to such an extent that fish are unable to survive in these conditions.

Some aquatic animals, however, have continued to thrive, especially in the far north. In Finland, where human population is low, the country's great many lakes provide an ideal habitat for beavers. The activities of this animal – chopping down trees and building dams and lodges – can flood hundreds of square kilometers of countryside. Beavers were hunted almost to extinction for their fur in the 19th century, though efforts have since been made to preserve and reintroduce the animals. In Norway and Sweden the European beaver was used for the experiment, but in Finland it was found that the Canadian beaver thrived better and the country now has both types of beaver. The experiment has been a great success; by 1980 the population of beavers in Sweden was estimated to have risen to some 40,000. The animals did so well that they succeeded in flooding the large town of Östersund in western Sweden a year or so later. Limited hunting of beavers has even had to be resumed in order to protect the human habitat.

The Atlantic salmon

Although it is a sea fish, the Atlantic salmon breeds in fresh water. For millions of years the fjords, rivers and streams of the Nordic countries have attracted vast numbers of salmon. Large individuals swim up the rivers in winter, smaller ones in summer. The marvel of the Atlantic salmon is that, after four years spent at sea, it returns thousands of kilometers to exactly the same stream where it hatched, in order to spawn. The key to how it finds its way is a remarkable sense of smell allied to an acute memory. Rocks, soil, vegetation and the presence of juvenile relatives give each stream its own distinctive flavor. The salmon literally "smell" their way home.

The females select spawning sites in the gravel where they are courted by brightly colored males as they lay their eggs. Acts of spawning continue for a fortnight, the salmon resting in deep holes in the riverbed between each tiring bout; finally they drift back to the sea. Mature salmon do not feed in fresh water, and they are so exhausted by their efforts that most males die before they reach the sea. Females seldom survive four spawnings.

The eggs hatch in April or May, and the young emerge from the gravel to feed on insect larvae and other invertebrates. The young salmon, or "parr", stay in fresh water for one to five years before migrating to the sea (as "smolt"), where they grow rapidly, reaching over 14 kg (30 lb) in just three years.

The salmon fishing industry

Angling is an inefficient way of catching salmon and it inflicts relatively little damage on fish stocks. Even netting the mouth of the estuary allows reasonable numbers of salmon to swim upstream to breed. Unfortunately, in the late 1950s a United States nuclear submarine reported large shoals of salmon under the ice of the Davis Strait, between Baffin Island and Greenland. This sparked off a rush of commercial fishermen to capitalize on the important discovery of where the fish spent their growing period at sea. Disaster followed as the Atlantic salmon population was decimated. Many North Atlantic netsmen lost their livelihood, and anglers their sport, as fewer and

Salmon leaping a waterfall (*below*) The urge to return to their spawning grounds is so strong that salmon can overcome formidable obstacles, aided in this case by their powerful tails. Sadly, such determination is often thwarted by dams and sluices.

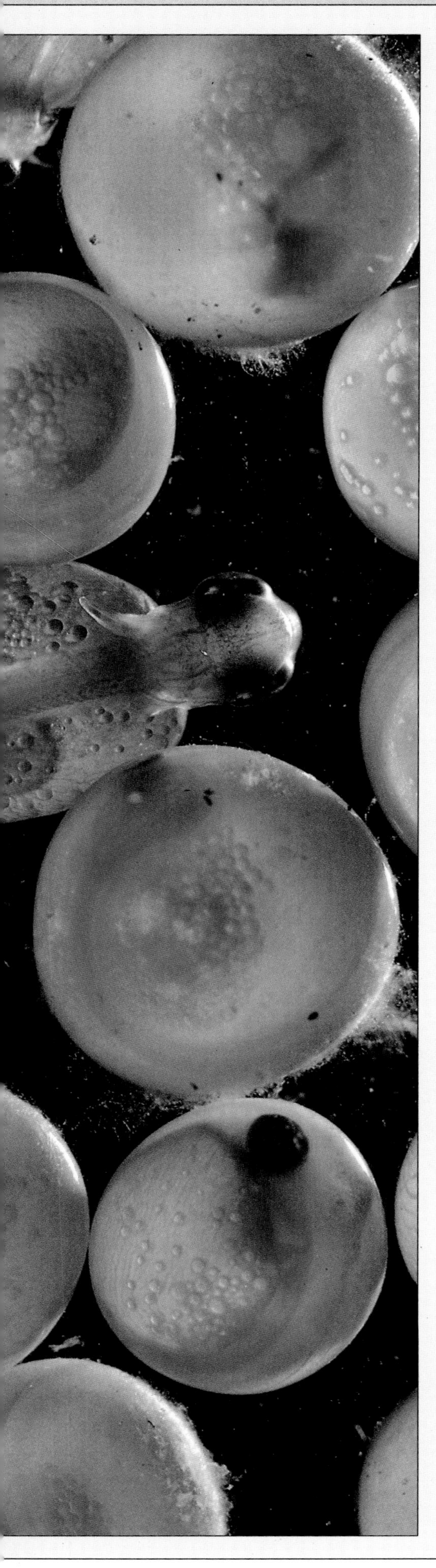

fewer salmon returned to spawn in the streams where they were born.

A recent threat to the salmon fishing industry has come from the accumulating effect of acid rain. Sweden is particularly susceptible to its effects because its rocks and soils are already naturally acidic rather than alkaline, so there is little buffering action.

Besides affecting organisms directly, the acids leach heavy metals from the soil. Aluminum that has been unlocked from the soil and leached into streams has a devastating effect on fish and their invertebrate food. In 1900 anglers took 30,000 kg (30 tons) of Atlantic salmon out of the seven main rivers of southern Sweden. Since 1970 nothing has been caught – the water is now too acid to support aquatic life.

The threat from salmon farming

Salmon farming has become a major industry, providing most of the salmon for the domestic markets of Europe. The fish are bred and reared in huge nets that are anchored in saltwater inlets and in sheltered bays. Carefully measured quantities of artificial food are scattered in the water by automatic feeding devices. Inevitably some of this food is swept out of the nets to fall on the seabed.

In the confines of these hatcheries diseases and parasites can spread very rapidly among the young salmon, and chemicals and antibiotics are added to the water in order to control them. These additives also leak into the surrounding water, and affect other marine life such as crustaceans and mollusks.

These farms are also a threat to the wild salmon. Each river has its own distinct genetic stock of salmon, finely adapted by evolution to the particular conditions of that river. Escaped farmed salmon interbreed with the wild ones, reducing the genetic variability of the species and also the ability of the fish to survive in particular rivers.

Hidden in the gravel (*left*) salmon eggs and alevins (newly hatched fry) are nourished by fat yolk sacs. They still have their external gills, but these will disappear as the yolk sac is absorbed.

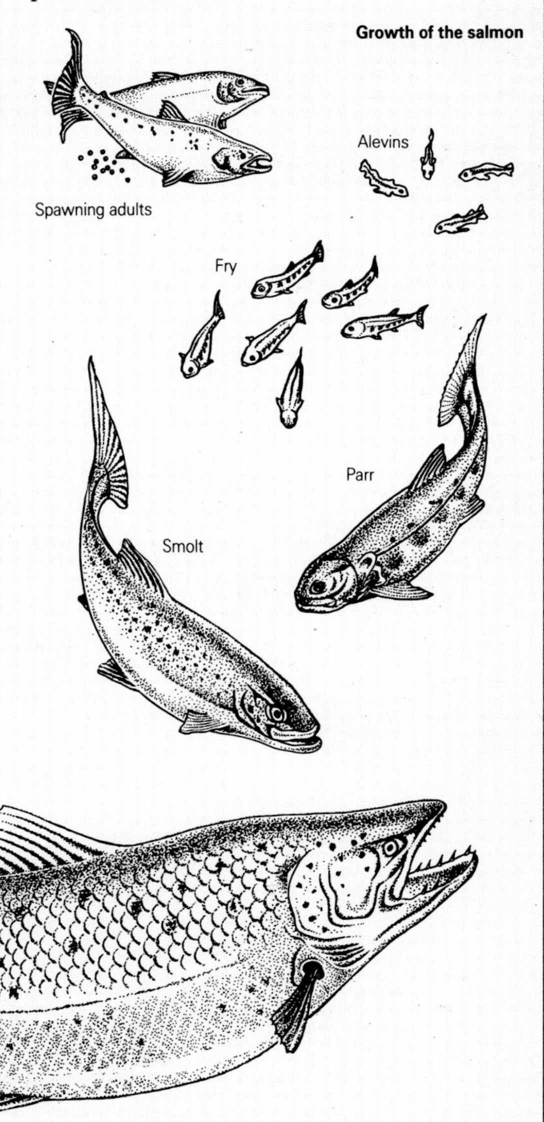

Life cycle of the salmon (*right*) Salmon deposit their eggs at 5–10 minute intervals between extended periods of rest. The eggs are nourished for the first few weeks of life by a yolk sac attached to the abdomen. As soon as its fins are more developed, the salmon fry's downstream journey begins. It develops distinctive markings and becomes known as a parr. In fresh water, salmon grow comparatively slowly. The parr's characteristic markings disappear when the salmon becomes a smolt and enters the sea where it remains for up to four years. As the adult fish return to the rivers of their birth and head upstream for the spawning grounds, the males' color heightens and their bottom jaw develops an upward hook, which impedes the temptation to feed. The rigorous journey and lengthy spawning process – about two weeks – considerably weaken the salmon and many die returning to the sea.

ISLAND ANIMALS

COLONIZATION AND CHANGE · ADAPT OR DIE · CROWDING OUT THE ANIMALS

The animals of the British Isles are essentially European in origin, but have been separated from the mainland for about 9,000 years; in the case of Ireland the period of isolation has been even longer. Virtually no part of the region remains free of human influence. Although some mammals such as foxes and badgers have survived in manmade environments, modern agriculture has taken its toll of other animal species. The largest native land mammal is the Red deer, living on moorland in the north and west; the biggest carnivores are the European river otter and European badger. The coasts are rich in wildlife; estuaries provide stopover points and wintering sites for shorebirds and waterfowl on their seasonal migrations; and colonies of seabirds nest and breed on steep cliffs and offshore islands.

COUNTRIES IN THE REGION

Ireland, United Kingdom

ENDEMISM AND DIVERSITY

Diversity Very low to low
Endemism Very low to low

SPECIES

	Total	Threatened	Extinct†
Mammals	51	1**	0
Birds	300*	3	1
Others	unknown	27	0

† *species extinct since 1600 - Great auk* (Alca impennis)
* *breeding and regular non-breeding species*

** Some other threatened cetaceans, for example, Fin whale *(Balaenoptera physalus)*, are occasionally reported in British waters

NOTABLE THREATENED ENDEMIC SPECIES

Mammals none
Birds none
Others Ivell's sea anemone *(Edwardsia ivellii)* (possibly extinct)

NOTABLE THREATENED NON-ENDEMIC SPECIES

Mammals Harbor porpoise *(Phocoena phocoena)*
Birds Red kite *(Milvus migrans)*, White-tailed sea eagle *(Haliaeetus albicilla – reintroduced)*, corncrake *(Crex crex)*
Others Kerry slug *(Geomalachus maculosus)*, Chequered skipper butterfly *(Carterocephalus palaemon)*, Freshwater pearl mussel *(Margaritifera margaritifera)*, Ladybird spider *(Eresus niger)*

DOMESTICATED ANIMALS (originating in region)

none

COLONIZATION AND CHANGE

At the peak of the last glaciation most of the British Isles was buried under a huge ice sheet, and was linked to the rest of Europe by a land bridge that is now covered by the North Sea. As the climate began to warm the ice melted and the sea rose, cutting off Ireland about 11,000 years ago. As a result, it contains fewer animal species than the rest of the region: there are no snakes in Ireland, for example, since the land bridge had disappeared before the climate had warmed sufficiently for them to move in. About 9,000 years ago the remaining land bridge between the British Isles and Europe was gradually flooded over and animals were no longer able to travel between the two.

With so brief a period of isolation, it is not surprising that the number of endemic species and subspecies is very small. A few races of small rodents on remote islands, the British red squirrel, the Irish stoat, the Scottish crossbill (both distinct species) and the Red grouse are among the few endemics. As the ice sheet retreated and the land warmed up, tundra was replaced by coniferous forest, and many species spread northward before the land bridges were severed, but a few were cut off in the higher cooler areas of Wales and Scotland, in the west and north of Britain. Ptarmigan, Black grouse and Snow buntings still breed in the Scottish highlands, and shorebirds such as greenshank, Golden plover and dunlin breed on upland bogs reminiscent of the treeless tundra marshes. The great flocks of ducks, geese and shorebirds that had nested in the region in cooler times now breed on the Arctic tundra, but they still fly great distances south in winter. They rest and overwinter in Britain's estuaries, which are rich in plankton and other marine life.

The waters of some of the larger northern lakes support relict populations of Arctic fish such as charr and grayling. The Arctic charr of Scotland's Loch Ness, whose deep waters remain very cold, has evolved from a seafaring species that ascends rivers to spawn. Trapped in the loch long ago, it now passes its entire life cycle in fresh water.

Brown trout are also nonmigratory in habit, and remain in lakes and rivers; however, they belong to the same species as Sea trout, which do migrate to the sea. Today, a new barrier to migration has appeared in the form of river pollution, which may be so bad in some rivers that migratory fish, notably salmon, cannot survive the journey to the sea.

A changing landscape

The lowland areas of Britain and Ireland were once covered in broadleaf deciduous woodland, but centuries of agriculture have created a mosaic of cultivated fields and sprawling human settlements. Larger mammals such as wolves and the Brown bear became extinct long ago, persecuted by farmers and shepherds. Other carnivores, such as the Pine marten, polecat and native wildcat, survive only in remote mountain areas, where they are safer from human predators than in their former lowland homes.

Here they mingle with animals more typical of the great northern coniferous forests of Europe – Crested tits, Coal tits and Long-eared owls. The capercaillie, which became extinct in the north during

Native predator *(above)* The Irish stoat or ermine is one of the few vertebrates endemic to the British Isles. A fierce predator, its slender body can slip into the burrows of smaller animals. It is distinguished from the mainland subspecies by its smaller size.

Woodland fox *(below)* A common predator of woodland and countryside throughout the region, the Red fox has also adapted well to living in cities. The female, or vixen, is smaller than the male and bears litters of four or five cubs.

the 18th century, has been successfully reintroduced, and large birds of prey such as the Peregrine falcon and Golden eagle are also occasionally observed. The Red deer, a woodland animal, has adapted to open country in the Scottish highlands and the high granite moorlands of the southwest peninsula.

Heathlands are limited to sandy, uncultivated soil in southern Britain. They are an important habitat for a number of declining species, such as the Dartford warbler, a bird at the northern limit of its range, and for reptiles and amphibians, particularly the rare and protected Sand lizard, which needs a warm, dry sandy environment in which to lay and incubate its eggs, and the Natterjack toad. Over 22,000 species of insect breed in the British Isles, including 55 species of butterfly and 40 species of dragonfly. Road embankments, churchyards, disused sand and gravel pits, gardens and garbage dumps all provide valuable havens for a rich variety of wildlife.

ADAPT OR DIE

The seasonally warm waters along the western coasts of the British Isles support a plentiful supply of fish. These provide a food source not only for the numerous seabirds that nest along the cliffs but also for seals and cetaceans. These range from the larger filter-feeding Fin whale, which lives on plankton and can be almost 20 m (65 ft) long, to the small Killer whale and the smaller Harbor porpoise, barely 2 m (6.5 ft) long. The large plankton-filtering Basking shark is also found off these coasts. Many migratory fish, such as herrings, have their spawning grounds near the North Sea coast. The Gray seal is found in colonies all around the coasts of the British Isles. Smaller numbers of the Common or Harbor seal are also present, and are most abundant on the northwest coast of Britain.

Coastal survival

Along the varied shores of these islands many small invertebrates extract their food from the sea. On rocky coasts sea anemones, barnacles, mussels, shrimps and worms take their food from the water, and crabs scavenge along the shore. When the tide is out some of these tiny marine animals have to survive exposure to the sun and wind and attack by predators such as gulls; they also have to withstand being buffeted by the waves as the tide advances and retreats.

Limpets clamp down hard on the rocks as the tide retreats, sometimes with such force that they wear cirular grooves on the rock surface; they use sticky mucus and a muscular foot to create a suction pad. This prevents them from being dislodged by the waves, and protects them both against desiccation and the prying beaks of predators. Although they move around, they always return to the same site to rest as the tide goes out. Barnacles remain permanently cemented to the rocks, but retreat between the plates of their shell, which close above them to keep out drying air and predators.

Mussels are fixed firmly to the rocks by tough threads, and also close their shells tight as the tide goes out. Sea anemones can survive out of the water only in very damp places; they have a mucus coating that helps to conserve moisture, and their waving tentacles attract, then paralyze, their prey. Many other seashore animals, such as crabs and shrimps, seek shelter in rocky crevices and below the thick curtains of seaweed in the rock pools.

On muddy shores many creatures live in burrows or bury themselves in the sediments. Ragworms have paddlelike appendages, which they beat rapidly to draw a stream of water, carrying both food and oxygen, through their U-shaped burrows. Cockles and razor shells lie hidden in the sand, but protrude siphons that draw water through their bodies. Burrowing crustaceans of the shoreline emerge from the sand and mud to feed as the tide washes over them, but must retreat before it returns to avoid being swept out to sea. Some species may possess a biological clock that warns them when the tide is about to return.

Learning from the birds

The various species of predator along the shoreline have evolved different feeding techniques to exploit the different kinds of food supply, and even within a single species opportunist individuals adopt new foraging techniques to make best use of available resources. For example, some members of a colony of Herring gulls may learn to locate crabs in the sand, some feed primarily on mussels or worms, and others scavenge for edible refuse at the local garbage tip. Foods may be available to the birds only at specific times of day, such as at low tide or when the rubbish is dumped. During the nesting season, when one parent must incubate the eggs, individual pairs of gulls may have complementary food preferences and foraging specializations, resulting in a more efficient use of feeding time. Fostering experiments have revealed that such behavior is not innate, but can be learned from the parent birds.

Inland birds show a similar ability to exploit new sources of food. When foil-capped milk bottles first began to be delivered to people's doorsteps many different birds, including Great tits, Blue tits, magpies, rooks, crows and jackdaws, quickly learned to tear holes in the bottle caps to reach the cream. The habit seems to have spread from various centers of origin, suggesting that individuals are able to adapt to environmental change through a process of learning.

Camouflage is an important adaptation to the environment evolved by many species. The Peppered moth is widely distributed in both the countryside and the towns. Originally most of the moths were pale in color and well camouflaged

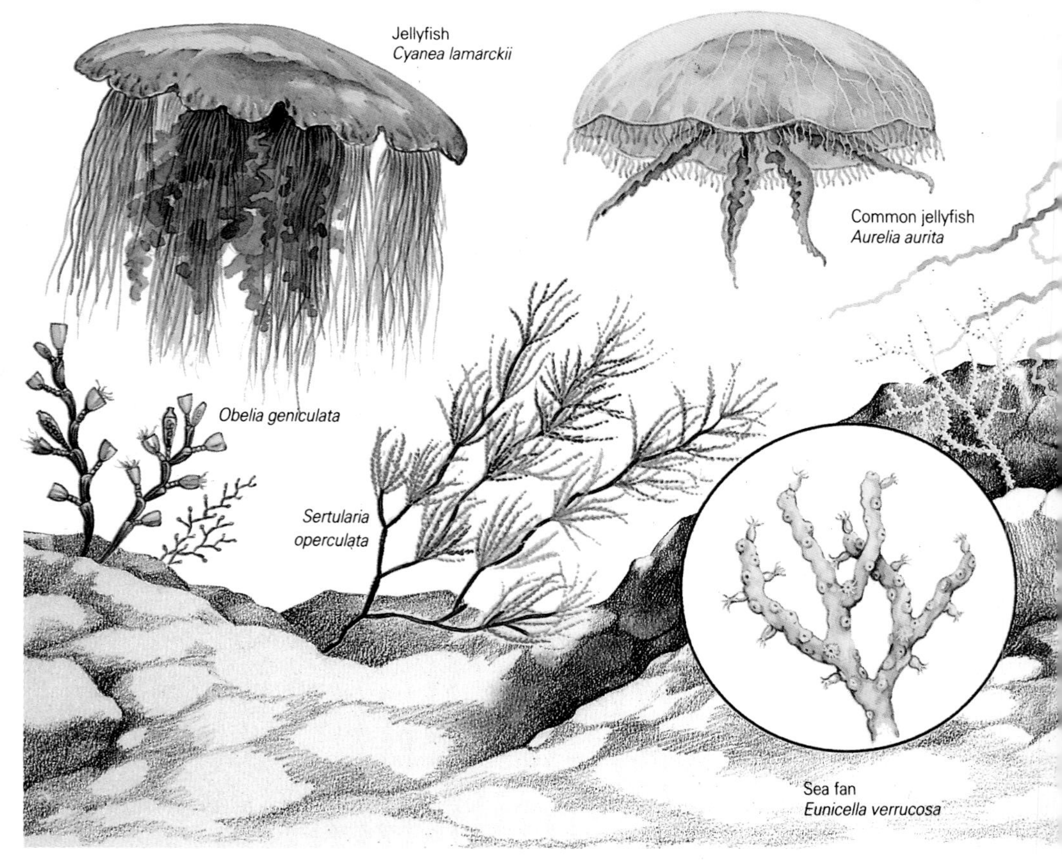

against their background, the lichen-covered, light-colored trunks of trees. In industrial areas, where smoke darkened the bark, the Peppered moth population gradually became darker and less conspicuous to predators – dark variants survived to reproduce in larger numbers than pale forms. With a return to cleaner air, the moths are now assuming their pale colors once more: they have adapted through a process of evolution to the changes in their habitat.

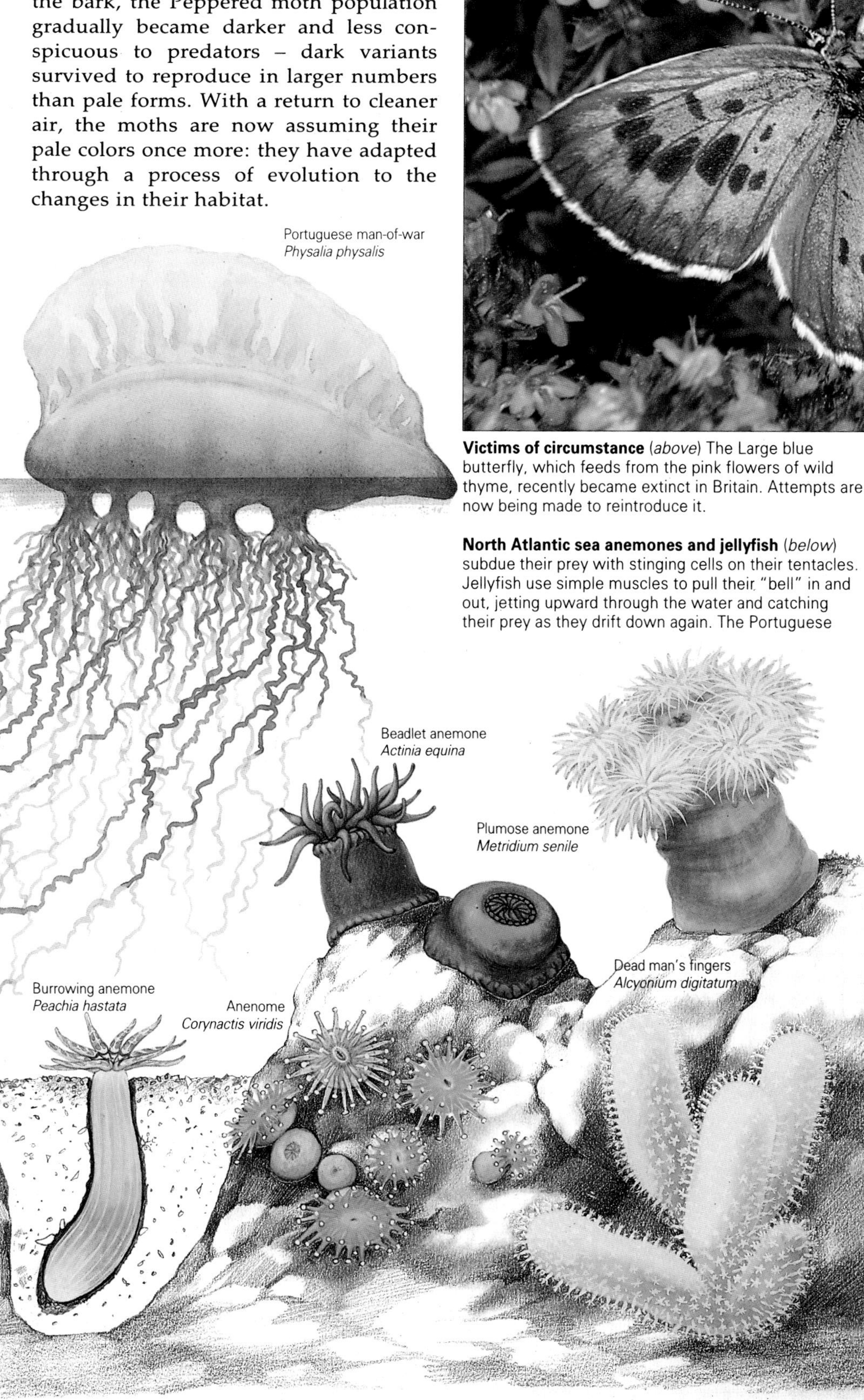

Victims of circumstance (*above*) The Large blue butterfly, which feeds from the pink flowers of wild thyme, recently became extinct in Britain. Attempts are now being made to reintroduce it.

North Atlantic sea anemones and jellyfish (*below*) subdue their prey with stinging cells on their tentacles. Jellyfish use simple muscles to pull their "bell" in and out, jetting upward through the water and catching their prey as they drift down again. The Portuguese man-of-war is a colony of individual organisms with powerful stings. The Beadlet anemone immobilizes its prey with stinging tentacles; the Plumose anemone has a sticky mucus net streaming over its feathery tentacles. A hydra, composed of a colony of polyps, seen also in close-up, and a much larger hydroid or sea fir. A Sea fan; when not feeding, the polyps retreat inside their cups. A burrowing anemone. The Jewel anemone and Dead man's fingers are also colonies of polyps.

THE DEMISE OF THE LARGE BLUE

Since 1979, the Large blue butterfly, once found in the south and southwest of England, has been considered extinct in Britain. The reasons for its demise are surprisingly complex. The newly hatched caterpillars feed on thyme, a plant of dry grasslands. In common with the larvae of other species of blue butterflies, they have the ability to produce a sugary secretion. After a while, the caterpillars drop from the plant and are picked up by a species of red ant, *Myrmica sabuleti*. They are carried to the ant nest, where they are "milked" for their secretion. Once underground, the caterpillars become exclusively carnivorous, feeding on ant grubs until pupation.

The ants that play such an essential role in the life history of the Large blue are unable to survive below a minimum temperature. When the disease myxomatosis almost wiped out the rabbit population of the British Isles in the 1950s, previously grazed turf grew taller. The ground temperature fell, creating an unacceptable environment for the ant which disappeared. So began the decline of the Large blue.

CROWDING OUT THE ANIMALS

More than two thousand years of agriculture, with accelerating industrial growth, have seen Britain's primeval forest and lowland swamps replaced by farmland and urban sprawl, with plantations of conifers and treeless rough grazing land in upland areas. This depletion of the region's natural habitats has, not surprisingly, had a serious effect on its animals. Insect life has been hard hit. Butterflies and dragonflies – victims of agricultural intensification, with the consequent loss of habitat and the effects of pesticides–are particularly threatened. Of the British Isles' total of 55 breeding species of butterflies, 10 are considered vulnerable or endangered, and at least another 13 have declined in numbers.

Many mammals are also becoming rare. The European river otter has disappeared from many parts of its former range. Bat numbers have also decreased, because their breeding and hibernation roosts in tall trees and old buildings have been destroyed, and their insect prey has been reduced by the use of pesticides. All bat species are now legally protected. Although most mammal species are declining, not all is gloom – some, such as the Wild cat and Pine marten, have recovered part of their original range.

A few animals have even benefited from close association with humans, taking advantage of new food sources and new sites for homes and dens. House mice, rats, Common woodpigeons, House sparrows and some seed-eating birds and rodents have profited by the change from woodland to fields and villages, and many animals have adapted to life in suburban and even urban environments.

Success with birds

An enormous effort in public education and the introduction of wildlife protection legislation has produced heartening results. More species of birds breed successfully in Britain now than they did at the turn of the century. Notable success stories include the natural return of the avocet and Common crane, an increase in numbers of the Red kite, and the reintroduction of the White-tailed sea eagle to its former northern haunts. Species such as

City dweller Robins are very popular birds in the British Isles. They thrive in urban gardens, nesting in a variety of unusual places from old boots to rusting motor cars. Their main enemies are domestic cats, magpies and the occasional kestrel.

Cetti's warbler and the Collared dove have also moved into the region of their own accord, the latter species exploiting the increased area of land given to grain crops. Along the coasts and extending considerable distances inland, gull colonies have expanded to take advantage of new food sources such as garbage tips. But on north and west coasts the numbers of several species of seabird have plummeted in recent years, a disaster thought to be partly caused by the commercial overfishing of coastal waters.

Afforestation, especially in its early stages, encourages the success of several bird species, particularly vole-hunting predators and birds that need ground cover. However, the planting of new upland plantations of conifers results in the loss of open moorland habitat and the birds which depend on it, such as the merlin, Red grouse and greenshank. The disappearance of lowland woodlands and hedgerows has caused many animals to be lost, although some hedgerow birds now thrive in suburban and city gardens.

A history of protection

Wild mammals and birds have always been valued for the food and sport they provide. Hunting sanctuaries established by the Anglo-Saxon and Norman kings a thousand years ago or more are today important as wildlife reserves. Game preserves and fox coverts have helped to maintain scattered woodland in the agricultural lowlands, and management for game birds has preserved the heather moorlands of many upland areas.

Fish, too, are considered valuable for food and sport but they are extremely sensitive to human activities. Salmon, for example, were once a common spectacle as they migrated from the sea to their breeding grounds upstream, but as a result of overfishing and pollution they have now disappeared from many rivers. Marine pollution is also increasing. In the late 1980s breeding seal colonies around the North Sea coasts were badly hit by a viral disease; a link has been suggested – but not proved – between exposure to pollutants and impairment of the seals' fertility and immune systems.

Farmers and foresters still have a very strong influence on land use and nature conservation. But even the most nature-loving landowners are constrained by the economic pressures that promote the use of modern intensive farming practices. Reliance on agricultural chemicals, as well as the use of large farm machines, which the striving for increased farming productivity makes necessary, are incompatible with management of the countryside to the benefit of wildlife.

There has, however, been a vast increase in public interest in and awareness of wildlife issues, as well as a growth in the number of voluntary organizations that are dedicated to land conservation. This may have come just in time to save the region's most endangered species, and to halt the decline of many others.

MAMMALS OF THE TREETOPS

The Red squirrel is a native mammal of the conifer forests of northern Europe, where it feeds on seeds and young shoots. As the forests retreated northward at the end of the last ice age the Red Squirrel spread into the expanding deciduous forests. In Britain, the felling of these woodlands had dramatically depleted this habitat by the late 19th century, leaving the Red squirrel populations severely fragmented. It was at this time that the Gray squirrel was introduced to Britain from North America for ornamental reasons. Since then the Red squirrel has disappeared from most of Britain's deciduous woodland, and is dominant only in its ancestral conifer forests in Scotland and on conifer plantations, where its habit of eating young shoots causes a good deal of damage.

It has been claimed that the Gray squirrel played an influential role in the decline of the Red squirrel, fighting and killing them and driving them out of their homes, but it is unlikely that direct aggression was the reason for their decline. Viral disease had already reduced the numbers of Red squirrels. This allowed the Gray squirrel to become rapidly established. A native of the mixed deciduous forests of North America, the Gray squirrel is well adapted to survive in Britain's woodlands. Larger and more robust than the Red squirrel, it can withstand periodic food shortages, storing enough fat to survive the winter cold; the Red squirrel has to forage for food at least every two days throughout the winter.

The introduced Gray squirrel, commonly found in parks and gardens, has become a serious pest in forestry plantations.

Non-native animals

Some of the most commonly seen animals in the British Isles are non-native species. The rabbit has had perhaps the greatest impact of all on the nature of the British countryside, and where they have been eliminated, thorny scrub invades the cropped grasslands that are the typical landscape of Britain's chalk and limestone hills. The rabbit was probably introduced in the late 12th century to be bred for food, but it did not become abundant in the wild until much later. Changes in agriculture, and the trapping and shooting of many of its natural predators, which also took farmers' livestock, were major factors in its spread.

Rabbits were themselves destroyed as pests – they compete with sheep for grazing, invade fields of root crops and nibble the tender young shoots of cereal crops. Despite attempts to control them, by the 1950s the rabbit population of Britain had risen to an incredible level of between 60 and 100 million animals, but this increase was not to last – in the early 1950s over 90 percent of the wild rabbit population was destroyed by an introduced virus, *Myxomatosis cuniculus*. Since then, rabbits have staged a recovery, but continuing outbreaks of the virus keep their numbers in check.

The rabbit is a good example of an introduced animal successfully colonizing an area where there is a vacant ecological niche. The rabbit was introduced onto mainland Europe north of the Pyrenees around 2,000 years ago and rapidly spread northward. It would probably have invaded the British Isles had they not been cut off by the English Channel.

Over the centuries, many animals have been deliberately introduced. The ferret, first brought here some 800 years ago for sport, and to catch rabbits for the cooking pot, began to hybridize with the native polecat. Also introduced for sport, the pheasant is an attractive addition to the region's wildlife, but gamekeeping to preserve the young birds for shooting has had a serious impact on the numbers of native predators.

Animals have also been introduced to ornament estates and parklands. Several species of introduced deer include the Sika deer from eastern Asia, the Chinese water deer and the muntjac (in many parts of England the commonest species of deer). Like the native Red deer, they

The South American coypu (*above*) is a large wetland rodent that was introduced to Britain to be farmed for its fur, where it became a pest in the wild. An expert swimmer, it spends most of its life in water, and undermines banks with its burrows.

The prolific rabbit (*below*) is famous for the rate at which it multiplies. It has modified the landscape of much of the region, grazing the vegetation to a short turf and promoting the growth of a rich variety of flowering plants.

damage young saplings by bark-stripping and browsing.

Accidental arrivals

Not all animals are deliberately introduced. The Ship (Black) rat and Common (Brown) rat arrived on ships. In the 14th century the flea of the Ship rat was the carrier of the Black Death, a bubonic plague that took a terrible toll of human life. The House mouse arrived in prehistoric times, as did the voles of the islands of Orkney and Guernsey.

Introduced species have often escaped from captivity to flourish in the wild. Mink were brought here to be farmed for their fur. In the wild, they breed by streams and rivers, taking fish and waterfowl. Coypu (large South American rodents reared for fur) also escaped from farms in eastern Britain, where they raided vegetable gardens and cereal crops and burrowed in the banks of dikes, causing flooding and destroying reed beds. A successful campaign of trapping appears to have eradicated this pest. Introduced Gray squirrels pose another threat to the environment. Their habit of stripping the bark from young trees, especially beech, oak and sycamore, does serious damage to woodlands.

Escapes from domesticity are commonplace – feral ponies, goats, sheep, cats and pigeons are all found. More decorative are the escaped Ring-necked parakeets that have established wild breeding populations in places. In other countries the parakeet is a serious pest on arable land; so far in Britain the cold winters have kept it in check.

Climate, habitat, availability of correct food, absence of predators and competition from local species are all important in determining whether an alien species will survive, or even reach pest proportions. Attempting to predict the outcome of an exotic introduction is risky; all too often, a thoughtless action will result in considerable cost to humans, native wildlife and the environment. Of the animals exported from Britain around the world, several have become serious pests in their new environments, including the rabbit in Australia and the House sparrow and starling in North America.

Introduced for sport The Ringneck pheasant is one of the most attractive additions to the region's wildlife. Many thickets and copses have been preserved to provide cover for the birds – and incidentally for other animals such as foxes, deer and songbirds.

Colonists of the cliffs

The breeding seabird colonies of the British Isles are of international importance. The region has some 11,000 km (6,800 mi) of varied coastline, together with an abundance of fish in the cold northern waters that provide ideal conditions for the birds. Cliff ledges are particularly suitable for nesting seabirds; they are safe from mammalian predators, and the updraft of air along the cliff face assists the birds in landing and taking off. However, they can be hazardous places for eggs and young, and many species have evolved adaptations to minimize the risks. Guillemots (murres) lay eggs that are pointed at one end; if disturbed, they roll around in a circle, rather than off the ledge. Kittiwakes make nests of seaweed that they cement to the ledge with guano (droppings) and mud. The young instinctively avoid the cliff edge, are relatively calm while begging for food, and do not jump around when exercising their wings.

Several species of seabird may share the same cliff, but will usually have different preferences for nest sites. Puffins and Storm petrels nest in burrows on the grassy upper slopes, while gannets fight for space on the cliff edge. Cormorants and shags nest near the base of cliffs, close to rocks on which they can stand to dry their wings after diving for fish. Guillemots huddle close together along narrow ledges, while razorbills nest in small groups on tiny ledges and buttresses. Fulmars repel predators by spitting foul-smelling oil.

Kittiwakes are cliff-nesting seabirds that spend the rest of the year outside the breeding season wandering the oceans. About half a million pairs now breed around the coasts of the British Isles.

ANIMALS OF WOODS AND MOUNTAINS

A VARIETY OF HABITATS · PREDATORS OF THE SKIES · THE CHANGING COUNTRYSIDE

Compared to its neighboring countries, France has a rich population of mammals and birds. The barrier of the English Channel prevented the spread of many species to the British Isles, and France's generally mild climate favored animals that are restricted by harsher climates to the east and north. The country is on a major north–south migration route for birds. Physical and climatic contrasts within the region have led to a great diversity of wildlife. The permanently snow-covered peaks of the Alps lie only a short distance from the arid warmth of the Mediterranean coast. The rocky shores of the west and northwest, lashed by Atlantic gales, contrast as habitats for wildlife with the high limestone plateaus and steep river gorges of the Massif Central, and also with the region's extensive woods and farmlands.

COUNTRIES IN THE REGION

Andorra, France, Monaco

ENDEMISM AND DIVERSITY

Diversity Low
Endemism Low

SPECIES

	Total	Threatened	Extinct†
Mammals	101	7	1
Birds	400*	5	0
Others	unknown	104	0

† *species extinct since 1600 - Sardinian pika* (Prolagus sardus)
* *breeding and regular non-breeding species*

NOTABLE THREATENED ENDEMIC SPECIES

Mammals none
Birds none
Others Corsican painted frog *(Discoglossus montalentii)*, asper *(Zingel asper)*

NOTABLE THREATENED NON-ENDEMIC SPECIES

Mammals Long-fingered bat *(Myotis capaccinii)*, Pond bat *(Myotis dasycneme)*, Mouse-eared bat *(Myotis myotis)*, Pyrenean desman *(Galemys pyrenaicus)*, European mink *(Mustela lutreola)*, Harbor porpoise *(Phocoena phocoena)*, Fin whale *(Balaenoptera physalus)*
Birds Red kite *(Milvus migrans)*, White-tailed sea eagle *(Haliaeetus albicilla)*, corncrake *(Crex crex)*, Little bustard *(Tetrax tetrax)*, Audouin's gull *(Larus audouinii)*
Others Corsican swallowtail butterfly *(Papilio hospiton)*, Quimper snail *(Elona quimperiana)*, Shining macromia dragonfly *(Macromia splendens)*, Longhorn beetle *(Cerambyx cerdo)*

DOMESTICATED ANIMALS (originating in region)

rabbit *(Oryctolagus cuniculus)*

A VARIETY OF HABITATS

During the last ice age vast glaciers held northern Europe in thrall, destroying all life. Although in France temperatures were doubtless much lower than they are today, no sheets of ice engulfed the region. Its wildlife escaped death by cold. The survivors of those times can be found in inaccessible mountain areas in the Alps and Pyrenees. Snow voles thrive above the treeline – up to 4,000 m (13,000 ft) on Mont Blanc in the Alps, where they can be seen jumping from rock to rock on sunny slopes. There are populations of other sure-footed mountain animals as well: chamois and the goat-like ibex, and mouflon – small wild sheep that survive on the steep wooded slopes by eating the bark of trees when snow covers the ground.

The remoter mountain forests afford refuges for large carnivores such as the Brown bear and lynx, both of which were hunted almost to extinction in the past. In recent years attempts have been made to rescue Europe's tiny population of lynx, which thrive best in woodland on precipitous slopes, by reintroducing them to former haunts in the Alps. To survive in winter lynx need at least a kilogram (2 lb) of flesh every day, and require extensive hunting grounds in order to catch enough prey. Hares form a large part of their diet, but they also take chamois.

The Brown bear population of the French Pyrenees has been reduced to as few as a dozen animals. Despite recent legislation to protect them, they are still

In search of wild waters (*left*) The European otter is becoming rare. Its numbers are declining due to water pollution, disturbance of waterways by boat traffic, urban and agricultural encroachment, and loss of den sites due to removal of trees.

persecuted by local farmers whose livestock may occasionally fall victim to the bears. The genet has a stronghold in the rocky, scrubby terrain in western France. With its long striped tail and spotted coat, it is an attractive animal, and many have been introduced into the area.

Forest mammals

Much of France was once covered in deciduous forest, but only a fifth of the country is wooded today. The Red squirrel thrives in the mixed woodlands of chestnut, beech and oak; polecats, stoats and weasels are common, and badgers prosper in areas where woodland alternates with pasture and cultivated ground. The Wild boar, too, can survive in cultivated country provided it can make its base in wild areas of swampy woodlands, reedbeds or forest. Together with Red and Roe deer, it has been actively conserved in France for centuries, reflecting the French enthusiasm for hunting. European otters, European mink and wildcats can still be found in remoter areas, although the expansion of agricultural land that has taken place in recent decades has destroyed extensive areas of previously wild habitat.

One animal that has managed to adapt to recent environmental change is the Beech marten. Like its relative the Pine marten, it lived in rural habitats until a few years ago. However, it is able to survive close to human habitations, and a recent study has shown that Beech martens are now living in built-up areas. The animals burrow through panelling and insulation in buildings, and they are regarded almost as a pest in some of the large French cities.

The variety of climates and habitats found within the borders of France means that the country is able to support at least 26 species of bat. Caves, holes in walls or trees, as well as the roofs of all kinds of buildings, provide shelter for nursery roosts and hibernation sites. Among other small mammals, France has nine species of shrew, and three species of dormice: the Fat dormouse, the Garden dormouse and the Hazel dormouse.

Wetland wealth

The lagoons and marshes of the west and south coasts include some of the most important wetland habitats in western Europe. These attract vast numbers of migrant shorebirds and waterfowl to overwinter, to rest and to breed. Many have been drained in recent years, but one of the most remarkable wetlands in the world – the Camargue, an area of lagoons and marshes in the Rhône Delta – has remained a wilderness because of the salinity of its otherwise rich soil. The area contains a spectacular variety of birdlife.

On the Etang de Vaccarès, the largest of the lagoons in the heart of the Camargue, over 5,000 flamingoes breed, making an unexpected splash of pink against the bright blue of the Mediterranean sky and water. Avocets and terns can also be found on the lagoons, while the short grass at their edges supports Short-toed and Crested larks, Kentish plovers, Spectacled warblers, Tawny pipits and pratincoles. Great gray shrikes and bee-eaters hunt over the surrounding dry vegetation, and Black kites, hoopoes, Penduline tits, Fan-tailed and Sardinian warblers, Scop's owl and Lesser gray shrike breed in disused drainage ditches. The dense woodland nearby attracts populations of spectacular Golden orioles, as well as Cetti's and Melodious warblers.

Predators beware (*below*) The hoopoe's young produce foul-smelling dung that the parents make no attempt to remove from the nest: the stench is thought to deter predators. The bird lives mainly on the ground, feeding on insects and lizards.

PREDATORS OF THE SKIES

A wide variety of birds of prey are found in France. The exposed mountain slopes of the Pyrenees and the open scrubby maquis of the south of France and Corsica are ideal habitats for airborne predators, and also support a wealth of small vertebrate prey – rabbits, hares, mice, voles and shrews. According to the habitat and the type of prey, species of predator have developed different hunting methods and adaptations, though all birds of prey are easily recognizable by strong hooked bills for tearing flesh and powerful talons for gripping their prey.

Osprey
Pandion haliaetus

Marsh harrier
Circus aeruginosus

A difference of style The osprey cuts a dramatic figure in the sky as it circles the water, searching for the gleam of a fish below the surface. It swoops rapidly, plunging beneath the water to emerge with a fish writhing in its talons. The Marsh harrier glides just above the grasses. Ruffs of feathers on each cheek act as sound collectors that pick up the slightest rustle made by its prey: small vertebrates such as birds, mammals, lizards and frogs.

Built for hunting

The large birds of prey that cruise the countryside – the buzzard, and Golden, Bonelli's and Booted eagles – have broad straightish wings for gliding long distances on high air currents, and a tail that can be broadly fanned as a rudder. The largest birds of prey – the vultures – have very long wings that enable them to glide long distances on warm air currents. Three species are found in remote areas in the Pyrenees – the Egyptian and Griffon vultures, and the lammergeier (Bearded vulture). These powerfully built birds need to cover extensive territory in order to find enough carcasses to satisfy their food requirements.

The eagles have thick, strong toes for gripping large prey such as hares, and may even be capable of breaking bones. Bonelli's and Booted eagles both prey on snakes and lizards, and have short toes for grasping their wriggling prey. The osprey, whose fish prey is even more slippery, has a series of tubercles (rough swellings) on the undersurfaces of its feet to improve its grip.

By contrast, kites, hobbies and merlins, which hunt small birds on the wing, have slender, long, widely spread toes with long, curving claws – improving their chances of catching prey. Their wings are narrow and curved to give them speed and maneuverability in the air. The sparrowhawk also preys on small birds, but its natural habitat is woodland, and it has shorter, rounder wings that make it easier for it to dodge among the branches. The peregrine has the most spectacular chase of all. It cruises at high altitude until it detects its prey of small birds on the wing. Folding its wings close against its side, it drops bulletlike through the sky at speeds of up to 350 kph (200 mph).

The kestrel has mastered the ability to hover above the ground, searching for prey. It hangs almost motionless in the air, its head remaining in exactly the same position despite the rapid beating of its wings. It uses its long, square-cut tail to angle itself carefully to compensate for the wind. Harriers prefer to hunt in grassland, cruising low over the ground to spot frogs, mice and other small creatures. They are rare among birds of prey in nesting on the ground, and their long legs enable them to keep a good lookout among the tall grasses.

Survival in a harsh climate

Even on the bleakest mountains, birds of prey can still find enough food in winter. The thick snow cover of winter provides small animals with an insulating blanket, because the temperature under the snow usually hovers around freezing, while the temperature in the open air may be more than 20°C below freezing (−4°F). The mountain hare, which molts to a white coat in winter, has particularly large paws that spread its weight over the snow. For safety it creates firm, well-defined runways across the dangerous loose snow. Active at night, the Mountain hare often burrows under the snow to sleep during the day.

The Alpine marmot, by contrast, sleeps through the whole winter at altitudes of 1,000 to 2,500 m (3,000 to 8,000 ft) living off reserves of body fat. Groups of marmots, huddled together to conserve heat, hibernate for six or seven months in a stopped-up chamber, insulated from the cold with layers of grass. In summer, they feed on grass and roots in the meadows above the treeline.

Among the mountain game birds the ptarmigan molts to winter white, which acts as camouflage in the snow. It lives on the high mountaintops of the Alps and Pyrenees, and scrapes an insulated hollow in the snow to shelter from extremely cold weather. In spring the male begins his spring molt to summer brown much later than the female, so that he becomes conspicuous against the greening landscape and is more likely to attract the attention of predators. This may reduce the chance of a predator discovering the female sitting on her eggs.

Kestrel
Falco tinnunculus

Peregrine falcon
Falco peregrinus

Master hunters (*above*) Both kestrels and Peregrine falcons have acute vision. The kestrel usually hunts at a lower level than the peregrine, and can hover almost motionless as it scans the ground far below for the movements of prey. The peregrine is the bird-world's fastest sky-diver; it plummets through the air, pulling up at the very last second to seize its prey.

HENRI FABRE AND THE INSECTS

Sérignan in southern France was the home of one of the pioneers of the study of animal behavior – Jean Henri Fabre (1823–1915). He was a schoolmaster who made many original observations and simple experiments on a number of insects. Although his conclusions about the intelligence of insects were often wrong, his observations were amazingly accurate and detailed. He spent days watching hunting wasps catching other insects to provision their nests, and concluded that their behavior, though simple and usually rigid, was sufficient to enable the insects to survive. One of his subjects was the bee-wolf (*Philanthus*), a solitary wasp that provisions its nest with honeybees. These it kills, unlike other solitary wasps, which merely paralyze their victims to provide food for their larvae. Curiously, the better armed and armored honeybee was incapable of aiming its sting effectively when attacked by the wasp, which has unerring aim, injecting venom into the bee's brain, causing instant death.

As Fabre put it, "How is it that the *Philanthus* has learned for purposes of attack what the bee has not learned for the purposes of defence? To this difficulty I see only one reply: the one knows without having learned and the other does not know, being incapable of learning." In other words, their behavior is instinctive. Fabre believed that instinct is the most important force in insect life, rejecting evolutionary theories as irrelevant.

Griffon vulture
Gyps fulvus

Booted eagle
Hieraaetus pennatus

Power in bill and claw (*above*) Vultures and eagles have powerful, hooked bills for tearing at flesh, and strong, curved talons for gripping prey or carcasses. The Griffon vulture has no feathers on its head and neck; possibly an adaptation to sticking its head deep inside carcasses to feed. Feathers would become too bloody. The Booted eagle feeds mainly on birds, but is also adept at catching snakes: its short toes with highly curved claws provide a fearsome grip.

THE CHANGING COUNTRYSIDE

In France – as in most of Europe – the main threat to wildlife is the loss of wild habitats to agriculture, and the increase in disturbance and pollution owing to the intensification of farming methods. The wetland habitats in particular have suffered greatly from the drainage of land to grow arable crops. Coastal wetlands are also affected by the increase in tourist resorts and marinas. Massive developments along the Mediterranean have driven out many species of marine wildlife, including the rare Mediterranean monk seal, not seen in France since 1980.

New migrating patterns

Migrating birds are extremely sensitive to changes in the environment; the failure of any stage in the dangerous business of international travel may result in the death of thousands of birds – perhaps a sizable proportion of the entire world population. Changes in the behavior of the Common crane are a measure of how alterations in land use have affected its traditional stopover sites.

A hundred years ago French agriculture was much less intensive than it is today, and much of the countryside remained undeveloped. Plenty of marshy areas provided sites for the migrating cranes on their long 5,000 km (3,000 mi) journeys between Scandinavia and northern Spain, and there were small fields in which they could feed. In the course of the 20th century nearly all these wetlands have been drained. The resting places have gone, and the birds must now fly huge distances to reach the few sanctuaries that remain. Lac du Der Chantecoq is their only major refuge in northeast France, and there are no stopover points between that and the fields of Les Landes on the southwest coast, a distance of about 650 km (400 mi).

The cranes must make the journey in a single flight. When they arrive they need somewhere secure to rest and feed before attempting to cross the Pyrenees. There are no nature reserves in this area; the cranes roost overnight within the well-guarded perimeters of a French military establishment, emerging each dawn to feed in the open fields. Physical changes in the landscape now dictate that the cranes migrate in large flocks of hundreds, and sometimes thousands, of

OIL SPILLS IN THE ENGLISH CHANNEL

The English Channel is one of the world's busiest seaways and a major route for huge tankers carrying hundreds of thousands of tonnes of oil. Oil spilled into the sea forms a film over the surface, preventing oxygen reaching the water below, asphyxiating fish and other marine life. It coats feathers so that birds cannot fly and, washed onto beaches, kills the invertebrates upon which the coastal food chain depends. In the last 25 years, human error and bad weather have caused two major oil disasters in the Channel.

In 1968 the *Torrey Canyon* went aground off the English coast during spring tides. The heavy sea spread the ship's cargo of oil over a huge area, contaminating both British and French coasts. Methods of dealing with oil spillage were not well developed at the time; some of the chemicals used to break up the oil did further damage to the marine environment. Thousands of birds and seals became oiled and most of the breeding puffins from the colonies of northern France were lost.

In 1978 one of the world's largest oil tankers, the *Amoco Cadiz*, went aground off the Brittany coast when her steering failed. Some 223,000 tonnes of crude oil spilled into the sea, and thousands of puffins and guillemots on passage from their northern breeding grounds in Scotland and the Shetland Islands died as a result. The puffin population has never reestablished itself after these two environmental disasters.

Only in the last few decades has conservation become an issue in France. The victims of the recently proposed scheme to dam the upper waters of the river Loire would have been migratory fish such as Atlantic salmon, River lampreys and eels. Trout and grayling would also have been affected, as well as all the bird and animal life dependent on the river and its natural environment. In the late 1980s, in the face of local opposition and an international outcry, the French government dropped the project.

Wetland wonder *(above)* Breeding colonies of Greater flamingoes are one of the great wildlife spectacles of the Camargue. The lagoons support a unique aquatic life that attracts many species of shorebirds.

Look but don't touch *(below)* The bright warning colors of the European Fire salamander advertise the noxious secretions that coat its skin. Unlike other salamanders, this species gives birth to fully formed, tadpolelike young.

birds. Climatic change may alter the migration behavior of the birds still further: warmer winters in France may mean that in future the cranes need not cross the Pyrenees at all.

Changes in farming can also affect wildlife in other ways. Hedgerows and copses on small farms in France provide refuges for many birds and animals that were once widespread in the deciduous woodland that covered the country. As these fragmented farms are amalgamated in the interests of greater productivity, such refuges disappear and the wildlife is squeezed out. Elsewhere, in mountainous areas, rural depopulation has helped animals to survive, but if too much farmland is allowed to revert to scrub, many meadow and grassland species will lose this habitat.

Conflicting interests

Reintroduction schemes have seen the return of many threatened species to protected areas, such as the lammergeier to the Tarn Gorge in the Cévennes and to parts of Brittany, beavers to the rivers of the Massif Central, and ibex and chamois to the Alps. But many rare and shy animals – mammals in particular – are vulnerable to the disturbance created by increasing numbers of visitors to France's national parks, and especially by the growth of the skiing industry in the Alps and Pyrenees. Although tourism is now a serious threat to the tranquility of many areas, the ski lifts that carry skiers up the mountains in winter give nature lovers improved access to see wildlife.

The birds of these mountain ranges are a great attraction. In the Alps, apart from Golden eagles, the birdwatcher may also be fortunate enough to catch sight of ptarmigans, Rock partridges, Hazel hens, Black grouse, Three-toed and Black woodpeckers, chough and Alpine chough, and Citril and Snow finches. Many of France's large birds of prey are found in the Pyrenees – as well as eagles and vultures there are Black and Red kites, hobbies, goshawks and peregrines.

Birdwatchers from all over Europe come to the Pyrenees for the spring and fall migration, when thousands of birds stream through the high mountain passes. Shooting rights throughout the mountains are rented out to hunters by the landowners, for the killing of migrant birds is a traditional activity in southern France. The World Wide Fund for Nature has purchased the rights on one valley so the birds may cross this area in peace. Cranes, storks and eagles are protected by law, but the noise made by hunters causes a great disturbance at a critical point on the migration route, and this affects all the birds.

Endangered delicacies

Gourmet's delight (*above*) Snails have been farmed since Roman times for eating. Napoleon's soldiers carried canned snails as emergency rations during their military expeditions. The standard ration was 1,000 snails per man per week.

The French are renowned as a nation of gourmets. One of the hallmarks of French cuisine is the regional variety of its dishes. This local flavor is derived from the age-old skill of thrifty housewives who make use of seasonal gluts of local produce. Wild boar, wild geese, pigeons, woodcock and plover are often used, along with a great variety of fish and shellfish. This culinary interest has helped to preserve some animals, such as deer and Wild boar; for others, it has been nothing short of disastrous.

Conservationists have been at the forefront of growing concern over the killing of millions of small birds migrating to and from Africa each spring and fall. The mass slaughter of songbirds destined for the pot is a traditional part of French rural life, especially in the south where historically the annual hunt was a way of supplementing the diet of poor people. Now dishes such as thrush pâté – a particular speciality of Provence and the Basse Alpes – are in demand as gourmet foods. In the past, the birds were usually caught in nets for family consumption. However, times have changed. Now restaurants demand wild delicacies for the menu. The killing has gone commercial. Shotguns are widely used, and indiscriminate hunting is a major threat.

Hundreds of millions of migrants – as many as one in six – die each year. In spite of national laws and European Community regulations even eagles, storks, buzzards, cranes and falcons fall victim. With the plight of the migrants already threatened by loss of habitat, the disappearance of staging posts, and by pesticide use, this slaughter – which is not confined to France, but takes place throughout southern Europe – contributes to a serious decline in the breeding birds of Europe.

Game birds fare better: pheasants, partridges and quails are raised rather like chickens, and are released into the game reserves perhaps a few weeks before the shooting starts; other species are protected during the breeding season. Of the larger mammals, chamois and mouflon are completely protected by law. Game reserves also protect Red, Roe and Fallow deer and Wild boar, whose hunting is regulated by open and closed seasons.

Of frogs and snails...

French cuisine is also noted for its use of frog legs, a delicacy it has helped to make world famous. Frog legs weigh up to one third of the total body weight, and are served deep fried, in soup or sautéed. Although large numbers of frogs are bred in frog "farms", French production is insufficient to meet demand, and frog legs are imported from Yugoslavia and Asia. Until recently the paddyfields of India and Bangladesh provided some 150 million frogs to the international French restaurant trade each year. However, malaria-bearing mosquitoes and crop pests proliferated in the absence of the frogs, which provided a natural control, and the Asian countries found they were spending more on imported insecticides than they made from the sale of frogs.

Snails are another well-known item in French cookery; snail farms date back to Roman times. The large Apple or Vine snail is the type generally served in restaurants, and is especially bred for the purpose. However, many country people collect the "petits gris" or Garden snail to serve with sauces of butter and garlic.

The lagoons of the west and south coasts are ideal for raising the shellfish, especially oysters, that feature in French cooking, while the seas above the shallow continental shelf provide mackerel, sardines, sole, turbot, sea bass and whiting, and various species of lobsters, including langoustine or scampi. Freshwater fish are also popular on the menu, especially in the Loire area – pike, pickerel and eel are probably the best known.

Conserved for hunting (*above*) The Red deer has been protected for hunting purposes for many centuries. This stag is challenging a rival with a series of loud roars during the breeding season – or rut – in the fall. If the challenge develops into a fight, the stags rush at each other, heads lowered, and clash antlers. They push against each other, each trying to gain the advantage of slope. Their sharp antlers can inflict fatal wounds.

National speciality (*right*) In France frogs were once much cheaper than meat, and frog vendors were a common sight in the streets. Now frog legs are something of a delicacy. In the wild the European edible frog spends its entire life in ponds. It hibernates on the floor of the pond in winter, emerging a blackish brown color, then molting to the more usual green.

WILDLIFE OF THE WETLANDS

ANIMALS IN A CHANGING LANDSCAPE · LIVING CLOSE TO PEOPLE · ANIMALS UNDER SIEGE

The wildlife of the Low Countries is of relatively recent origin; it is a region of young landscapes, shaped by the meltwater deposits of the last glaciation some 25,000 years ago, which created a mosaic of marshes, estuaries, dunes and gentle hills. Most of the wildlife migrated into the region as the land warmed, and is similar to that of neighboring countries. It underwent little change until recent centuries when – as the Low Countries became Europe's most populated and agriculturally productive region – many species, especially large mammals, became extinct. Other species have been attracted into new habitats – fields, cities, heaths, ditches and canals – that resulted from the intensification of human activity. The region is renowned for the vast numbers of migratory shorebirds that visit the coastal wetlands.

COUNTRIES IN THE REGION

Belgium, Luxembourg, Netherlands

ENDEMISM AND DIVERSITY

Diversity Very low to low
Endemism Very low

SPECIES

	Total	Threatened	Extinct†
Mammals	48	2	0
Birds	300*	2	0
Others	unknown	26	0

† species extinct since 1600
** breeding and regular non-breeding species*

NOTABLE THREATENED ENDEMIC SPECIES

Mammals none
Birds none
Others Zuiderzee doribella sea slug *(Doridella batava)*

NOTABLE THREATENED NON-ENDEMIC SPECIES

Mammals Pond bat *(Myotis dasycneme)*, Harbor porpoise *(Phocoena phocoena)*
Birds Red kite *(Milvus migrans)*, corncrake *(Crex crex)*
Others Rosalia longhorn beetle *(Rosalia alpina)*, Large copper butterfly *(Lycaena dispar)*, Edible snail *(Helix pomatia)*, Freshwater pearl mussel *(Margaritifera margaritifera)*

DOMESTICATED ANIMALS (originating in region)

none

ANIMALS IN A CHANGING LANDSCAPE

From the Atlantic coast to the dense coniferous forests of the Ardennes in the southeast, the Low Countries contain a considerable variety of landscapes, supporting wide a diversity of wildlife. Harbor seals find a refuge all along the coast, while the extensive river forelands, dunes and tidal mudflats are inhabited by a profusion of terrestrial and marine life. The belt of offshore dune-covered islands, tidal mudflats and marshes that comprise the Waddenzee in the north of the region is undoubtedly the single most important wetland in Western Europe. It is a staging post and wintering ground for migratory waterbirds, including up to one million shorebirds, and is an important nursery for many fish and crustaceans. The intertidal mudflats of the Schelde estuary are also of global importance as stopover areas for shorebirds as well as ducks, geese, terns, and gulls. There are also significant breeding populations of other birds in these wetlands, including one of the largest populations of avocets in the world. Other important breeding birds include spoonbills, eiders, Sandwich and Little terns.

Fringing the Atlantic Ocean are extensive gorse-covered dunes. These have a temperate climate all the year round and are inhabited by Brown hares, Roe deer and Red foxes, as well as a host of smaller mammals, lizards and snakes. In the fairly recent past, the land behind the coastline consisted mostly of inhospitable stretches of marshy plains with lowlying peatlands. Since the 18th century these have been drained to make fertile polder land for agriculture. In consequence, much of the original wildlife has disappeared, though where patches of marshland and woodland have been protected, diverse species of animals continue to thrive, among them rarities such as the Common otter, Beech marten, spoonbill, Purple heron and Grass snake.

Woodland habitats

Farther to the east, the flat polders gradually give way to a hilly countryside, formed as the ice sheets retreated from the Netherlands. As the temperature of the land rose, these lowland plains and hills became covered with broadleaf forest, and animals typical of woodland

Solitary hunter (*left*) A Gray heron waits for fish to surface in a ditch. Wetland habitats dominate the landscape of the Netherlands, attracting a wide range of shorebirds and waterfowl. Both the government and the people are keen conservationists: over 50 percent of the people belong to some kind of conservation agency.

Fleet of foot (*right*) The Brown hare relies on speed to escape from predators, often outrunning its pursuers by zigzagging to-and-fro to throw them off balance. Initially it may try to hide, crouching low in the vegetation, but if the predator comes too close, it will leap from its cover and flee; it can cover the first 4.5 m (15 ft) in one bound.

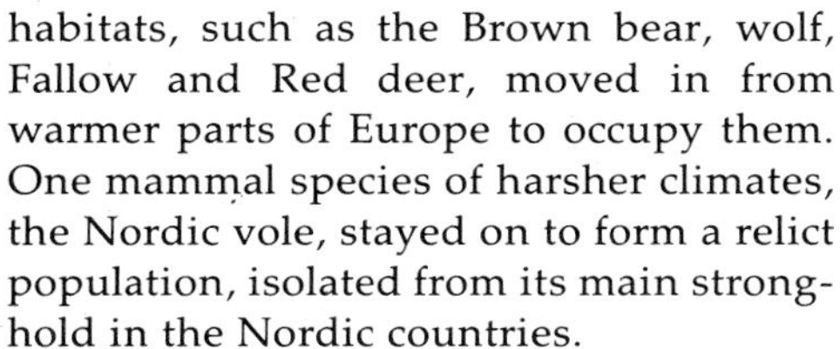

habitats, such as the Brown bear, wolf, Fallow and Red deer, moved in from warmer parts of Europe to occupy them. One mammal species of harsher climates, the Nordic vole, stayed on to form a relict population, isolated from its main stronghold in the Nordic countries.

Over the centuries, the clearance of much of this land for agriculture put increasing pressure on the wildlife. Many species of animal once common in the region are now extinct, either because of loss of habitat or as a result of direct persecution. Today, the immense, well-managed coniferous forests of the Ardennes in southern Belgium and Luxembourg, the northernmost extension of the mountain systems of Central Europe, provide refuges for large mammals such as Red and Fallow deer, Wild boar and wildcat. Elsewhere, however, the human alteration of the landscape has created a mosaic of swampy meadows, pastureland, peat bogs and moorland that each have their own distinctive wildlife. For example, areas of cleared sandy soil now support heathland, with scattered junipers and occasional patches of conifers; this is inhabited by rare mammals such as Pine martens and polecats. Such heaths also attract a great variety of invertebrates, especially digger wasps, beetles, bees and spiders.

During the last few decades, as a result of climatic change, a number of bird species have colonized the region. These include eiders, Black-necked grebes, crossbills, Red polls and Penduline tits. Improving environmental conditions are also resulting in the natural recolonization and expansion of recently lost species such as the goshawk.

LIVING CLOSE TO PEOPLE

The Low Countries' high density of population, and the intensification of its agriculture, have had mixed benefits for the region's wildlife. Birds have generally proved to be more successful than many mammals in adapting to manmade habitats. In stark contrast to other animal groups, the number of new bird species that have colonized the region has exceeded the number of species that have been lost during this century.The creation of open fields and meadows attracted birds from the eastern European steppe and tundra such as Black grouse, godwits, ruffs and redshanks. A more recent development has been the dramatic expansion of the Collared dove's range in the region. At the turn of the century this member of the pigeon family was found in Europe only in the southeast. It then spread westward, and began to breed in the southernmost part of the Low Countries in the 1950s. It is now found throughout the whole region.

In recent years the region has become the main wintering grounds for many thousands of geese, including the entire northwestern European population of the Brent goose, and over one million waterfowl. They are attracted here by the plentiful pasture and mild climate. Strict hunting restrictions ensure their safety.

Many birds survive through their ability to adapt to the ever-changing environment. Species like the House sparrow, swift and Collared dove have accepted the concrete jungle of cities as their new habitat. However, not all species that have shown great affinity with people in

Diving and dabbling ducks The shelduck is a bird of coasts and estuaries, where it feeds mainly on mollusks, foraging in the mud or dabbling in the shallow water. The pintail feeds on water plants, supplementing its diet with small invertebrates, which it obtains by up-ending itself in the water; it feeds mainly at night. The Tufted duck is a bird of freshwater lakes and marshes. A diving duck, it has a varied diet of small fish, amphibians and invertebrates, supplemented by grasses, water plants and even berries.

the past have survived the changing conditions of today: the White stork, the Netherlands' most popular and best-loved bird species, is now very rare – by 1988 the natural breeding population had dwindled to only two pairs. In the wild White storks build their nests at the top of trees. With the development of Europe's farming landscape, they began to build them on rooftops and church steeples. Stork pairs return year after year to the same nest to breed. In the Netherlands, as elsewhere in Europe, they have long given pleasure to the local people, who often construct special cartwheel platforms to attract the storks to breed. But despite this patronage and protection, stork populations have declined in western Europe. This is partly the result of changing farming practices, but also stems from alterations to the habitat in their wintering grounds in Africa.

Wetland colonies

The richness and diversity of bird life in the wetlands of the Low Countries, whether natural or manmade, is largely due to the successful partitioning of resources. Among the species exploiting these feeding grounds there is a wide variety of food preferences, as of feeding techniques. Many shorebirds and gulls are adaptable, varying their feeding habits according to the food available.

The mud of estuaries and tidal flats is rich in invertebrates hidden in burrows just below the surface at low tide – marine worms, crabs, cockles, razor shells and other mollusks. Sandy shores offer rather less food, but the wetter areas lower down the shore are profitable feeding grounds. A wide range of shorebirds exploit this food source, but in different ways. They detect their prey by touch rather than by sight. Their bills probe to different depths, and so competition between the

THE MYSTERIOUS EUROPEAN EEL

The European eel is a culinary delicacy in the Low Countries, and is commercially fished in inland waters. However, the details of its life are shrouded in mystery. It is a freshwater species for most of its life, but the eel spawns in the Sargasso Sea in the western Atlantic Ocean, near Bermuda. The secrets of its reproductive biology were unraveled only in 1904 when it was realized that elvers (previously thought to be a separate species) were the young of the mature European eel.

The hatchlings take two years to make the journey back across the Atlantic, a distance of about 6,000 km (3,700 mi). At this stage they are tiny, transparent and leaflike, but as they reach the coast they develop fins and become slimmer and more eel-like, and are known as elvers. They enter the rivers of western Europe in their millions, and the Dutch government allows the opening of outlet and drainage sluices during certain times of the year to ease their passage. Their physiology undergoes a dramatic change as they adapt from salt to fresh water. It takes them seven years to mature and develop sexual organs.

At this stage they gradually alter in form again. They develop large eyes that allow them to live at deep sea levels, and thick blackish skin. Their metallic sheen changes from gold to silver, their gut degenerates, and they will never feed again as they begin the journey back to the Sargasso Sea. No returning adult eels have ever been caught beyond the continental shelf – presumably they travel at great depths.

Slithering overland The widespread European eel spends much of its adult life in fresh water, but it can also travel over damp ground on its migration route toward the sea.

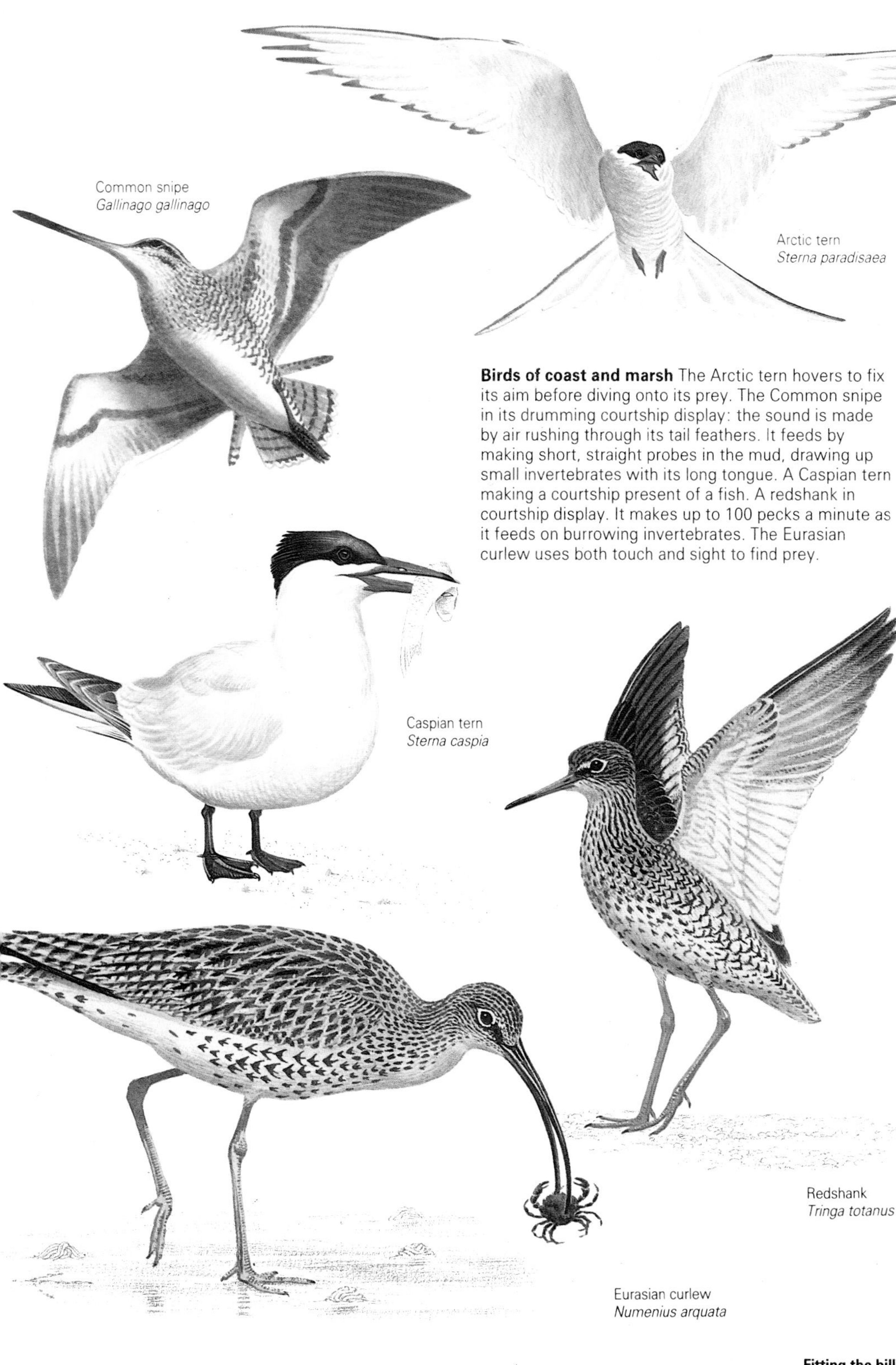

Birds of coast and marsh The Arctic tern hovers to fix its aim before diving onto its prey. The Common snipe in its drumming courtship display: the sound is made by air rushing through its tail feathers. It feeds by making short, straight probes in the mud, drawing up small invertebrates with its long tongue. A Caspian tern making a courtship present of a fish. A redshank in courtship display. It makes up to 100 pecks a minute as it feeds on burrowing invertebrates. The Eurasian curlew uses both touch and sight to find prey.

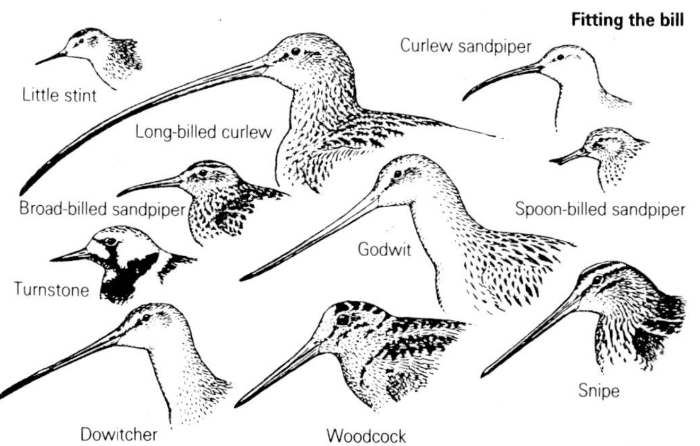

Shorebirds' bills (*right*) are adapted to a variety of feeding habits. The Little stint picks at surface prey; the Long-billed curlew and Curlew sandpiper probe farther for deep-burrowing shellfish and marine worms. The Broad-billed sandpiper's bill copes with larger prey, such as mollusks. The Spoon-billed sandpiper feeds by swinging its bill in the water. The Ruddy turnstone uses its bill to turn over stones, exposing prey. The godwit probes deep into wet mud. The bills of dowitchers, woodcocks and snipes have many touch sensitive organs.

species is reduced. The sanderling, for example, picks food from the surface of the beach, scuttling rapidly over the wet sand near the breaking waves. The avocet also feeds in the wettest mud, slicing through soft, almost liquid mud as it swings its head from side to side.

Larger prey is exploited by birds with more powerful bills. The oystercatcher can tackle cockles and mussels by hammering the shells open, while the turnstone has a neat method of turning over small stones and other objects to get at invertebrates hiding beneath. Gulls are more omnivorous feeders, scavenging for scraps on garbage dumps and in gardens, as well as hunting along the shorelines. They will drop mollusks onto rocks to smash open their shells, and will also dive for fish. Large gulls will also take bigger prey, including the eggs and fledglings of nesting seabirds.

Most ducks and geese are herbivorous, grazing on the beds of eel grass, and on the saltmarshes and fields farther inland. The proximity of the sea affords food for fish-eating ducks and terns, who are attracted to the sand banks and island dunes where they can nest out of reach of foxes and other predators.

Natural managers

All wetland habitats are in a natural state of transition to a drier habitat. However, the animals that exploit their resources may help to slow down the process. The Graylag goose, for example, can consume up to 30 percent of its body weight through its daily intake of reeds and grasses. In spring it keeps down the growth of sprouting reeds, thus helping to keep open stretches of shallow water such as those of the Oostvaardersplassen at the southern end of the IJsselmeer. If these were to become covered over, the lake would lose significance for wildlife. In addition, the geese fertilize the lake, enhancing opportunities for freshwater fish and amphibians. Thus they play a key role in sustaining the abundant wildlife of this important wetland site.

On their own, however, the geese would not be able to prevent the area from becoming overgrown, first with reeds and then with willows and alders. Since 1984 the managers of the Oostvaardersplassen reserve have introduced primitive breeds of cattle and horses to graze the established reed borders of the lake all year round.

ANIMALS UNDER SIEGE

The assault on the natural environment of the Low Countries as modern farming methods and technology bring even the most marginal land into cultivation puts wildlife under severe pressure. With only 4 percent of land left as wilderness – most of it consisting of the extensive intertidal flats of the Waddenzee – it comes as no surprise that four mammal species have disappeared from the region in recent years and that a further 26 are in imminent danger of local extinction. For other animal groups the situation is even more disturbing. Of the 75 species of butterfly, 15 are known to have been lost and a further 49 (or 82 percent) are in serious danger of extinction.

Protection schemes

The Dutch government was a pioneer in the establishment of protected areas. It was instrumental in setting up the National Red List Publication, which gave details of not only threatened mammal species, but also amphibians, butterflies, plants and even fungi. The publication provided a sad catalog of extinctions, but also set out an action plan for future conservation and management.

The sincerity of the government's concern to protect the little that remains of its formerly rich natural heritage is possibly best reflected by its efforts in respect of its reptile and amphibian populations. The Netherlands is one of the few countries that have accorded strict legal protection to all reptile and amphibian species, and special sanctuaries, some as small as 1 ha (2.47 acres), have been set up to preserve the breeding sites of species such as the Midwife toad, the Fire-bellied toad and the European common spadefoot toad. Underpasses have even been built beneath the busiest roads to allow these creatures to migrate safely to their breeding ponds.

In this densely populated region, the size of the threat to wildlife by humans is enormous. Drainage and river management, but more importantly environmental pollution, have had a far-reaching impact on protected wilderness areas. Pollution of the soil from industrial effluent and vehicle exhaust is beyond control. High phosphate, nitrate and potassium levels in the soil not only threaten drinking water supplies, but also damage the wetlands. The Dutch government, in its National Environmental Plan of 1989 called for a 70 to 90 percent reduction in all environmental pollution. The implementation of this plan had far-reaching implications for all sectors of the economy, including agriculture and industry. The per capita expenditure on conservation related programs in the Netherlands was already one of the highest in the world. The cost of realizing the plan represented up to 3.5 percent of the gross national product.

New spaces for wildlife

Despite the dimensions of the problem, there have been some successes. New wildlife habitat continues to be created as a byproduct of land reclamation. The important Oostvaardersplassen wetland area was developed in less than 20 years after the closing of the Flevopolder, the last of the large polders constructed in the IJsselmeer in 1967. Now officially designated as a nature reserve, this 5,000 ha (12,500 acre) wilderness area of shallow freshwater lakes, bordered by extensive reed and willow forests, is typical of the region's recently created wetlands. Within only a short time of its creation a number of mammals had moved into the reserve – first, various species of mice and rats, then predators such as stoats, weasels, Red foxes and polecats, and large herbivores such as the Roe deer. The reserve provides breeding, staging and

Safety in numbers The Sandwich tern is one of the many species of seabirds that breed along the coast of the Low Countries. Synchronized breeding reduces the toll on numbers from predation. The terns may fly over 75 km (46 mi) from the colony on foraging trips.

wintering sites for a great many bird species. It now contains breeding colonies of cormorants and spoonbills; other species that breed there include the bittern, Little bittern, Night heron, Purple heron, Great and Little egret, Short-eared owl, Bearded reedling and Graylag goose.

Habitat creation to help the reintroduction of species is high on the agenda of the Dutch government. In 1988, after 160 years' absence, three beaver pairs were released in the Biesbosch National Park, following similar successful release programs in Germany. If they are successful, these animals should have a positive influence on the water management of this tidal area – part of the Rhine delta – enriching it for other wildlife.

At about the time this program was being put into effect, a team of landscape architects, ecologists and hydrologists drew up a scheme (known as Plan Stork) for managing a 40,000 ha (100,000 acre) stretch of meadows and flats between the Rhine, Waal and Meuse rivers. All agricultural activities will be halted in the outer dikes in order to allow widely differing habitats to become established, in the hope of attracting species such as the Black stork, osprey and European badger back to the area.

STREETWISE ANIMALS

At dawn in the city center thousands of animals – rats, mice, foxes and even Beech martens – scour the streets in search of edible litter. Most of these scavengers will be gone before the first people arrive for work.

Although most city-dwelling animals are also found in the suburbs and countryside, some of them – particularly birds such as pigeons, starlings and sparrows – seem to do better in an urban environment. They have adapted to living off the leftovers of human consumption and are faced with fewer predators. Many birds in city parks and suburban gardens are actively fed by people, who may also erect nesting boxes. The garbage dumps are fertile breeding grounds for a multitude of insects, attracting large numbers of birds. Predatory birds such as kestrels and peregrines have even been known to take up residence on window ledges high above the streets. The cities may also harbor a few exotic birds, such as parakeets that have escaped from captivity. The temperature of the cities is on average 2°C (3.6°F) warmer than the surrounding areas. This enables these tropical birds to survive, and may also reduce winter mortality among other city-dwelling animals.

The increased wildlife population is a considerable burden on many municipal authorities. Starlings (which roost in huge numbers in city trees and on buildings) and pigeons have become a real pest. Their droppings are unsightly, and may damage the fabric of the monuments and ancient buildings of these cultural centers.

The hedgehog's response to danger is to curl up in a ball relying on its spines to protect it – not always the best policy.

The North Sea fishing grounds

The North Sea – the arm of the Atlantic Ocean running between the eastern coast of the British Isles and the European mainland – is a relatively shallow sea. Near the Dutch coastline it ranges in depth from 30 m (100 ft) in the south to 100 m (330 ft) in the north. Large amounts of mud and silt – the source of minerals and other nutrients – are constantly being deposited by the large river estuaries of the Rhine, Schelde and Elbe. The shallowness of the North Sea, combined with the influence of the warm North Atlantic Drift and with strong offshore currents, which help to oxygenate the water and disseminate nutrients, creates an ideal environment for marine life.

Among the swarms of plankton living in the waters of the North Sea are large numbers of a minute crustacean: the copepods of the genus *Calanus*. These are eaten by herring and by many other plankton-feeding fish, which in turn support larger predators. As well as herring, the North Sea contains haddock, skate, plaice and cod (the herring's main predator), and there are a multitude of jellyfish, bivalves, squid, shrimps, crabs and starfish. Fish-eating seabirds abound – gannets, terns, gulls and, nearer to the coasts, cormorants, guillemots, and other diving birds. Fin whales and porpoises are visitors to the area, and Harbor seals are often seen around the Waddenzee and off the coast of Britain.

The pattern of currents affects the distribution of nutrients, and consequently of the plankton and the fish that follow them. Fish travel considerable distances during the different stages of their life cycle. Herring, for example, spawn on shallow gravel banks, and the young fry swim up to the plankton to feed. As they grow they gather in shoals in shallow coastal waters and estuaries. After six months they leave for the open sea where they mature. About five years later they make the return journey back to the spawning grounds.

Other species, too, have their particular spawning grounds. Plaice, which are bottom-dwelling flatfish, spawn at the northern end of the English Channel. The young, which at this stage are symmetrical in form with an eye on each side of the head, drift northeast into the North Sea and scatter across a series of nursery grounds along the Dutch and Danish coasts. As they grow they move to deeper water and a metamorphosis occurs: one eye migrates across the head to lie alongside the other, and the mouth twists into the same plane as the eyes. The fish do not have a swim bladder and sink to the seabed, where they will remain, lying on their blind side and frequently burying themselves in the sand. The larger the fish, the deeper the water they will choose to inhabit.

The shallow waters of the Waddenzee, saturated with sunlight and enriched by fertile mud from the large Dutch rivers, are the single most important spawning ground and nursery area for the fish of the North Sea. Almost 60 percent of the North Sea's Brown shrimp, more than 50 percent of its sole, 80 percent of its plaice and all of its herring populations are supported in the Waddenzee at some point during their life cycle.

These concentrations of fish have encouraged a lucrative fishing industry to develop on both sides of the North Sea. However, since the early 1960s overfishing, in response to increased demand and aided by improved methods of shoal detection, has led to a sharp decline in production, from 4 million tonnes to less than 1 million tonnes. For most of the commercial species a quota system is in force to limit annual catches, and the fishing grounds and seasons are strictly controlled in an attempt to build up stocks to their previous levels.

Pollution of the North Sea is posing an even more serious threat. The rivers pump out industrial chemical wastes, including nondegradable pesticides, and there is dumping of highly toxic waste from tankers, including incineration vessels. The disposal of low level radioactive wastes from mainly landbased nuclear plants results in high levels of radioactivity in fish. The International North Sea Conference has imposed strict regulations concerning sea pollution, including restrictions on the dumping of hazardous industrial waste and sewage sludge, and in 1987 a task force was set up to coordinate research into the ecology of the North Sea and its conservation. Public awareness of the problem was heightened by the crisis that overtook the North Sea seal population in 1988, when a viral infection known as phocine distemper killed over 170,000 seals. There is concern that pollution of the North Sea is affecting seal fertility, and may also have an effect on their immune system.

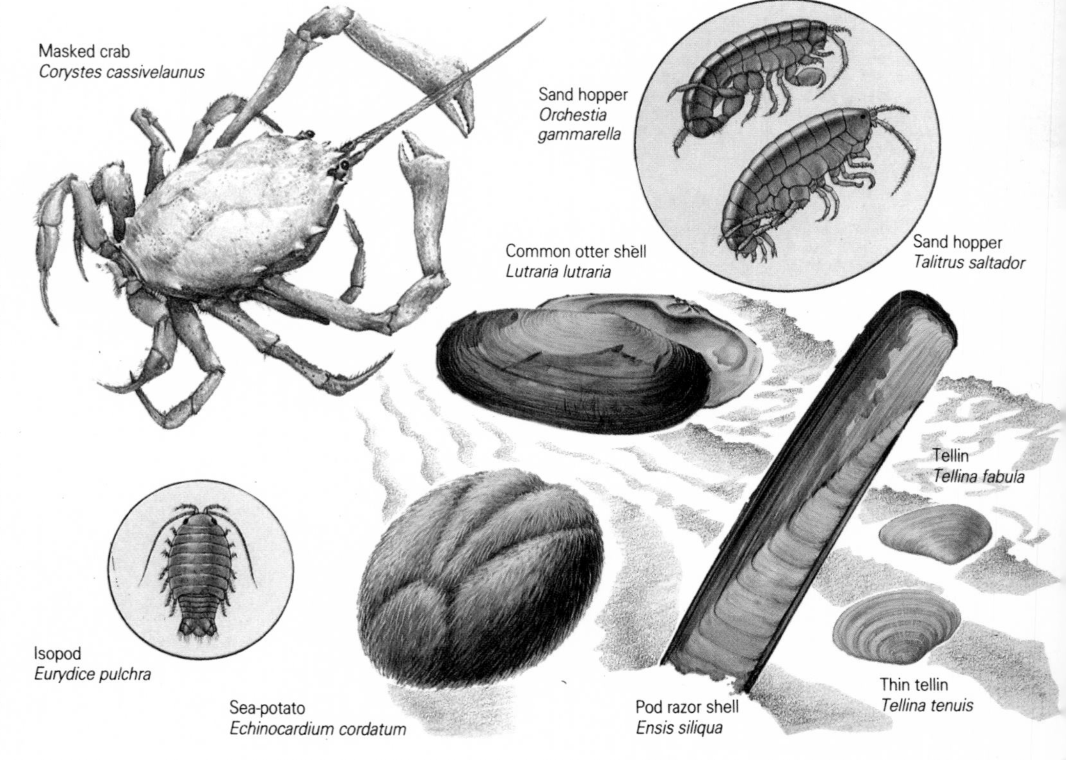

Crustaceans, mollusks and echinoderms (*below*) of a sandy shore in northwestern Europe. Within the debris of the strand line: sand hoppers such as *Orchestia gammarella* and *Talitrus saltator*. On the upper shore: Masked crab (*Corystes cassivelaunus*); and an isopod *Eurydice pulchra*. On the lower shore and continuous shallow water: the Common otter shell (*Lutraria lutraria*); Pod razor shell (*Ensis siliqua*); Thin tellin (*Tellina tenuis*); another tellin, *Tellinai fabula*, and the Sea potato (*Echinocardium cordatum*).

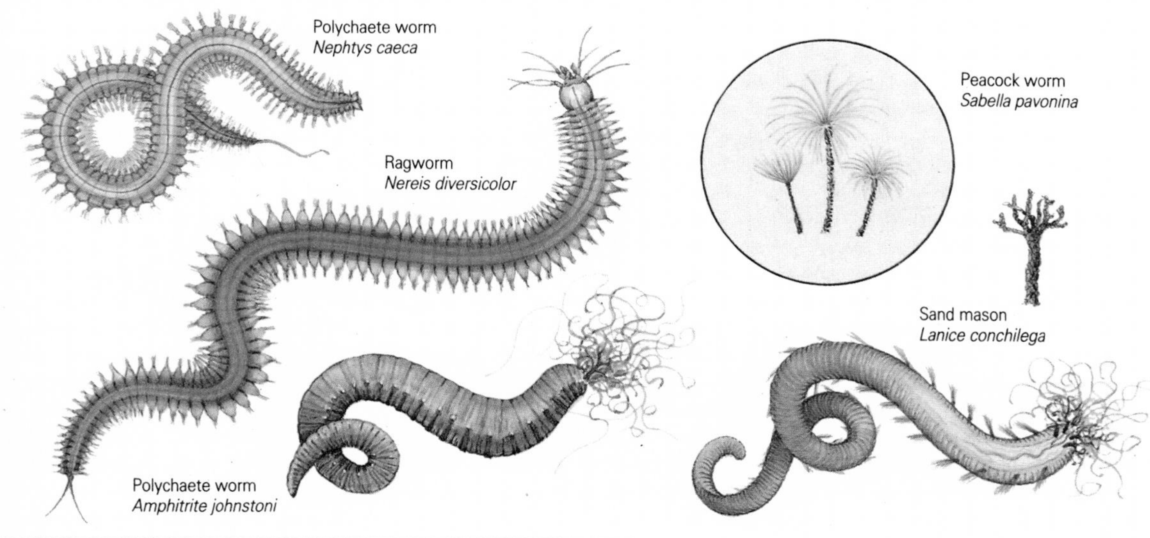

The Harbor seal (*above*) has several breeding beaches around the North Sea where it is continuously under threat from various forms of pollution. The seal's large eyes and long whiskers are adaptations for living in semidarkness underwater.

Wormlike animals (*left*) typical of a sandy shore in northwestern Europe. On the upper shore, a polychaete worm, *Nephtys caeca*. On the mid-shore: the ragworm (*Nereis diversicolor*) and another polychaete worm, *Amphitrite johnstoni*, which make burrows in the sand. The Sand mason (*Lanice conchilega*), with its branching cast of sand grains, which it makes as it burrows; and Peacock worms (*Sabella pavonia*), which construct tubes that are visible above the sand.

Meadow birds of the Netherlands

The manmade polders and associated meadows and pastureland of the Netherlands – which in many ways resemble the Siberian tundra – have been colonized by large numbers of tundra birds such as the ruff, lapwing, Black-tailed godwit and redshank. The population of ruff and Black-tailed godwits is particularly significant; the 100,000 or so pairs represent almost 90 percent of the European godwit population. The polders are popular courting territory for the ruff, the males (ruffs) and females (reeves) gathering at special display grounds called leks. Here, the males parade around, erecting the striking ear tufts and ruffs of feathers around their necks and competing for the best places on the lekking ground. Within the lek each male has a "residence", a bare patch of earth some 30 cm (12 in) in diameter, on which he displays. Positions in the center of the lek are highly prized and attract the most females. Immature "satellite" males with white ruffs are confined to the periphery of the lek where they are tolerated by the resident males because they help to attract reeves.

Modern farming techniques have drastically changed the management of the polders. The lowering of water levels, early and repeated mowing, and the increase in cattle have caused many meadow species – including the godwit and ruff – to suffer a sharp decline in their populations. Of the 49 species listed as highly endangered in the Netherlands, almost one third are meadow birds.

Courtship display A ruff at the peak of breeding condition. The relatively dark color of his feather ruff indicates that he is one of the dominant, mature males of the lek – the birds' display ground.

WHERE EUROPE MEETS AFRICA

A LAND RICH IN SPECIES · ANIMALS OF THE OPEN COUNTRY · AN AGE-OLD CONFLICT

The wildlife of the Iberian Peninsula is a unique blend of species from both Europe and Africa. There are warmth-loving creatures such as Egyptian mongooses, Hermann's tortoises and Cattle egrets, as well as species that are typical of more northern climates. These inhabit the forests and high peaks of the mountain areas and include Pine martens, Crested tits, capercaillies and ptarmigans. The most inaccessible of these forests are a last refuge for some of Europe's rarest mammals such as Brown bears, Gray wolves, wildcats and lynx. The extensive grasslands, maquis scrubland and farmland of the peninsula's high, dry interior provide open hunting country for a wide range of birds of prey, including the Booted eagle, the magnificent Spanish imperial eagle and four species of vulture.

COUNTRIES IN THE REGION

Portugal, Spain

ENDEMISM AND DIVERSITY

Diversity Low to medium
Endemism Low to medium (Canary Islands and Madeira)

SPECIES

	Total	Threatened	Extinct†
Mammals	85	8	0
Birds	350*	17	1
Others	unknown	67	0

† *species extinct since 1600 - Canarian Black oystercatcher* (Haematopus meadewaldoi)
* *breeding and regular non-breeding species*

NOTABLE THREATENED ENDEMIC SPECIES

Mammals Pardel lynx (*Felis pardina–Felis lynx pardina*)
Birds Spanish imperial eagle *(Aquila adalberti)*, Dark-tailed laurel pigeon *(Columba bollii)*, freira *(Pterodroma madeira)*
Others Hierro giant lizard *(Gallotia simonyi)*, Majorcan midwife toad *(Alytes muletensis)*, Valencia toothcarp *(Valencia hispanica)*, Madeiran land snails *(Leiostyla* spp.)

NOTABLE THREATENED NON-ENDEMIC SPECIES

Mammals Pyrenean desman *(Galemys pyrenaicus)*, Long-fingered bat (*Myotis capaccinii*), Mouse-eared bat (*Myotis myotis*), Gray wolf *(Canis lupus)*, European mink *(Mustela lutreola)*, Mediterranean monk seal *(Monachus monachus)*, Harbor porpoise (*Phocoena phocoena*)
Birds White-headed duck *(Oxyura leucocephala)*, Black vulture *(Aegypius monachus)*, Great bustard *(Otis tarda)*, Audouin's gull *(Larus audouinii)*
Others Herman's tortoise *(Testudo hermani)*, Spanish moon moth *(Graellsia isabelae)*

DOMESTICATED ANIMALS (originating in region)

rabbit *(Oryctolagus cuniculus)*, canary *(Serinus canarius)*

A LAND RICH IN SPECIES

Iberia's vast range of habitats contributes to the richness of its animal species. Behind a long and varied coastline the land rises to a large central plateau – the Meseta – while to the north and south are high mountains with beech, oak and pine forests, alpine meadows and bare peaks. Great rivers wind across the land to end in deltas whose wetlands support a wealth of bird life.

The barriers of sea and mountain that isolate the Iberian Peninsula from Africa on the one hand, and from the rest of Europe on the other, have not always been in existence. About 100 million years ago animals typical both of the northern tundra and of the tropics were able to migrate freely into the peninsula. Once it had become cut off from its neighbors by the alterations in the sea level, however, there proved to be no escape for those species unable to cross mountains or oceans. As a result, a unique Afro-European blend of species evolved in the peninsula, making it the richest of all European regions in terms of its huge diversity of animal life. This is seen in the remarkably large number of species that can be found here.

Among the invertebrates, for example, some 70 percent of all European dragonfly species are found here, in addition to more than 200 species of butterfly and numerous subgroups, or races, that are found only in the peninsula or on the adjacent islands. The vertebrates show a similar level of diversity: the number of bird species recorded in the region is over 350 (about three-quarters of them breed here regularly); there are 135 indigenous species or subspecies of mammal, 24 species of amphibian, 56 species of reptile and 56 species of continental or migratory fish. About 30 percent of species in each class are unique to Iberia.

Such species, known as endemics, are often restricted to particular mountains or valleys, or to certain islands. For example, the Iberian rock lizard is confined to four widely separated areas in the Pyrenees, the Cantabrian Mountains and the central ranges. Valverde's lizard is restricted to mountain ranges in the southeast, while the endemic Iberian viper is found in those of the north. Other remarkable reptiles are confined to the sand dunes of the coast. The endemic Spiny-footed lizard has long spreading toes that enable it to traverse the shifting sands with ease, while the burrowing Beriaga's skink, also endemic, has greatly reduced limbs to assist its way of life.

Rare magnificence The Spanish imperial eagle is distinguished from the more common Eurasian race by its pale shoulders. The bird is highly endangered – only about 100 pairs remain – and, though protected by law, it is still persecuted by farmers.

The agile ibex The Spanish ibex moves to high mountain pastures in the summer; in winter it seeks out steep slopes where the vegetation remains relatively free of snow. Several populations of the Spanish ibex are now in danger of extinction.

Island specialties

The Iberian Atlantic islands of the Azores, Madeira and the Canaries are all largely volcanic in origin. This means that they were colonized by animals with the power of flight or the ability to swim long distances; or by animals that were transported accidentally in masses of floating vegetation or carried on upper air currents. The indigenous wildlife of these Atlantic islands is thus rather limited; for example, the nine islands that make up the Azores archipelago, lying some 1,360 km (850 mi) off the coast of Portugal, have virtually no native vertebrates other than birds and marine creatures. Where terrestrial animals have become successfully established, however, their isolation from their mainland relatives has resulted in the evolution of many endemic species and subspecies.

The Canary Islands, lying further to the south off the coast of Africa, contain many such species of bird. These include the Blue chaffinch and the Long-toed and Laurel pigeons, which inhabit the islands' evergreen forests (now reduced to mere remnants of their former extent), the Canary Islands chat and the canary itself. Berthelot's pipit and the Plain swift are found only in the Canaries and Madeira. The Canaries mouse-ear bat, the Madeira bat and the Canaries shrew are also confined to these islands, as are distinct species of butterfly such as the Canary blue, the Azores grayling and the Madeiran speckled wood. Madeira boasts many endemic land snails, each species confined to a separate *barranco*, or gully, in the volcanic slopes of the island.

Giantism – whereby isolated species evolve larger forms in the absence of predators or competitors for food – is encountered among island communities throughout the world. A regional example of the phenomenon is the Hierro giant lizard of the Canaries, which is now an endangered species.

The Balearic Islands in the Mediterranean, once the northern end of the mountain ranges of the southeast peninsula until they were cut off by rising sea levels, also contain a number of endemic species. The best known is the Majorcan midwife toad, only discovered as a living animal in 1980, though previously known from fossil evidence. There are also two endemic species, and numerous races, of lizard. Formentera boasts a large endemic race of Oak dormouse, and Minorca has a score of individuals of an endemic subspecies of Pine marten. Of seabirds, the islands support a distinct race of Cory's shearwater.

ANIMALS OF THE OPEN COUNTRY

The arid plains of the Meseta have their own distinctive animals. Hiding places are rarely available in such open habitats, so cryptic coloration, or camouflage, is an essential feature of ground-nesting birds such as Black-bellied and Pin-tailed sandgrouse (unknown elsewhere in Europe), Stone curlews and many species of larks.

All but a handful of western Europe's Great bustards are found in Iberia. With their long legs and marvellous camouflage, they are typical steppe-dwellers, able to move swiftly through the high grass, stretching their heads on tall necks to watch out for predators. The region also supports the majority of western Europe's Black and Griffon vultures. These large carrion-eating birds have long wings that enable them to soar high on warm air currents to scan large areas for prey. Their bald heads have evolved to allow the birds to delve deeply into animal carcasses without the need for lengthy preening afterwards.

Four species of shrikes – the Great and Lesser gray, Woodchat and Red-backed shrikes – are found in Iberia. Fierce little predators, they scan the ground from favorite perches for their prey of small vertebrates, large insects and occasionally fish. They have the large head and eyes of a typical predator, with short, heavy, hooked bills and powerful legs, feet and claws to seize their prey. Shrikes store some of their food in larders – thorn bushes on which the carcasses of their victims lie speared for up to a month before being consumed.

The *dehesa* is cork or holm oak parkland, the ground between the trees being grazed or cultivated in 10 year rotation. Fragments of the original Mediterranean woodland remain in places, and these pockets support a wide range of colorful birds: Golden orioles, Great spotted cuckoos, Woodchat shrikes, and rollers, bee-eaters and hoopoes. Although the Iberian population of the Black stork, which roosts in the trees of the *dehesa*, has been reduced to less than 100 pairs, it is still the most important breeding population in western Europe. Most of the remaining pairs of Spanish imperial eagles also breed here. They are a distinct race whose nearest relatives are found in southeast Europe, and are perhaps the most endangered of all Iberian birds: the world population of these eagles numbers only approximately 100 pairs.

Living at height

Mountains dominate the landscape of the region. Black and Middle-spotted woodpeckers, wallcreepers, lammergeiers and capercaillies are specialities of the northern mountain ranges. The Pyrenean subspecies of ptarmigan is one of the few birds to remain in the high mountains all year round. In winter its normally gray-brown plumage is exchanged for a coat of pure white, its feet develop broad tufts of feathers for walking on the snow, and in harsh weather it digs itself into a snowdrift and becomes torpid until more favorable conditions return.

One of the most magnificent mammals of the region is the ibex, distinct races of which are found in the Pyrenees, the Sierra Nevada, the Sierras of Cazorla and Segura and the Sierra de Gredos. The legendary agility of the ibex, which can leap from great heights, is attributed to its cloven hooves, the two halves of which separate on landing to provide a firm grip. Another agile mountain animal, the chamois, is capable of upward leaps of some 4 m (13 ft). It is widespread in the northern mountains, especially in the Picos de Europa.

In the high mountains winters are long and the growing season is short. Some animals have adapted their breeding cycles to compensate. The Alpine newt, which is found in the Cantabrian Mountains and the Pyrenees, attains sexual maturity in the larval stage: the final stage

Short-toed eagle *Circaetus gallicus*

Great gray shrike *Lanius excubitor*

Red deer *Cervus elephas*

of metamorphosis to the adult form does not occur, a type of behavior known as neoteny. Many endemic species of butterfly – Forester's furry blue and Lefebvre's ringlet in the north, Zapater's ringlet and Oberthur's anomalous blue in the central sierras, and the Nevada blue in the Sierra Nevada – respond to the short growing season by producing only one brood of young each year.

Coasts and wetlands

Iberia's extensive coastline encompasses precipitous cliffs, sweeping sandy bays, stretches of dune and shingle and some of Europe's most extensive and unspoiled coastal saltmarshes, all offering varied habitats for birds. Many are now being lost to encroaching tourist developments and largescale drainage and hydroelectric power schemes.

Narrow rocky ledges on the Islas Cíes off the northwestern coast support the only colonies of Iberian guillemots in the world. Western Europe's main breeding populations of the Collared pratincole, Marbled teal, Squacco heron and Purple gallinule are found in the great coastal marshlands of the peninsula. On the flat marshlands of the Odiel river in the southwest, 300 pairs of spoonbills, which usually nest in trees, rear their young among the saltmarsh grasses.

As many as 30,000 Common cranes – an extraordinarily high number for western Europe – overwinter in the west of Spain every year, while the only stable breeding colony of Greater flamingoes outside southern France is found at Fuente de Piedra, a large saline lake in Andalusia in the south, where up to 12,000 birds may nest in good years. The freshwater lagoons at Cordoba and Cadiz in the same area are home to over 200 White-headed ducks. This declining European species is remarkable for the extremely large size of its eggs; the large, precocious fledglings are able to dive for food within only a few minutes of hatching.

The Guadalquivir river of southern Spain is one of the few remaining spawning sites of the European sturgeon; other endangered species of fish include the Valencia toothcarp and its close relative *Aphanius iberus*, both of which are confined to a handful of rivers on the eastern Iberian coast.

Azure-winged magpie
Cyanopica cyana

Crested lark
Galerida cristata

Animals of the sand dunes The Coto Doñana, an extensive area of dunes and pine trees, is rich in wildlife. The Short-toed eagle feeds mainly on snakes. The Great gray shrike impales its prey on thorns, producing a gruesome larder. The Crested lark is very agile on the ground. The Azure-winged magpie occurs only in Iberia and eastern Asia. The Red deer emerges from the shelter of the trees to feed at dusk and dawn. The venomous Montpellier snake preys on small mammals, birds, lizards and other snakes. The Wild boar forages at dusk for nuts, mushrooms, roots, worms and snails.

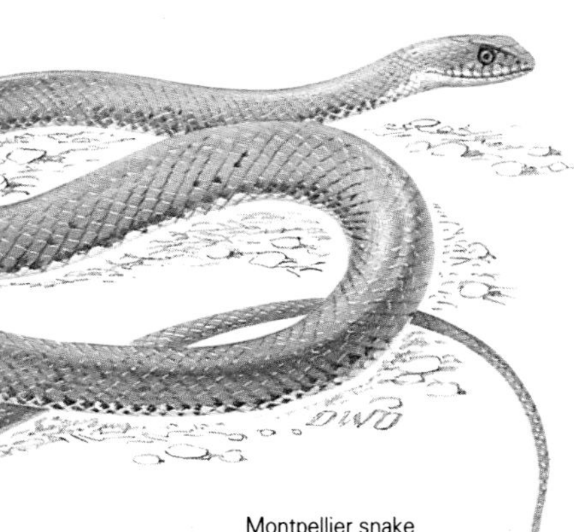

Montpellier snake
Malpolon monspessulanus

Wild boar
Sus scrofa

AN AGE-OLD CONFLICT

Livestock rearing is an important source of income for farmers in many parts of the Iberian Peninsula. For centuries they have sought to protect their flocks and herds from the predation of large carnivores. The population of the Iberian wolf, an endemic subspecies, is currently thought to be in excess of 2,000 individuals, about one-fifth of which live in Portugal. It is believed that at least 30 percent of these wolves are killed annually in organized hunts, by illegal shooting or by the setting of poisoned baits. Such is the supposed economic threat to the farmer that when Spain ratified a major international convention regarding the protection of animals in 1986, the wolf was excluded from its clauses. In Portugal wolves may still be hunted from October to February.

A QUESTION OF TIMING

Eleonora's falcon breeds only around the Mediterranean and on the Canary Islands, spending the winter in Madagascar and nearby islands in the Indian Ocean. Unusually for a bird of prey it is highly gregarious, nesting in noisy cliffside colonies of up to 70 pairs. Those that are sited along the Iberian coastline contain some 500 pairs altogether, representing a substantial proportion of the world population.

Although the birds return to the Mediterranean at the end of April each year, mating does not take place for several months and the two or three eggs are not laid until early August, hatching some 28 days later. In spring the adult falcons feed mainly on insects, which they trap when in flight, but the delayed reproductive cycle of these elegant, long-winged raptors is perfectly timed to take advantage of the southerly migration of small birds at the end of the summer. Young inexperienced songbirds making their first Mediterranean crossing are easy prey for the adult falcons, which often form loose phalanxes offshore to coordinate their hunting efforts. The abundant food supply ensures that the falcon chicks grow rapidly. They have made their first flights by the beginning of October and will have learnt to hunt before embarking on the long journey via the Red Sea and the African coast to their wintering quarters in the tropics.

The Pardel lynx, a subspecies unique to Iberia and valued for its superb pelt, was once widespread throughout the peninsula, favoring forested areas in the lowlands. Centuries of hunting, as well as indiscriminate persecution by farmers and the destruction of its natural habitat, have driven it into less favorable terrain – the rugged central sierras and the marshlands of the Guadalquivir, with each area supporting a population of approximately 100 individuals.

The fate of the Brown bear provides a good illustration of many of the problems that beset Iberia's large predators. The Spanish subspecies is one of the smallest bears in the world. Fewer than 100 individuals remain, most of them in Asturias, in the Cantabrian Mountains. A small population of some 20 bears farther east is cut off from the main population by a network of major north–south road and rail crossings; these bears are in grave danger of declining even further due to lack of genetic variability.

In spite of strict national and international legislation to protect it, the numbers of Brown bear are dwindling. Many fall victim to poachers; others are killed by poisoned baits put out by farmers for other animals (farmers are now compensated for livestock losses inflicted by Brown bears). In addition, the bears' natural habitat – upland beech and oak forests – is being exploited for timber at an alarming rate. Other threats include opencast mining, reservoir construction that threatens to cut the bears off from certain seasonal food sources, and the promotion of tourism. Ironically this uses the wildlife of the area – the bears in particular – as an attraction, but the influx of visitors may disturb the bears so much that their breeding is inhibited.

Successes and failures

Both Spain and Portugal are signatories of all the major international conventions concerned with the protection of animal species. In recent years Spanish progress in wildlife conservation has outstripped that of its poorer neighbor, with comprehensive national and regional legislation dedicated to the conservation of both species and their habitats. As a result of these efforts, assisted by a comprehensive network of protected spaces, species such as the rare Crested coot and Andalusian hemipode have recently begun to breed again in the marshes of the Guadalquivir. In addition, Black-shouldered kites, unique to Iberia and northwest Africa, and Azure-winged magpies, found only in Iberia and the Orient, are increasing in the southern part of the region.

Legislation is not always sufficient. Despite stringent regulations governing

Gibraltar apes (*above*) The small colony of Barbary macaques in Gibraltar is a major tourist attraction. It is not known whether they have always lived here, or whether they were introduced from North Africa long ago, perhaps by the Phoenicians or the Romans.

Fighting bears (*left*) European Brown bears threaten each other in the snow. The remote mountain forests of Iberia provide a refuge for some of Europe's rarest large mammals. The Spanish population of the Brown bear is seriously threatened by habitat disturbance.

A male midwife (*right*) The Midwife toad has gained its name because the male carries the eggs wrapped around his legs for safety until they hatch. The young then swim off when he next enters the water. These nocturnal creatures make a sound like chiming bells.

THE THREAT TO THE PYRENEAN DESMAN

Pyrenean desmans are primitive, semi-aquatic insectivores, closely related to moles. They have dense underfur and long, water-repellent guard hairs. The ears and nostrils can close on submersion; the tail, longer than the body, is flattened toward the end to act as a rudder, and the strong, clawed toes are linked by membranes to provide greater swimming power. But the desman's most distinctive feature is its long, trumpet-shaped snout, covered in tactile whiskers, which is highly mobile and used for searching out aquatic invertebrates on the riverbed.

These creatures are confined to upland rivers, streams and lakes in the Pyrenees, the Cantabrian Mountains and the central Iberian sierras. Never far from water, studies suggest that their whole lives may be spent within a single 200 m (650 ft) stretch of river. Restricted both in geographic range and habitat, the desman is highly vulnerable to human activities, particularly in the form of water pollution, which destroys its natural food supply.

More and more of the places that the desman inhabits are being lost to hydroelectric schemes and irrigation projects; reforestation of hillsides with plantations of pines leads to the acidification of soil and waters in these areas. In recent years a further threat has manifested itself in the form of feral American mink that have escaped from fur farms in the area.

collection, reptile populations are being threatened by the collection of certain species for the pet trade. These include Hermann's and Spur-thighed tortoises, Mediterranean chameleons and Stripe-necked terrapins. The continued use of poisoned baits, banned by law, also threatens Iberia's rare birds of prey with elimination. The lammergeier, or Bearded vulture, once widespread in the peninsula, is today more or less confined to the Pyrenees, where its population has been reduced to at most 45 pairs. The Spanish imperial eagle has also declined to precariously low numbers, and neither population can afford the further losses that illegal hunting incurs.

In the farming country of the *dehesa* the intensification of agricultural methods, including the use of largescale machines, with consequent uprooting of trees, and of artificial fertilizers and pesticides, is threatening the unique animal life of these habitats with extinction. Especially vulnerable are the large birds that nest in the tops of ancient evergreen oaks – Black vultures, Spanish imperial eagles and Black storks – and mammals such as the genet and Pardel lynx, which rely on the *dehesa* trees for cover and prey.

Unwelcome invasions

The animal communities of the Atlantic islands evolved in the absence of predators, and many of the animals thus lack the normal defense mechanisms of related species on the mainland. The influx of non-native animals such as rats and cats, together with the loss of their habitats to tourism and farming, have wreaked havoc among them. The Canarian black oystercatcher has not been seen for more than 20 years and only a handful of Hierro giant lizards remain. Similarly, in the Balearic Islands, the Ibiza beech marten has probably become extinct in the last decade, and the Ibiza genet is endangered by disturbance from tourism.

On the mainland, the thoughtless introduction of trout into glacial lakes in the Cantabrian Mountains and the Pyrenees has resulted in a decline in the numbers of native amphibians, particularly the Pyrenean brook salamander, since the fish prey on both young and adults. The introduction of mouflon, a wild sheep, into the former game reserve of Cazorla y Segura in the southeast has driven the native ibex into a small area among the highest peaks. The situation is even more precarious for the Pyrenean ibex. This subspecies of the Spanish ibex is in real danger of disappearing for good, having been reduced to less than 30 individuals as a result of poaching and of competition for pasture from domestic livestock.

Evolution in action

Eleven thousand years ago the sea level of the Mediterranean was 30 m (100 ft) lower than it is today, and almost all the Balearic Islands were linked to each other and to the Iberian Peninsula by land bridges or by very shallow water. As the waters rose, their animal communities lost contact with each other, and today many animal species exist in isolated populations scattered throughout the islands. The genetic characteristics of the populations have gradually changed over time to adjust to the local environment, producing a range of distinct subspecies (races) or even species. This process, known as adaptive radiation, still continues. A good example is provided by the four species of lacertid lizards (wall and rock lizards) found on these islands.

Two of these species occur elsewhere in the Mediterranean: the scarce Moroccan rock lizard is native to northwest Africa, but was introduced to Minorca in 1968; Minorca is also home to the Italian wall lizard, a species found on several Italian islands. The other two are endemic to the Balearic Islands: the Ibiza wall lizard is present on Ibiza, Formentera and their surrounding islets, while Lilford's wall lizard occurs on Majorca, Minorca and nearby islands. Both these lizards are robust, versatile species, well suited to tolerate hostile environments with little vegetation, which makes them superbly adapted to small offshore islets that are little more than barren rock and where insects (the normal diet of small lacertids) are in short supply. Both species supplement their diet with vegetable matter.

There are numerous races of both species, each generally belonging to one islet or group of islets. More than 50 distinct populations of the Ibiza wall lizard are known, including about 25 genetically discrete subspecies or races. Extreme examples are the navy-blue lizards with turquoise bellies of the Illa Murada subspecies and the melanic (black) race of Ses Margelides. The 20 or more accepted subspecies of Lilford's wall lizard include the large, short-limbed melanic populations of Imperial and the yellow-bellied race of Dragonera.

Gulls, hedgehogs, genets and introduced species such as rats are the main predators of these lizards, although they are unlikely to reduce their populations significantly. People, on the other hand, are a more serious threat. The capture of the Ibiza wall lizard for the purpose of scientific study in the 1950s offered an easy source of income to fishermen. In the absence of natural predators, the lizards displayed no fear and were caught in their thousands. At the same time, specific races were unwittingly dispersed from island to island, producing a loss of genetic integrity and the subsequent extinction of several subspecies.

There are additional threats. In order to improve navigation around the main islands of the archipelago, whole islets have been dynamited out of existence; yet others are regularly used for target practice by the warships of the Spanish navy. More recently, the expanding tourist industry has led to an increase in pleasure boats, with disturbance and the introduction of domestic animals an obvious outcome. Of 43 islet populations of the Ibiza wall lizard described by researchers in the early 1980s, no fewer than 18 were considered to be highly endangered or already extinct as a result of genetic mixing; only 10 were found to be thriving. It is likely that the Lilford's wall lizard is in the same jeopardy.

A sharp-toothed predator (*right*) The Common genet is one of several animals that prey on the lizards of the Balearics. It has not been established whether the genet is native, or whether it was introduced, perhaps by fishermen.

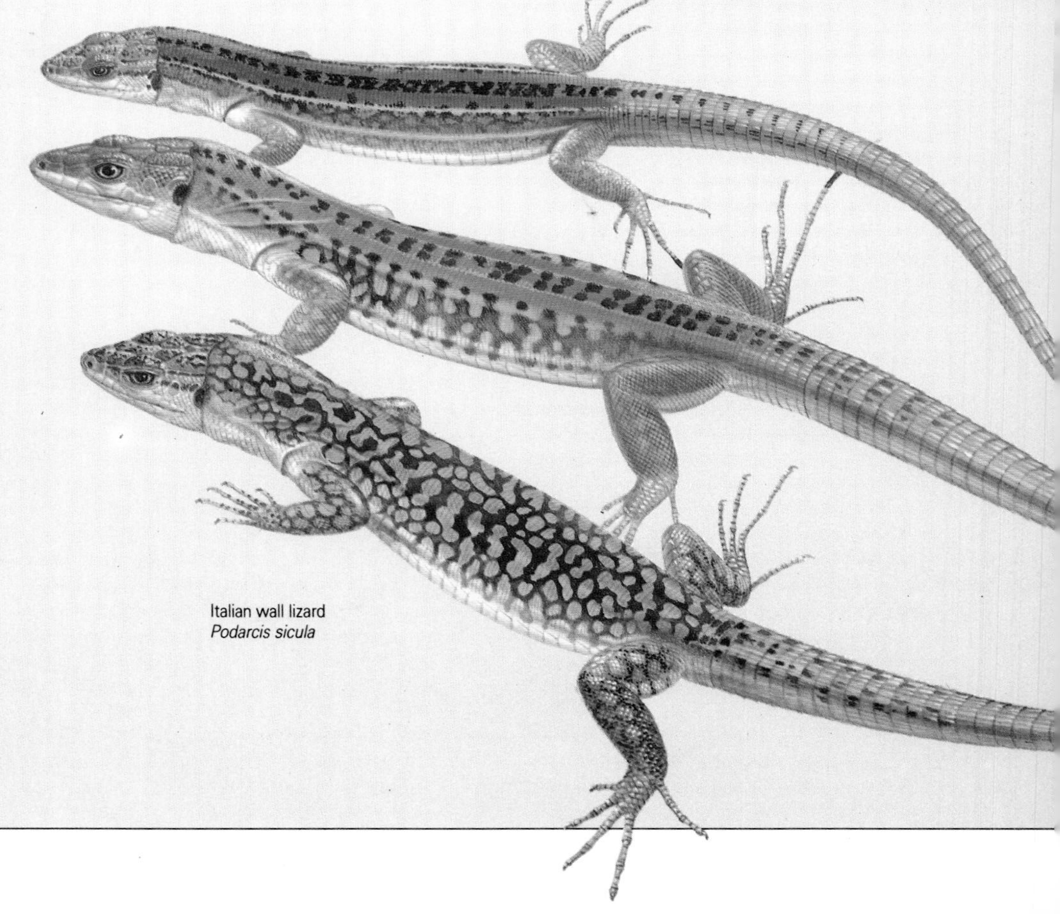

Italian wall lizard
Podarcis sicula

Balearic lizard diversity As the Balearic Islands became cut off from the mainland by the rising waters of the Mediterranean Sea, their lizard populations continued to evolve in isolation, producing many new species and subspecies. The Italian wall lizard has a distribution extending from the Balearics and mainland Spain to the Italian islands and Turkey. Lilford's wall lizard is found on Majorca, Ibiza and nearby islands. Unusually for a lizard, this species consumes a high proportion of vegetable matter in its diet. It is distinguished by the rather thick tail. The Ibiza wall lizard, whose tail averages twice its body length, lives on Ibiza, Formentera and the surrounding islets. It, too, has a varied diet, including vegetation, marine crustaceans and even its own offspring when food is scarce. The Moroccan rock lizard is not native to the region, but was introduced to Minorca from northwest Africa in 1968.

ANIMALS OF DRYLANDS AND ISLANDS

VARIED ISLAND LIFE · LIVING HIGH AND DRY · A LONG RELATIONSHIP WITH PEOPLE

Although the animals of Italy and Greece are generally typical of those that have adapted to the relatively arid environment of the Mediterranean, both countries are home to a surprisingly large number of species. The Italian peninsula, for example, has a diversity of habitats, ranging from the high peaks of the Alps in the north to the deserts of Calabria in the southernmost tip, each with their own characteristic wildlife. The remote mountains of Greece provide refuges for some of Europe's rarer animals, such as wolves and vultures. On the myriad islands of the region several species, in particular lizards, have evolved in isolation to form numerous subspecies. Many birds pass through Italy and Greece during their migrations between Europe and northern Africa; some stay for their breeding season.

COUNTRIES IN THE REGION

Cyprus, Greece, Italy, Malta, San Marino, Vatican City

ENDEMISM AND DIVERSITY

Diversity Low
Endemism Low to medium (particularly Islands)

SPECIES

	Total	Threatened	Extinct†
Mammals	88	4	1
Birds	400*	12	0
Others	unknown	49	0

† *species extinct since 1600 - Sardinian pika* (Prolagus sardus)
* *breeding and regular non-breeding species*

NOTABLE THREATENED ENDEMIC SPECIES

Mammals none
Birds none
Others Milos viper *(Vipera schwiezeri)*, Sardinian cave salamander *(Speleomantes genei)*

NOTABLE THREATENED NON-ENDEMIC SPECIES

Mammals Long-fingered bat (*Myotis capaccinii*), Mouse-eared bat (*Myotis myotis*), Gray wolf *(Canis lupus)*, Mediterranean monk seal *(Monachus monachus)*
Birds Dalmatian pelican *(Pelecanus crispus)*, Pygmy cormorant *(Halietor pygmeus)*, Little bustard *(Tetrax tetrax)*, Audouin's gull *(Larus audouinii)*, Lesser kestrel (*Falco naumanni*), Imperial eagle (*Aquila heliaca*), Black vulture (*Aegypius monachus*)
Others Loggerhead turtle *(Caretta caretta)*, olm *(Proteus anguinus)*, Corsican swallowtail butterfly *(Papilio hospiton)*

DOMESTICATED ANIMALS (originating in region)

goose *(Anser anser)*

VARIED ISLAND LIFE

A major factor in accounting for Italy and Greece's diversity of animal life is the presence of so many islands. The region contains some of the largest islands in the Mediterranean, notably Crete, Rhodes and Sardinia, as well as hundreds of smaller ones, very often with their own range of endemic species and subspecies. (Sicily, which is separated from Italy only by the narrow straits of Messina, shows a much lower variation in species.) Some remarkable island specializations have developed. For example, more than 70 subspecies have evolved from the three main species of Mediterranean lizard, one of which has adapted to feeding on insect parasites living in the nests of island gulls in the Adriatic Sea.

Sardinia – which is separated from the Italian mainland by the Tyrrhenian Sea – has a characteristic range of island wildlife. It shares some species with the mainland, and others with Corsica and mainland France, as well as having a number of endemic species. Within historic times Sardinia was inhabited by a species of pika (a relative of the rabbit), which may have survived until the late 18th century on the tiny island of Tavolara, off northeastern Sardinia. The mouflon (a wild sheep) has long been known to inhabit Sardinia, Corsica and Cyprus, but its occurrence here may not be natural. It is quite likely to derive from an early domesticated sheep brought here by people before sheep akin to modern varieties were developed.

A species of brook salamander found only in mountain streams in Sardinia is closely related to two other species, one confined to the Pyrenees, which straddle the border between France and Spain, and the other to Corsica. This suggests that a common ancestor colonized these areas when they were connected by land bridges. The distribution of the five species of cave salamander remains unexplained: one species lives in the cool caves of Sardinia, and another in northern Italy – but the three other close relatives are found, surprisingly, only in California in the southwestern United States.

Sea mammals and birds

Until the 1950s and 1960s the coasts of Italy (particularly Sardinia) and Greece were the stronghold of one of Europe's

rarest mammals: the Mediterranean monk seal. This species (its name comes from the Greek word *monakhos,* meaning solitary) once ranged throughout the Mediterranean and the Black Sea and westward out into the Atlantic. The monk seal is now on the verge of extinction; one of the few viable populations survives in the northern Sporades islands, off the east coast of Greece.

Dolphins are far more common, and are often seen in the Mediterranean. Their sociable habits and complex methods of communication have long been of fascination to humans. Thought to symbolize joy, they were given special protection by the ancient Greeks: killing them was an offence punishable by death.

Together with many of the Greek islands, the coasts of Sardinia provide the nesting grounds for Eleonora's falcon. This species nests at the end of summer, and its newly fledged young feed on the abundant supply of migrating songbirds flying south to overwinter in Africa. Audouin's gull, one of Europe's rarest birds, also breeds in coastal Sardinia and in many of the Aegean islands.

Until recently the isolation and sheer number of the islands and beaches in the Mediterranean have protected much of their wildlife. However, the rapid growth of tourism, with the dumping of raw sewage and refuse into the sea, is now destroying the habitats on which these animals are dependent. The draining of wetlands and the widespread introduction of more intensive methods of agriculture are also adding to the problem.

A vocal amphibian (*above*) The continuous trill of the male Green toad is a common sound in Italy and Greece. This is one of the few toads that is capable of spawning in brackish as well as fresh water.

Small but sure-footed (*left*) The chamois is quite at home on steep cliffs. The elastic soles of its hooves grip the rock surface; the outer rims of the hooves are firmer and help to prevent slipping.

LIVING HIGH AND DRY

The countries of the Mediterranean have a relatively arid environment, characterized by olive groves, vineyards and other farmlands, interspersed with open scrub on hillsides. Only the occasional remnant survives of the forests that existed thousands of years ago.

Several species of bird are well adapted to this dry and scrubby landscape. In the past they benefited from the spread of human activities, but modern agricultural practices may now be harming them. Many species of warbler flourish in gardens, orchards, olive groves and scrub: in Greece the Olive-tree and Ruppell's olivaceous warblers; in Italy the Dartford, Garden, Melodious and Spectacled warblers; and in both countries the whitethroat and blackcap, Fan-tailed and Bonnelli's, Sardinian, Orphean and Subalpine warblers.

Less than 50 years ago vultures and other scavengers were still widespread in southern Europe. The Egyptian vulture survives in southern Italy and is still widespread in Greece. Here it shares parts of its range with the Griffon vulture (which just survives on Sardinia as well), the Black vulture and Europe's rarest vulture, the lammergeier or Bearded vulture, also found on Crete.

A place for reptiles

Many reptiles thrive in the dry environment of Italy and Greece. Far from being the cold-blooded animals they are sometimes held to be, reptiles can control their blood temperature and are generally active only after they have warmed themselves. They regulate their temperature by moving in and out of the sunshine. In addition, lizards and snakes flatten their bodies to increase the area exposed to the sun. If the temperature becomes too hot they retreat to the cool of their burrows and become nocturnal, or even go into a state of torpor for the summer, a condition called estivation. Among the reptiles of the region are the geckos, which have adapted particularly well to existing in a human environment in the cracks of walls and pavements. Geckos are mainly nocturnal, with large eyes and a vertical pupil. They can walk up vertical surfaces and in some cases run upside down across ceilings. With the exception of Kotschyi's gecko – a lizardlike species found in Greece, Crete and many of the Aegean and Ionian islands, in adjacent Balkan countries and in a small area of southwestern Italy – all the European species have adhesive toepads: the most highly developed are those of the Moorish gecko, which can run up and down apparently smooth walls. Even Kotschyi's gecko can hang upside down from overhanging rocks. To protect these sensitive and delicate pads geckos often curl up their toes when resting.

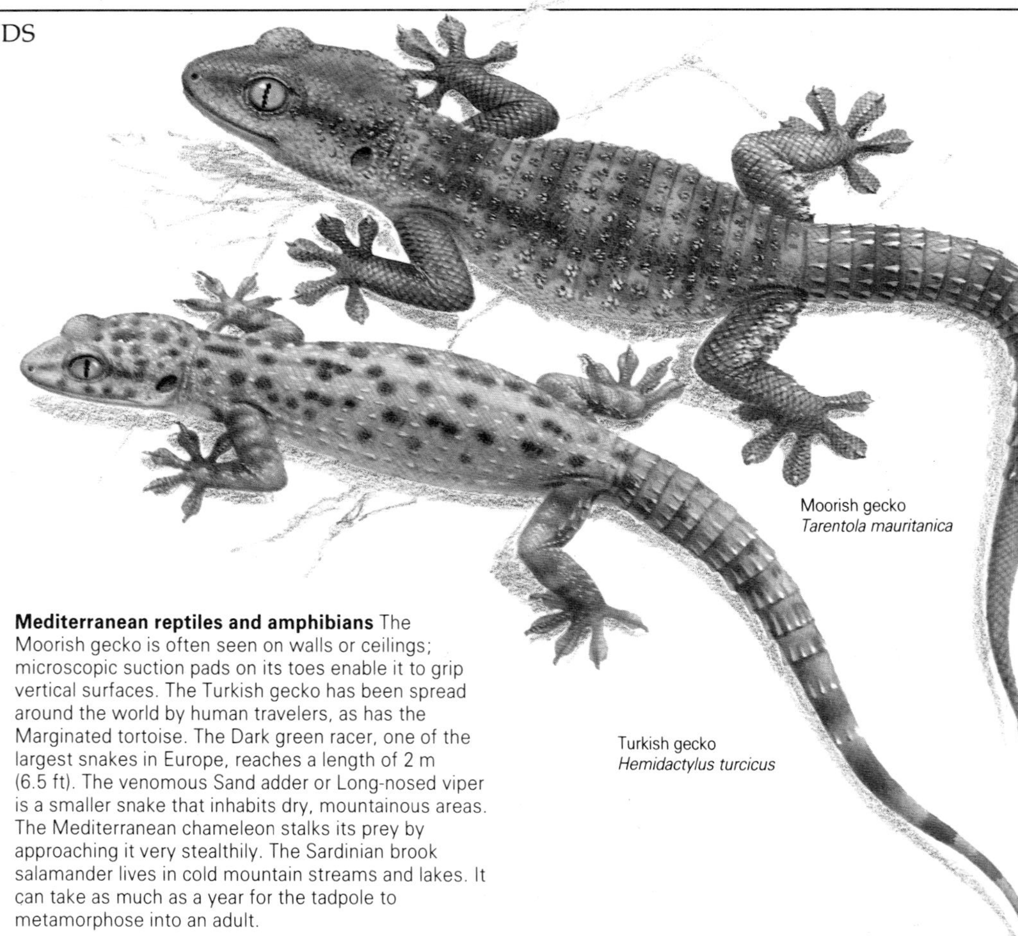

Mediterranean reptiles and amphibians The Moorish gecko is often seen on walls or ceilings; microscopic suction pads on its toes enable it to grip vertical surfaces. The Turkish gecko has been spread around the world by human travelers, as has the Marginated tortoise. The Dark green racer, one of the largest snakes in Europe, reaches a length of 2 m (6.5 ft). The venomous Sand adder or Long-nosed viper is a smaller snake that inhabits dry, mountainous areas. The Mediterranean chameleon stalks its prey by approaching it very stealthily. The Sardinian brook salamander lives in cold mountain streams and lakes. It can take as much as a year for the tadpole to metamorphose into an adult.

Adaptable insects

The scarab or dung beetle is found throughout Mediterranean Europe; it was considered sacred by the ancient Egyptians. This 30-mm (1-in) long beetle makes a fist-sized ball of cattle or goat's dung, which it rolls along the ground held between its hindlegs. Having found a suitable spot, the scarab buries the ball and feeds on it underground. In late summer the male and female use the dung to construct an underground nest into which the female lays an egg; once it has hatched, the larva feeds on the dung until the spring, when it emerges as an adult beetle.

The antlion, similar to the lacewing, is another insect of warm climates. Its larvae are voracious predators. The female lays

SPONGES

Sponges are among the simplest known forms of multicellular animal life. Some of them consist of just a hollow tube of cells communicating to the outside by numerous tiny holes or pores. There is a single larger opening through which water is drawn by the beating of the tiny hairlike flagella that line the tube; microscopic particles of food are then extracted from the water. The familiar bath sponge is the fibrous skeleton of a more elaborate sponge, but it functions on the same basic principle.

The Greek islands have been the center of the sponge-diving industry for hundreds of years. The sponges are gathered from as deep as 200 m (650 ft). In the past the collection was difficult, requiring great skill and endurance on the part of the divers, but it is now carried out by trawlers or by divers that are equipped with underwater breathing apparatus. Both of these methods can devastate sponge populations.

However, a method has now been devised so that sponges can be commercially farmed. A living sponge is first cut into twelve or more pieces, and then each piece is weighted to anchor it to the seabed; all will develop into separate sponges. Living sponges are dark brown; after collection they are cleaned, bleached and trimmed to the desired size and shape.

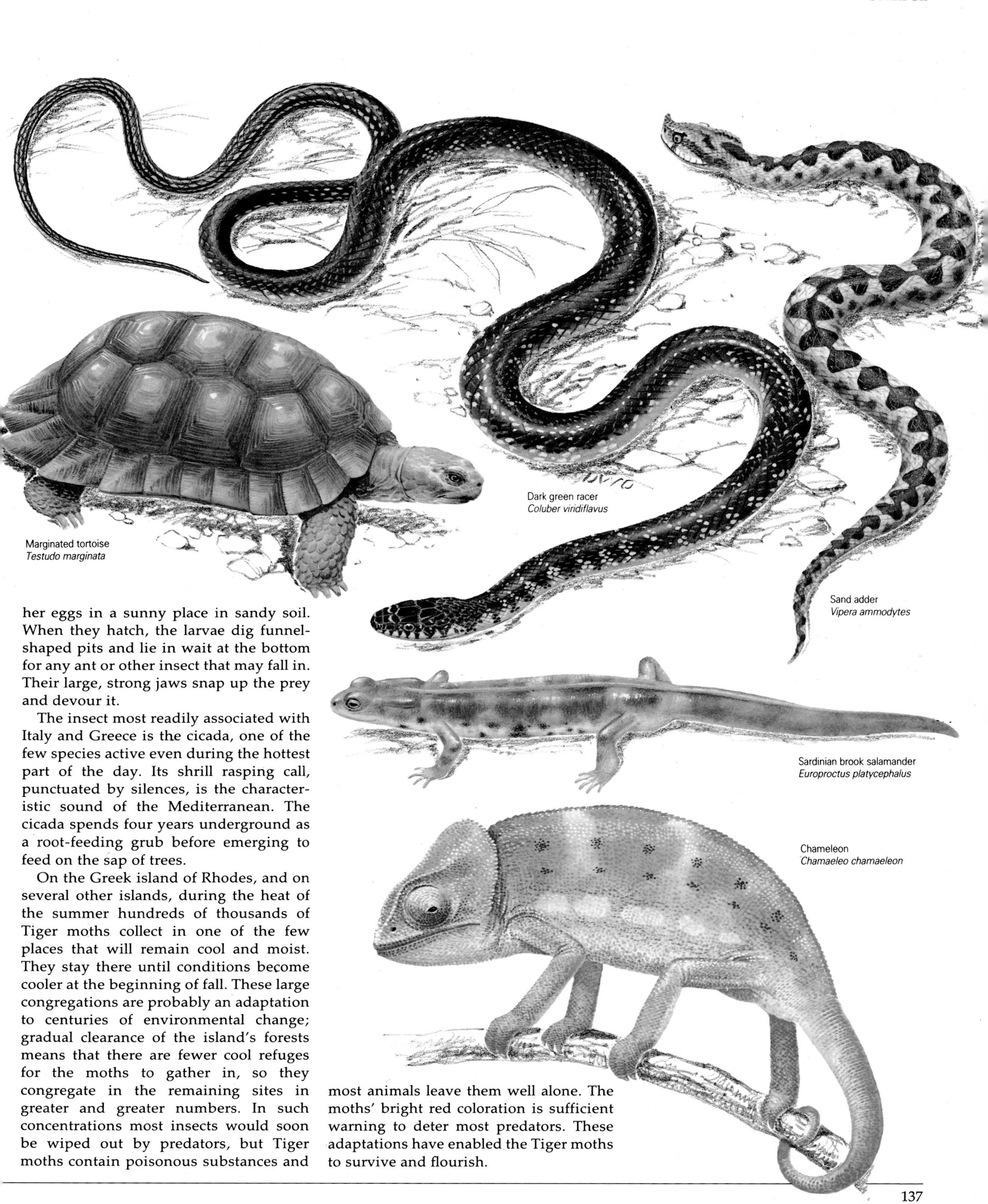

Dark green racer
Coluber viridiflavus

Sand adder
Vipera ammodytes

Marginated tortoise
Testudo marginata

Sardinian brook salamander
Europroctus platycephalus

Chameleon
Chamaeleo chamaeleon

her eggs in a sunny place in sandy soil. When they hatch, the larvae dig funnel-shaped pits and lie in wait at the bottom for any ant or other insect that may fall in. Their large, strong jaws snap up the prey and devour it.

The insect most readily associated with Italy and Greece is the cicada, one of the few species active even during the hottest part of the day. Its shrill rasping call, punctuated by silences, is the characteristic sound of the Mediterranean. The cicada spends four years underground as a root-feeding grub before emerging to feed on the sap of trees.

On the Greek island of Rhodes, and on several other islands, during the heat of the summer hundreds of thousands of Tiger moths collect in one of the few places that will remain cool and moist. They stay there until conditions become cooler at the beginning of fall. These large congregations are probably an adaptation to centuries of environmental change; gradual clearance of the island's forests means that there are fewer cool refuges for the moths to gather in, so they congregate in the remaining sites in greater and greater numbers. In such concentrations most insects would soon be wiped out by predators, but Tiger moths contain poisonous substances and most animals leave them well alone. The moths' bright red coloration is sufficient warning to deter most predators. These adaptations have enabled the Tiger moths to survive and flourish.

AN ANIMAL SLAUGHTER

Many birds that breed in the northern parts of Europe migrate south to winter in the Mediterranean basin, or even farther south in Africa. For centuries people have exploited these migrant birds for food, easily trapping or hunting them as they paused for rest and shelter on peninsulas, capes and islands throughout the region. In Italy, Malta and other parts of southern Europe the slaughter of migrant birds still takes place today – hundreds of millions are killed each year. Freedom to shoot robins, blackbirds, owls, Golden orioles, Turtle doves, warblers – in fact, anything that flies – is a long standing tradition in many Mediterranean countries, dating back to a time when it was a way of supplementing monotonous subsistence diets.

Today the birds are killed simply for sport and as a delicacy. A referendum on hunting held in Italy in 1990 failed because the pro-hunters abstained from the vote; the low turnout meant that the ballot was invalidated. Migrant birds are already subjected to many environmental hazards; continued hunting on a large scale can only have a serious impact on many species, especially birds of prey.

A LONG RELATIONSHIP WITH PEOPLE

Records of the region's wildlife have been left to us by classical Greek and Roman writers on natural history, such as Aristotle (384–322 BC) and Pliny the Elder (23–79 AD): we know, for example, that wolves, lions, bears, ibex, monk seals, pelicans, ibises, ibex, eagles and vultures were all once abundant. Within historic times, however, the distribution of these and other species of the region has been radically altered, both directly and indirectly, by their interaction with people. Many still survive, but only in the more remote mountains and swamps that are unsuitable for farming or that have not been swallowed up by development.

Fellow travelers

In addition, people have played a part in introducing new species, and altering the distribution of others. Domestic animals are the most obvious example, but tortoises, chameleons, frogs and geckos

Silencing the songbirds *(above)* A nightingale lies trapped on a stick coated with birdlime, a popular device for catching birds in Italy and Greece. Hundreds of millions of migrant birds are killed every year in Mediterranean countries, sometimes for food but more often for sport.

have also all been transported around the Mediterranean, sometimes deliberately, but more often accidentally. People have carried tortoises with them on their travels since time immemorial. The reasons for this are not always clear; it is most likely that the animals were used for food and sold as pets for children. Consequently the Marginated tortoise, whose range was originally probably restricted to mainland Greece, is now found on Sardinia, as well as on several islands in the Aegean.

The Mediterranean chameleon, a North African species that is found in Crete, Sicily and Malta and at the tip of the Peloponnese on mainland Greece, is also likely to have been introduced by people, as is the agama (another kind of lizard). Agamas are native to North Africa and Asia and are now present on Corfu, some

European range was confined to Greece and the Balkan countries, where it may have been introduced by the Ottoman Turks. Then in the 1930s, apparently in response to a genetic alteration, it proved successful in adapting to habitat changes caused by the intensification of farming methods, and quickly spread northward and westward until it had colonized most of Europe by the 1960s.

Vultures and other birds that scavenge the flesh of dead animals for food have benefited from living close to humans. Without extensive flocks of domesticated sheep, goats and cattle, some of which inevitably stray and die, and without the refuse produced by human communities, their populations would doubtless have been smaller. During the 20th century, however, pesticides and poisons have taken their toll of these birds, as have changing patterns of livestock husbandry and the disposal of waste. Feeding platforms on which animal carcasses that are unsuitable for human consumption are set out may have to become a part of the scenery in nature reserves in the future. Without measures such as this, lammergeiers (Bearded vultures), and Black and Griffon vultures will have only a slender chance of surviving.

Fattening up for winter *(above)* The Edible dormouse can almost double its weight in preparation for its winter sleep by feeding on nuts, berries and fruit.

Brassica bugs mating *(left)* These colorful insects will take advantage of agriculture to feed off the juices of cabbages and other crops. They derive their bright colors from various pigments in the plants on which they feed.

Aegean islands and around Salonika in northern Greece.

Some species have spread in the wake of farming and are found only in habitats heavily modified by human activity. They include the Blunt-nosed viper (western Cyclades, Greece) and the Algerian whip snake (Malta). Scientific evidence from the sites of temples on many of the islands suggests that the introduction of the Blunt-nosed viper may have been connected with early religious practices.

The region has served as a stepping stone for species extending their range from Asia into Europe. Rats and mice probably traveled this route, and more recently the Collared dove has spread into almost all of Europe from Greece. Until the beginning of the 20th century its

An animal success story

Not all aspects of people's relationships with wildlife are harmful. The growing awareness of our destructive impact on nature began to develop in the mid-19th century, leading to today's conservation movement. An early example of animal protection took place in Gran Paradiso in the Italian Alps, where the Alpine ibex, a mountain goat, was preserved for hunting purposes. Were it not for this status the ibex would by now be extinct.

The ibex's horns were a prized hunting trophy. They were held to have peculiar properties and were used for a variety of purposes: goblets of ibex horn were supposed to detect poisons, shavings were used to cure hysteria, and ibex blood was held to prevent the formation of kidney stones. Ibex were hunted and poached by Europeans just as ruthlessly as rhinos are today, for virtually the same reasons and with the same ultimate result. They were hunted to extinction everywhere in the Alps except in Gran Paradiso. Here their numbers grew and stabilized, and ibex have now been reintroduced into many other parts of their former range.

A superb hunting machine

The Italian wolf (*above*) One of the few remaining populations of the once widespread Gray wolf survives in the Abruzzo National Park in Italy. All wolves have acute senses, using sight, smell and hearing for hunting and communication.

Wolves were once the most widespread large mammal, apart from humans, ranging over most of Europe, North America and Asia. However, by the 19th century they had disappeared from most of the countries of western Europe – the victims of persecution and habitat destruction – and today are restricted in Europe to a few large forests in Eastern Europe and the Soviet Union, and to some isolated mountain refuges around the Mediterranean. About 100 individuals are protected in the Abruzzo National Park in central Italy.

Wolves are the largest members of the dog family. They reach a height of 80 cm (32 in) at the shoulder, have long legs and run on their toes. They take prey by stalking it and making a short dash for the kill, but are equally capable of loping for several kilometers across country, tiring the animal that is being pursued. The Gray wolf is by far the most numerous of the two surviving species, and is divided into many subspecies (up to 32 have been described).

Running with the pack

Wolves mostly live in social groups known as packs. The nucleus of the pack is the breeding pair: wolves usually mate for life, and only the leading male and female in the pack breed. The other adults of the family group take part in hunting expeditions and help in caring for the pups. Wolves have strong social behavior and communicate within the pack by scent, posture and facial expressions similar to those of domestic dogs.

Mating takes place in late winter. The babies are born nine weeks later, usually in an underground den. At first blind, helpless and dependent on the mother's milk, they emerge from the den at a month old. They are fed on food regurgitated by their parents and adult helpers, and can be large enough to travel with the pack at 3–5 months if food is ample.

Hunting, like rearing young, is a cooperative venture. Wolves work together in driving and ambushing their prey. They may take turns at leading the chase, following the twists and turns of the prey, while others take short cuts to conserve energy. When they catch up with a prey animal that is capable of fighting back, they surround it and take turns at darting in to attack. These strategies allow them to catch large animals they could not tackle alone.

A pack of wolves needs a large area in order to find sufficient food. Packs commonly range over more than 100 sq km (40 sq mi), but where food is scarce the range may be ten times as large. Other wolves are excluded from the pack's territory. They may be attacked, but are usually detered by "keep out" notices left in the form of urine scent marks at the edges of the pack's domain. The pack may also howl to proclaim ownership.

The size of the pack depends on the type of terrain being hunted over and the availability of prey. Where large prey are hunted, packs may number up to 20 individuals, but smaller prey such as deer can be tackled by packs of about 7 members. If natural prey is scarce, wolves may have to survive by scavenging human garbage and taking domestic livestock, and will travel alone.

The cooperative behavior of the wolf is unable to save it when it comes into close contact with people. Wolves menace humans extremely rarely, but they are often persecuted by farmers protecting their livestock. One of the most serious threats to wolves in Italy comes from inbreeding with dogs. They will survive only if they are granted sufficient space, free from persecution, to roam in packs.

On the alert (*above*) Cooperative hunting in packs enables wolves to capture large prey such as deer and sheep. During the chase a wolf can maintain a speed of up to 70 kph (43 mph) and can cover 5 m (16 ft) in a single leap. In many areas where it lives its survival is precarious, and though it once inhabited plains and open woodland, it has now sought refuge in large forests, on mountains or in marshy areas, where there is less risk of human persecution.

Wolf hierarchy Wolves have a strict order of dominance within the pack, including the pups. A subordinate wolf keeps its tail lower and its ears back as it greets a more dominant member of the pack nose to nose. A subordinate pup reacts non-aggressively as it is grabbed by the jaws of another in play. With a stiff-legged shove one pup tries to exert authority over another of the same age. From late fall to late winter, when breeding starts, such interactions between members of the pack become more frequent. Only one female usually mates, and females may fight aggressively for that right.

LIVING CLOSE TO PEOPLE

SURVIVAL AND DIVERSITY · BIRDS OF TEMPERATE FOREST · WINNERS AND LOSERS

Central Europe supports an impressive number of mammals, birds and fish, as well as countless invertebrates. Its geographical position – between the highlands of the Alps and the lowlands bordering the North Sea and Baltic, and in the transitional zone between the maritime climate of the northwest and the harsher continental climate of the east – makes it a mixing ground for species. Few of its animals are unique to the region; many reinvaded the area from the southeast and southwest at the end of the last ice age as the land began to warm up. Virtually none of the landscape is unaffected by human activity, and many species have been lost, together with their habitats. Others have migrated into the region, taking advantage of new food supplies on farming land and around human settlements.

COUNTRIES IN THE REGION

Austria, Liechtenstein, Switzerland, Germany

ENDEMISM AND DIVERSITY

Diversity Very low to low
Endemism Very low

SPECIES

	Total	Threatened	Extinct†
Mammals	83	2	1
Birds	440*	6	0
Others	unknown	63	0

† *species extinct since 1600 - Bavarian pine vole* (Pitymys bavarius)
* *breeding and regular non-breeding species*

NOTABLE THREATENED ENDEMIC SPECIES

Mammals none
Birds none
Others Goesswalds's inquiline ant *(Leptothorax goesswaldi)*

NOTABLE THREATENED NON-ENDEMIC SPECIES

Mammals Pond bat *(Myotis dasycneme)*, Mouse-eared bat *(Myotis myotis)*
Birds Red kite *(Milvus milvus)*, White-tailed sea eagle *(Haliaeetus albicilla)*, corncrake *(Crex crex)*, Great bustard *(Otis tarda)*, Aquatic warbler *(Acrocephalus paludicola)*, Lesser kestrel *(Falco naumanni)*
Others Danube salmon *(Hucho hucho)*, Great peacock moth *(Saturnia pyri)*, Scharlachkafer beetle *(Cucujus cinnaberinus)*, Ravoux's slavemaker ant *(Epimyrma ravouxi)*

DOMESTICATED ANIMALS (originating in region)

none

SURVIVAL AND DIVERSITY

Altogether some 40,000 animal species inhabit Central Europe. These include 78 species of land mammal, 280 breeding bird species, another 160 species of birds that visit or winter here regularly, and about half of Europe's fish species. This diversity of wildlife has survived despite the substantial deterioration in habitats that has taken place during this century, in particular the widespread replacement of the broadleaf deciduous forests that originally covered the region's lowlands and low mountainous areas by agricultural land and conifer plantations. Attempts have recently been made to increase the range of species by reintroducing widely diverse animals such as the beaver and the lynx.

Alpine wildlife

Rising to more than 4,000 m (13,000 ft), the Alps have a rich variety of animal life, including several species that are found nowhere else (endemic species). Two members of the goat family live in the mountains. The chamois, well adapted to steep cliffs and high pastures, is widespread, but the ibex, once almost hunted to extinction, today survives only in the Swiss National Park and a few other areas where it has been reintroduced. In the high meadows lives the Mountain hare, an isolated relic from a wider distribution during the cooler climate of the ice age. The pale-coated Snow vole is also found here, as well as the Black salamander, which hibernates in winter in the disused burrows of Alpine marmots.

There is a distinctive range of alpine birds, including species such as the Golden eagle, the lammergeier or Bearded vulture (now being reintroduced to the central Alps) and the raven. These birds are increasing in numbers after a long period of persecution. Among the smaller birds are the Snow finch and the wallcreeper that glean prey from insect-and spider-rich cliffs and walls; the chough and Alpine chough are masters of flight, able to conquer the windiest of mountaintops. There are also species, such as the Black redstart, which are more typical of lowland towns and villages.

The isolated position of individual Alpine valleys, and the severe climate, have encouraged the evolution of local subspecies. For example, a different version of the attractive Apollo butterfly is found in nearly every remote valley of the Alps and the lower mountains.

Animals of forest and coast

The extensive deciduous forests covering the low mountains of central Germany are home to many species typical of the belt of broadleaf woodland that once covered most of the central European lowlands. This area is a stronghold of both the Red kite and the Eagle owl. Deer and Wild boar roam the forest floor.

Many natural woodlands have survived along the banks of major rivers, where annual floods prevent other forms of land use. These riverine woodlands are extremely rich in animal species, including specialists such as the River warbler and colonies of the European tree frog. Oxbow

In full voice (*left*) A cock capercaillie displays to a potential mate with an accompaniment of sounds: clicks, gulps, pops and grindings. During the spring breeding season, when he attempts to gather a harem of up to a dozen females, the cock fiercely defends his territory against all invaders. The capercaillie is the largest species of grouse, up to 1 m (3 ft) long. Its numbers have been drastically reduced by the loss of its natural coniferous habitat.

A rare sight (*below*) The European lynx is a shy, nocturnal predator that has been hunted almost to extinction for its fur, though it has recently been reintroduced to Switzerland. Its survival depends on the continued existence of large tracts of forest and mountainside to provide sufficient prey – mainly hares, birds (including capercaillies) and occasionally Roe deer.

lakes support a profusion of dragonflies, and clouds of butterflies gather on the wet ground, where they extract minerals from the soil. The riverine forests along the Danube in northern Austria are famous for their large stock of Red deer, and also for a celebrated rarity: the Great horned beetle, now very rare due to the clearance of old growth forests. Near the Austrian border with Czechoslovakia is the only woodland-breeding colony of the White stork in central Europe.

One of Central Europe's most important sites for wildlife is the shallow stretch of the North Sea between the East Frisian Islands and the mainland, which is exceptionally rich in birdlife. Millions of geese, ducks and shorebirds use these waters as a refueling place on their migrations to and from their Arctic breeding grounds.

Three-toed woodpecker
Picoides tridactylus

BIRDS OF TEMPERATE FORESTS

Where there is a significant supply of decaying or dying trees, the remaining forests of Central Europe are the strongholds of many varieties of woodpecker, including the Black woodpecker – the largest and most impressive member of the woodpecker family in Europe – as well as some of the rarer species such as the White-backed woodpecker. Most species of woodpecker feed mainly on insects, spiders and grubs for which they drill with their powerful beaks in the crevices of tree bark. They play an important role in the ecology of the woodland, controlling the numbers of bark- and wood-boring insects and breaking up huge quantities of decaying wood.

Woodland diet

The Green woodpecker and the Gray-headed woodpecker prefer open woodland or, especially the Green, gardens and parks with mature trees. These woodpeckers depend to a great degree on Meadow ants for food, and their decline is closely linked to a decrease in these ants. This has been caused by the lusher growth of vegetation in the meadows and on the margins of the forest due to enrichment of the land by nutrients from agricultural fertilizers and sewage.

The Great spotted woodpecker is more adaptable in its feeding habits. It will catch insects in flight and makes so-called "anvils" into which it wedges pine cones to peck out the fat-rich seeds; in winter it may eat up to 1,000 pine seeds every day, the contents of 50 cones. It also sucks sap, damaging the trees by cutting a ring of bark around the trunks.

The wryneck, a member of a primitive subfamily of woodpecker, obtains its main food (various kinds of ant) by picking them from the surface of the wood with its tongue, as its bill is too weak to dig into wood. It has a habit of perching on branches rather than clinging to tree trunks, and often feeds on the ground. Its name derives from the snake-like twisting of its neck when threatened by predators in the nest.

The nest cavities excavated by woodpeckers are used by many other species. Woodpeckers usually make a new hole each year; the old ones are sometimes taken over by nuthatches, which cement up the entrance with mud to prevent larger birds displacing them. Tits also use these holes; when the entrance becomes even larger as the wood rots, redstarts may move in, and eventually even Stock doves and Tawny owls.

Many woodland birds, including tits, jays, nutcrackers and nuthatches, store nuts to supplement their diet in winter. The European jay is an important disperser of acorns, which inevitably helps to spread oak trees over a broader area. Many other woodland birds such as finches, crossbills and nutcrackers are seasonal migrants, and congregate in large nomadic flocks in fall and winter.

European woodpeckers *(above)* A bird of coniferous forests, the Three-toed woodpecker is the only European species without any red markings. The larger Green woodpecker lives in deciduous woodlands and parks, where its loud laughing call reveals its presence. In addition to insect larvae, it also feeds on the ground on ants, berries and plants. The Great spotted woodpecker is the commonest European species, and is renowned for its courtship chases in which the pair race round branches and tree trunks. The wryneck lives in deciduous woodlands and has a distinctive, highly camouflaged plumage. In behavior and appearance it is more like a perching bird than a woodpecker, preferring to perch on branches than creep up tree trunks. It feeds mainly on insects and spiders.

One of Europe's largest owls *(left)* The Eagle owl takes a variety of prey, from rabbits and hares to lizards, fish and frogs. Hunting and habitat destruction are causing its numbers to decline in many places.

DESIGNED FOR DRILLING

Woodpeckers are highly specialized so that they can extract invertebrate prey from the crevices of wood or bark. They have short powerful legs, and their toes are arranged so that the second and third toes are directed forward, and the fourth and first backward, in order to grip the trunk or branch. The toes are armed with very sharp, curved claws that act like grappling hooks. The straight stiff feathers of the tail prop the bird against the trunk. The central tail feathers are retained until the remaining feathers have been shed and regrown, ensuring that there are always enough fully grown feathers to act as a prop for the bird. The skull is strengthened, and muscles cradle the brain so that it can withstand the shocks of repeated hammering; the Black woodpecker, for instance, may strike up to 1,200 blows a day.

The woodpecker's long tongue is impressively designed to help it capture its insect prey. The bones and associated elastic tissues that support the tongue are anchored in the right side of the upper mandible or under the eye socket and loop over the brain to enter the mouth at the back of the head. When these bones move forward, the tongue is extended – in some species, such as the Green woodpecker, it may protrude up to 10 cm (4 in) beyond the bill. The tongue's sharp, horny tip spears ants, grubs and other insects, which are then entrapped in a coating of sticky mucus.

WINNERS AND LOSERS

The increase in Central Europe's human population and the spread of agriculture have created new sources of grain and other foodstuffs for wild animals. Cities, too, provide food in the form of garbage dumps and the nutrient-enriched waters of sewage farms, while artificial reservoirs support rich populations of aquatic invertebrates. As a result, many species have become more abundant in recent years. The Beech or Stone marten is an opportunist feeder that thrives near human habitation: it is found in towns and villages throughout the region and has even been known to damage cars by chewing rubber and plastic tubing in the engine. The raccoon, which was introduced before World War II from North America to northern Hessen in central Germany, is also expanding its range.

The region's bird life has seen even greater changes. Some 20 species have established new breeding sites in the past 50 years. One of the most prolific is the Black-headed gull, now a common breeding bird with more than 50,000 pairs on inland waters. The gulls winter in both towns and cities; between 20,000 and 30,000 gulls spend the winter in Munich, in southern Germany, more than 800 km (500 mi) from any coast. Diving ducks such as the Common pochard and Tufted

duck have also increased in numbers, wintering in their thousands on lakes and reservoirs. Near Munich is one of Europe's three major molting grounds; more than 200,000 ducks gather there in the summer to molt their flight feathers on an artificial reservoir that is part of the city's sewage treatment system.

Towns and villages have surprisingly abundant animal life. Munich boasts 99 species of breeding bird, and Vienna, in northeast Austria, has even more. In Munich 350 different moth species have been trapped in the grounds of the Nymphenburg palace. Peregrines have been introduced in Cologne and Göttingen, in west central Germany, in order to keep the large populations of feral pigeons in check. Urban populations of Alpine redpolls have also increased. These birds have adopted the ornamental conifers of city parks and gardens as a replacement for their native alpine habitats. Resident tit populations benefit from extensive winter feeding by city dwellers.

Following the plow

Many animals invaded the region from the steppe grasslands of southeastern Europe, taking advantage of the opportunities presented by the clearance of land for agriculture. Species such as the Brown hare and the skylark, which extended their ranges into Central Europe long ago, are representatives of the native steppe fauna; so, too, are most of the locust species and many of the butterflies and ground beetles. These animals of the steppes were already adapted to eating grassland seeds, so they quickly colonized the new grain-growing lands.

From the eastern steppes came the slow-moving Green toad, which was very common around the Neusiedler Lake in eastern Austria before the expansion of vineyards around the lake's shores. The European souslik (a member of the squirrel family) reaches the westernmost limit of its range here, and the Common hamster is also abundant.

Some species have profited so much from human activity that their numbers have reached pest proportions. The ever-expanding vole population now causes substantial damage to crops. As a result of their proliferation, populations of the Common buzzard and kestrel, which feed on voles, have also increased. The Collared dove has dramatically extended its range throughout the region in just half a

TO MIX OR NOT TO MIX?

Invasions of animal species in response to climatic and other changes have occurred frequently in Central Europe. During the 10,000 years since the last ice age many plants and animals reinvaded the region from two dispersal centers, in the southeast and the southwest. These two invasion fronts met in the area of Germany alongside the Alps, and many species still show transitional states there. For instance, the Long-tailed tit, which came in a white-headed race from the east and a striped race from the west, exists in Central Europe as a hybrid population with different percentages of white-headed, striped and hybrid birds. By contrast, where the eastern subspecies of the Carrion crow – the Hooded crow – meets the completely black western subspecies, the two mix only to a very limited degree. The river Elbe in northern Germany forms a stable border between the two distinct subspecies.

The mechanisms by which such differences are maintained can be unexpected. For example, the eastern and western subspecies of the House mouse meet just outside Munich; each subspecies is well adapted to its own particular set of internal parasites but not to those carried by the other one. The combination of two sets of parasites selects strongly against hybridization and therefore keeps the populations sharply differentiated.

Disappearing visitor *(above)* A White stork nest was once a common sight in Europe's villages. Today, the bird is in decline due to the loss of its wetland feeding grounds, both in Europe and on its migration routes to Africa and the Middle East.

Skilled mountaineer *(left)* The ibex was almost exterminated by hunters earlier this century. Animals from a small surviving herd in northern Italy have now been successfully introduced to several parts of the Alps, notably the Berchtesgaden and Swiss national parks.

Master of camouflage *(right)* Clinging on with suckerlike toes, the European tree frog is able to change color to match its background, turning silvery-gray when on lichen-covered bark. At its northern limit in Central Europe, it hibernates in winter.

century, following the expansion in the cultivation of maize. Originally it bred only in southwestern Europe.

Habitat fragmentation

While some animals have gained from the spread of human activity, the loss of natural habitat has been devastating for others. Almost half the species found on Central Europe's arable land have become less numerous as a result of the intensification of agriculture in recent decades. Species adapted to nutrient-poor conditions have declined.

The habitats that have suffered most are heaths, upland bogs, meadows, rivers and lakes, and fields that were formerly cultivated by traditional methods. Species that need hot, dry conditions at ground level have been badly affected since the general enrichment of the land with fertilizers has led to thicker, taller vegetation shading the ground below. This has caused many butterfly species to decline. Game birds, such as partridges and pheasants, which depend upon insects to feed their young, have also suffered from increasing use of pesticides.

Among the few birds that have adapted to modern agricultural practices is the lapwing. Originally restricted to seaside bogs and swampy meadows, it has evolved an extensive field-breeding population, behavior also adopted by another shorebird, the Black-tailed godwit. In contrast, curlews and redshanks are confined to wet meadows and coastal regions in the breeding season.

Many of the region's natural forests have been replaced by artificial plantations. Their lower diversity of plant species and lack of mixed mature and sapling trees adversely affects wildlife. With fewer shrubs and forest glades, the populations of insectivorous birds that nest and feed in the understory have diminished. Similarly, the absence of decaying wood and dying trees in the plantations prevents woodpeckers living there and has led to the decline of many wood-boring beetles. The capercaillie (a kind of grouse) has suffered a severe decline through the loss of its natural woodland habitat.

Habitat fragmentation has been particularly damaging to the larger woodland mammals. The Red deer is now restricted to isolated forests in Central Europe, but its much smaller relative, the Roe deer, has adapted better to change. Its numbers are thriving because increases in amounts of cultivated land and deliberate winter feeding by hunters have augmented its food supply.

Aquatic wildlife

In their natural state the many lakes of the Alpine area were low in nutrient content. As the human population has risen and agriculture expanded, however, they have become enriched with nutrients from sewage and farm waste, encouraging the growth of algae and aquatic invertebrates. As a result increasing numbers of ducks stop over or winter on these lakes. Over recent years dense populations of the small Zebra mussel have built up in many of the Swiss lakes and in Lake Constance, where Austria, Germany and Switzerland meet. Diving ducks, swans and coots feed on the mussels and in turn attract wintering White-tailed sea eagles.

The wildlife of the region's rivers has survived less well. Originally each river system had its own fish species or subspecies, but modern fishery practices have blurred these distinctions. Migrations of salmon up the Rhine used to fill the river, while vast swarms of mayflies covered the banks with thick mats of dying insects. Pollution has wiped out much of this wildlife.

Water quality has improved in recent years. However, recovery of the invertebrates and fish will be impossible while most rivers remain artificially channeled, preventing the buildup of islands, sandbanks and shallows.

Animal construction builders

Beavers, once common across much of North America, Europe and some parts of northern Asia, have been severely reduced by human persecution and large-scale destruction of their habitats. They were exterminated in Switzerland, Austria and western Germany in the second half of the 19th century, leaving only a handful of populations in western Europe, Scandinavia, on the lower Rhône in France, and on the river Elbe in eastern Germany. In recent years there have been successful attempts to reintroduce the beaver to some of its former haunts in Central Europe. Some individuals were taken from the Rhône to Switzerland in the 1950s, and others were reintroduced to Austria and southern Germany in the 1970s and 1980s; there are now more than 350 living in the headwaters and oxbow lakes of the Danube. Meanwhile the population on the Elbe has increased to about 2,500 and some of these animals have recently been moved to the Hessen area in central Germany.

Beavers are well known for their building activities. They modify their environment considerably, scraping and digging earth to make canals, which link their ponds to their feeding areas. This allows them to swim in comparative safety from predators, rather than moving overland.

At home on land and in the water (*right*) The beaver has waterproof fur, and its tail not only acts as a rudder but also provides propulsion as it is flexed up and down when swimming. When alarmed, a beaver will slap its tail on the water surface to warn the others. Although the beaver is the second heaviest rodent in the world – occasionally weighing over 30 kg (66 lb) – it is extremely agile in water.

Snug inside the lodge (*left*), the beaver is well protected from predators. The kits are born in the lodge in late spring, and at birth they already have a full coat of fur, their eyes are open, and they are able to move around. They can swim after just a few hours, but are too buoyant to swim down the underwater exit passage. Inside the lodge it is dark, yet the beavers regularly rise at sunset to feed, retiring to the lodge at sunrise. It has been shown that beavers time their activities by an internal biological clock.

A beaver dam (*below*) In deep rivers or lakes, beavers dig burrows in the banks, but in shallower waters they may significantly alter the landscape. Using their sharp front teeth, they fell trees to dam rivers, creating ponds and lakes where they build their lodges. The underwater entrances to the lodges deter predators, and allow the beavers to hunt below the ice in winter. Beavers sleep and raise their young on a dry platform inside the lodge. Extra mud and sticks are added to the outside of the lodge in fall; the frozen mud provides insulation and is an extra protection against predators. Small vents in the roof provide ventilation.

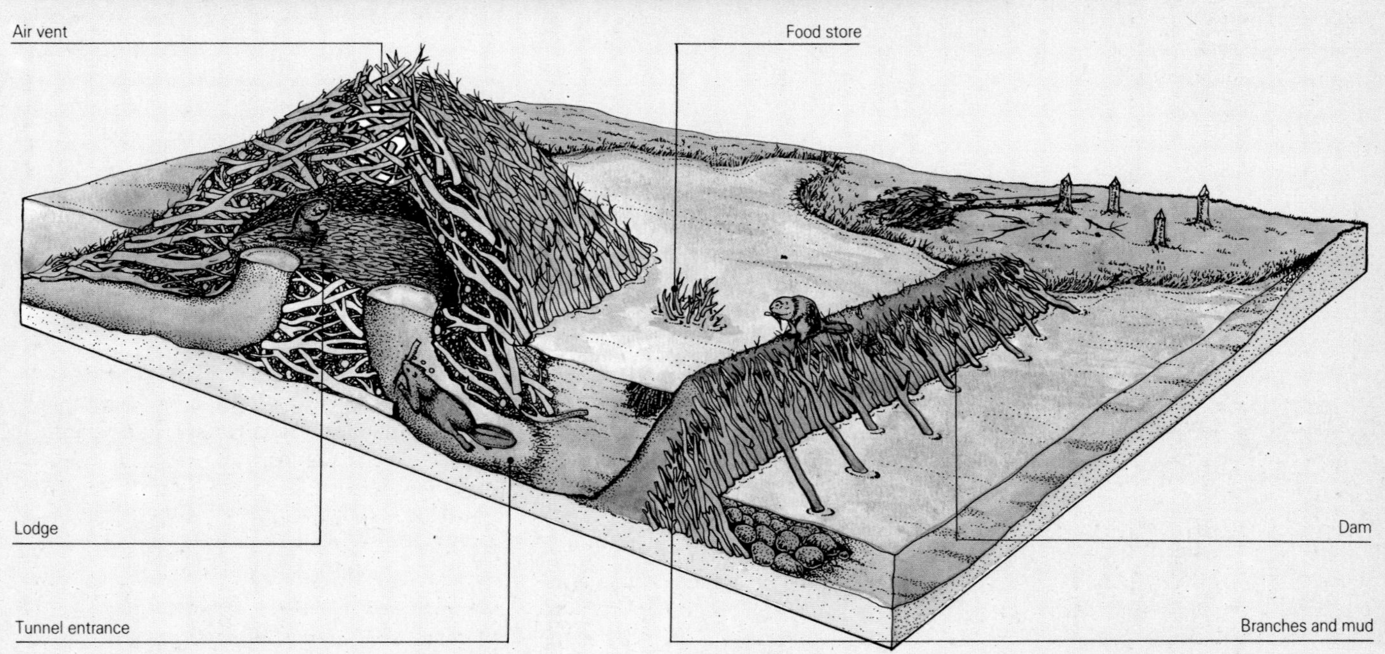

They dam streams, using sticks, stones and mud. Beaver dams may be enormous structures up to 3 m (10 ft) high and 100 m (330 ft) long. The dam creates a pond or lake behind, which enlarges the area they have to swim in, and also allows them to build a safe home in the middle of the expanse of water.

Beavers' homes, or "lodges", are large conical piles of sticks and mud. They begin by digging a burrow underwater with their forepaws; this extends upward toward the surface and is covered with sticks and mud. A tunnel entrance leads in from below the water to a living chamber hollowed out within the mound above water. Few predators are likely to find their way in here. It makes a snug winter home.

Family animals

A beaver lodge typically contains an adult pair, together with young from the previous year or two. All generations help to build the lodge and keep it in good repair. Only one litter is born a year, containing from one to five babies, known as kits. They are covered in fur at birth and are able to swim within a few hours, but beavers take a long time to mature and may not breed until they are three or more years old.

Beavers are herbivores, feeding in the spring and summer on the softer parts of plants. In the fall they switch to the woody stems of trees such as aspen or willow. The beaver's huge chisel-shaped incisor teeth can cut down small trees with ease. They store the stems underwater near the lodge, and use the bark for food during the winter.

Beavers have many adaptations for their semiaquatic life, including webbed feet and a large scaly tail that can be used for steering or as a propeller. Ears and nose close tight on diving, and the back of the tongue blocks the throat. The lips shut tight behind the incisors so a beaver can use its teeth or carry a stick underwater. Their bodies are torpedo-shaped to help them thrust through the water, and their fur is thick and waterproof. This was the reason they were so extensively hunted in the past – their fur was much valued, especially for making hats. "Castoreum", a musty-smelling reddish secretion from glands in the genital region, was also prized by the perfumery trade. Scent-marking of territory by adult males is particularly intense in the spring.

ANIMAL RICHES AT RISK

REFUGES FOR WILDLIFE · BIRDS OF THE WETLANDS · THE HUMAN IMPACT

Eastern Europe's wide range of habitats, which include shorelines, woodlands, mountains and vast, treeless steppes, contain large numbers of animals; its remote forests shelter some of Europe's rarest large mammals. Threading its way through much of the region is the mighty river Danube and its numerous tributaries, along which are dotted some of Europe's finest wetlands; the river itself contains several unique species of fish. The Danube ends in a magnificent delta, which is a paradise for wildlife, especially breeding birds. Beneath the ground, particularly in Yugoslavia and Romania, are extensive cave systems inhabited by bats and a range of highly specialized animals. Unfortunately, serious industrial pollution together with habitat loss is taking its toll of animal life throughout the region.

Expert fishers (*above*) The Great white pelican and many other wetland birds breed by the thousand in the region's extensive marshes. These birds are increasingly threatened by drainage and pollution of their habitats, particularly in the Danube delta.

COUNTRIES IN THE REGION

Albania, Bulgaria, Czechoslovakia, Hungary, Poland, Romania, Yugoslavia

ENDEMISM AND DIVERSITY

Diversity Low
Endemism Low

SPECIES

	Total	Threatened	Extinct†
Mammals	97	5	1
Birds	400*	13	0
Others	unknown	31	0

† *species extinct since 1600 - tarpan* (Equus gmelini)
* *breeding and regular non-breeding species*

NOTABLE THREATENED ENDEMIC SPECIES

Mammals none
Birds none
Others Dalmatian barbelgudgeon *(Aulopyge hugeli)*, asprete fish *(Romanichthys valsanicola)*, Kocevje subterranean spider *(Troglohyphantes gracilis)*

NOTABLE THREATENED NON-ENDEMIC SPECIES

Mammals European bison *(Bison bonasus)*, Gray wolf *(Canis lupus)*, European mink *(Mustela lutreola)*, Long-fingered bat (*Myotis capaccinii*), Mouse-eared bat (*Myotis myotis*)
Birds Dalmatian pelican *(Pelecanus crispus)*, White-headed duck *(Oxyura leucocephala)*, Lesser kestrel *(Falco naumanni)*, Great bustard *(Otis tarda)*, Aquatic warbler *(Acrocephalus paludicola)*
Others olm *(Proteus anguinus)*, Danube salmon *(Hucho hucho)*, Scarce fritillary butterfly *(Euphydryas maturna)*

DOMESTICATED ANIMALS (originating in region)

none

REFUGES FOR WILDLIFE

Many animals that have disappeared from other parts of the continent inhabit the mountains and forests of Eastern Europe. Like many mountainous areas in Europe, the Carpathian Mountains – extending from southern Poland to northeastern Romania – provide a refuge for some of the larger mammals that are now rare or extinct in most lowland areas. Among the animals that survive here, relatively safe from human encroachment and habitat destruction, are the Brown bear, Gray wolf and lynx.

The forests of the region include the Bialowieza in northeastern Poland, the largest tract of natural forest left in Europe. Its marshland borders give way to mixed stretches of hornbeam, pine, spruce and a wide variety of shrubs, and it is inhabited by European bison, elk (moose), lynx, Brown bears, various owls, woodpeckers and flycatchers. The beech forests of the Carpathians shelter deer, as well as nutcrackers, woodpeckers and other woodland birds.

A number of species are confined to alpine habitats above the treeline, having been isolated there at the end of the last ice age. Among them is the marmot, a thickly furred, burrowing squirrel that lives in the Tatra Mountains (northern Carpathians); another is the Snow vole, which lives in scattered populations at altitudes of up to 2,000 m (6,500 ft). An isolated population of the Northern birch mouse is also found in the region at high altitude. Montandon's newt is the only amphibian confined to these mountains; unlike most other newts, it will breed in a wide range of habitats, including lakes, ditches and even puddles and polluted water near human habitation, but prefers clear, cold, acid pools.

Sheltered in the forest (*right*) Large mammals – such as this Wild boar, as well as deer, Brown bears, wolves and European bison – still survive in protected areas of the region's mixed forests. Elk (moose) also retreat to the forest in winter.

The Carpathians have proved a barrier to the spread of many mammal species; for example, the Greater molerat (a burrowing rodent), is found only to the east of the range, and the Pygmy field mouse only to the west. The Common hamster and Steppe polecat are established on the lowlands that lie on either side of the mountains, but they do not live on the mountains themselves.

Animals of the steppes

The grassland steppes of eastern Europe are rich in insect life and support large populations of rodents such as hamsters and voles. The open nature of the grasslands means that these animals are easy hunting both for airborne predators such as eagles and owls and for terrestrial carnivores such as foxes.

One of the world's heaviest flying birds, the Great bustard, is a summer visitor to these plains; it likes to breed in open country and is well adapted to the treeless landscapes of eastern Germany, Hungary and other parts of Eastern Europe. However, its numbers started to decline in the 19th century with the plowing of the steppes. The increasing use of agricultural machinery and, later, the widespread adoption of pesticides, which have affected all the grassland birds of prey, have now lowered the bird's numbers to precarious levels.

Wetlands and waterways

The Danube, one of Europe's great rivers, flows eastward from the Black Forest in southwest Germany to the Black Sea. Sadly, pollution and overfishing have drastically reduced the numbers of its many species of fish. The sturgeon, a large fish of ancient ancestry that lives for up to 100 years, is now endangered – despite the fact that a female may lay as many as 2.4 million eggs. These are prized and eaten as caviar. Another commercially important species is the wels; this massive catfish reaches a weight of more than 200 kg (440 lb).

There are also many fish unique to the Danube, such as a herring species that has adapted to the freshwater environment. Other endemic fish are the Danube lamprey and Vladykov's lamprey, members of a primitive, jawless fish family. The huchen, which is a large relative of the trout, is confined to the Danube's upper tributaries, and grows to a length of 1.5m (5 ft) and a weight of more than 50 kg (110 lb). Three species of gudgeon and two of loach are unique to the Danube system, as well as the Romanian bullhead perch, a distinctive-looking species that was not discovered until 1957.

As the fish species of the Danube have declined, so have the populations of birds dependent upon them for food. The delta and the huge floodplains of the lower reaches of the river were, until very recently, the breeding grounds of the largest concentrations of wetland birds in Europe. Tens of thousands of geese, including most of the world's population of the Red-breasted goose, used to winter in these marshes. But loss of habitat due to human disturbance, canalization and drainage, as well as overfishing and pollution, have done irreparable damage to the region's wetland wildlife.

BIRDS OF THE WETLANDS

The Danube Delta is a classic example of a delta, with the river splitting into three main channels as it spreads out toward the Black Sea, and engulfing a vast area of some 435,000 ha (107,500 acres). Many smaller channels help slow down the river's massive flow, while its load of silt and other nutrients is trapped in the huge retaining net of reed beds, islands and weed-choked channels. The assembly of bird species found breeding here is probably second to none in Europe, though the diversity of breeding birds in the similar wetland environments of Czechoslovakia, Hungary and Yugoslavia is also very impressive.

The different bird species are adapted to exploit different parts of the wetland ecosystem. In the more open waters of the delta two species of pelican occur: the Great white and the Dalmatian. These pelicans feed in groups, driving the shoals of fish into the shallows where they scoop them up into their voluminous throat pouches. The fish are quickly swallowed so that they enter the crop, at the bird's center of gravity, allowing the pelican to maintain its balance while flying. Common cormorants and Pygmy cormorants also fish here. They chase their prey underwater but, unlike most waterbirds, they lack the oil glands to keep their plumage waterproof for long periods. After they have caught a fish, they return to land to dry their feathers, standing for extended periods with their wings outstretched.

Hidden among the corn (*above*) The young of a Montagu's harrier eagerly await their next meal from the safety of their nest. The harrier's long, narrow wings enable it to glide close to the ground in search of its prey of small mammals, birds and reptiles.

Animals of the lakes (*left and right*) The abundant insects that populate the lakeside support warblers such as the blackcap and the Sedge warbler. The European water vole tunnels into the banks close to grazing areas of damp turf. Lakes which have water plants with leaves that protrude above the surface support mayflies, the Common frog and the Gray moorhen. Submerged plants which are rooted in the bed of the lake provide ideal conditions for the water boatman or backswimmer, the Great pond snail, diving ducks such as the Tufted duck, scavenging flatworms and predatory fish such as perch.

LIVING IN THE DARK

Some of the most spectacular cave formations in the world are found in Eastern Europe. Their stable temperature and humidity attract a wealth of animals. The horseshoe bats are the most visible, hanging free from the cave roof or walls. Individual bat species use different parts of caves for hibernating and roosting; some, such as the Mouse-eared bat, prefer the warmer depths, while Long-eared bats roost in colder areas near the entrance. Birds also nest near cave entrances – these include the Rock dove (ancestor of the urban feral pigeon), the Red-rumped swallow, jackdaw and the Crag martin.

Other animals spend their entire lives in the depths of the caves. Many invertebrates, such as beetles, springtails, centipedes and segmented worms, feed on the detritus and feces of the bats. They, in turn, are food for other small animals: Cave spiders, crickets and millipedes. Most of these creatures are colorless and blind; some have long antennae and a greatly developed sense of touch. Perhaps the most highly adapted cave dweller is the proteus or olm of northern Yugoslavia, a salamander that lacks eyes. Whitish in color, with pink feathery gills, it retains the aquatic larval form all its life.

Among the most visible of the wetland birds are the herons and egrets, largest of which are the Gray and Purple herons. Both spend most of their time standing motionless among the reeds, waiting for an unwary fish, frog or even small bird to pass by, so that they can snatch at it with rapierlike speed. In common with some other herons and storks, they sometimes walk slowly through the shallows, their wings outstretched like a parasol to provide shade so that they are able to see into the water.

Spoonbills also wade in shallow water to feed, using their spatulate bill to sift particles of food, such as small invertebrates, from mud and water. By contrast, the Glossy ibis uses its long, downward-curving bill to probe in the mud and among herbage for food.

Sounds of the reed bed

The dense reed beds of the delta, with reeds growing up to 4 m (13 ft) high, often form huge floating islands where crakes, rails and several species of warbler nest; so, too, does the Little bittern, which is the smallest of the herons. Like all the bitterns, it makes a deep resonant booming call during the breeding season. The low frequencies carry over a long distance in a habitat where it is often difficult to see through the thick wall of reeds.

Other birds of these swamplands have striking calls that are able to carry through the thick vegetation. One of the most noticeable is the powerful song of the Great reed warbler, which incorporates froglike calls. The Bearded reedling, which uses a penetrating "ping" call to communicate, is a species almost entirely confined to reed beds; it builds its nest attached to the reed stalks. For most of the year it is insectivorous, but in winter it feeds on seeds as well. The Penduline tit is also confined to marshes, but it builds its nest in trees such as willows. The nest forms a distinctive woven globe at the end of a branch, usually hanging over water to protect it from predators.

Wetland hunters

The presence of waterbirds and a wide variety of amphibians, fish and insects attract a variety of predatory birds to the delta. In recent years, however, their numbers have declined due to agricultural and industrial pollution.

Ospreys are on the edge of their breeding range here, but are frequent visitors on passage. A large bird, the osprey is supremely adapted for catching fish. It is well camouflaged – dark brown above and white beneath – and its unusually strong feet absorb the first shock of the water as it dives for fish. It is equipped with long, sharp claws and horny-spined toes that help it to grip its slippery prey.

The massive White-tailed eagle is the Old World counterpart of the Bald eagle of North America. Until very recently it was still fairly common in the delta and other parts of the Danube. Although it may hunt fish like the osprey, it is also a scavenger, often perching on a branch by the river's edge waiting for a dead fish or the carcass of an animal to float past.

Harriers are characteristic marshland predators that crisscross marshes with their slow, flapping flight. They feed mainly on small mammals and birds, reptiles and insects, and have specialized hearing that allows them to locate their prey in thick vegetation. Marsh, Pallid and Montagu's harriers breed in or near the Danube delta in summer, and the Hen harrier visits in winter. Among the

Spectacular courtship The Great bustard's courtship display transforms the ordinarily dull brown bird into a magnificent spectacle of shimmering white feathers. The swollen throat pouch amplifies the moans that accompany the display.

smaller birds of prey found here is the Red-footed falcon, perhaps the most gregarious of all the falcons. It nests in old crows' and herons' nests and is largely insectivorous.

Adding a brilliant splash of color to its surroundings is the kingfisher. It feeds almost exclusively on small fish and waits on a perch until it spots its prey; then it plunges powerfully into the water to seize the fish, and returns with it to the perch where it batters its victim against the branch and swallows it head first.

Nesting in the drier areas close to the marshes are the Woodchat and the Lesser gray and Red-backed shrike. These are strikingly marked songbirds that prey on insects and small animals up to the size of small lizards and frogs. Shrikes are well known for their habit of storing prey by impaling it on thorns. The delta's rich variety of insects, including mosquitoes, also provide nourishment for myriads of small birds that are not specialized for a wetland environment. Swallows and dozens of other species feed in these vital refueling grounds during migration.

THE HUMAN IMPACT

Much of Eastern Europe survived in a virtually primeval state until the industrial era. The marshes and floodplains of the Danube river system were hunted and fished but remained mostly undrained; malaria, carried by mosquitos, and other diseases discouraged human settlement. Although European bison had long been driven from the steppes and confined to a few forested areas, tarpans (wild horses) were present on these grasslands until the 19th century, as were the saiga (an antelope) and the Bobak marmot. However, with the mechanization of agriculture and the plowing of vast areas of the steppes, came habitat fragmentation and the extinction of many species.

Hunters and protectors

Naturalists began to visit Eastern Europe in the late 19th century. Most of them shot what they observed, including many bird species now threatened by extinction. In the first decade of the 20th century a pioneer bird photographer compiled a valuable photographic record of the then abundant bird life of Albania, Hungary, Romania and Yugoslavia.

By the 1960s the Black grouse was on the brink of extinction in the Carpathians. Yet, as recently as the 1920s, the older inhabitants in some mountain areas could recall when these birds were so common that the cocks "lekked" or displayed around the villages. Today, the remaining populations are strictly protected, but their decline may partly be due to the long-term effects of climatic change and is thus irreversible.

Great and Little bustards continued to be hunted in many parts of Eastern Europe long after their numbers were seen to be falling. Even when protection was afforded, it was often only partial. In Romania, for instance, male birds were unprotected until the late 1960s, although there may already have been fewer than 1,000 pairs left.

While Eastern Europe's strong tradition of hunting is responsible in part for the rarity of some species, it may have helped protect others. Brown bears, for example, are relatively common in some parts of Romania and Yugoslavia where hunting under license provides revenue for their management and conservation. In the mountains of Romania game species such as the Carpathian Red deer (a particularly large variety with massive antlers), Roe deer, Wild boar and chamois thrive, despite being hunted.

Forest camouflage The Roe deer fawn's white spots help to conceal its outline in the dappled sun and shade of the forest. A fawn is often left alone for long periods by its mother, and it instinctively freezes at the first sign of danger.

In many hunting reserves, as well as in nature reserves, state forests and other protected areas, species such as lynx, almost extinct elsewhere in Europe, survive. The lynx is one of Europe's largest predators. It feeds mostly on hares and rodents, but also takes birds and sizable mammals such as young Roe deer and chamois. At the end of World War II it was believed that as few as 500 or 600 individuals remained in the Carpathian forests of Romania. This decline was probably due to the indiscriminate use of strychnine in baits put out for wolves,

that were eaten by lynx as well. Subsequent protection has led to an increase, and lynx are apparently abundant again in many of the forests.

Farther north the elk – an animal of damp deciduous forest where it browses on alders, willows and birches, and grazes on aquatic plants and grasses – has gradually extended its range southward into the region, with a little human help. After World War II only a few survived in a single population in Poland; this was augmented by a new population of animals introduced from the Soviet Union and released into the wild in 1956. Throughout the next two decades the elk spread rapidly from here and also from the Soviet Union, reaching Czechoslovakia by the 1970s. Elk are large, long-legged deer, able to cover considerable distances in the course of a single night. They will probably continue to colonize most of the remaining suitable habitat in Eastern Europe.

Two introduced species of fur-bearing mammal have now colonized Eastern Europe. They are the muskrat (a large aquatic vole) from North America, and the Raccoon dog (so called because of its raccoonlike appearance); it originated in eastern Asia and was introduced into European Russia in the 1930s. Having extended its range westward, the Raccoon dog is now widespread and locally abundant in the region. Its fast spread is explained by its ability to hibernate during the harsh winter, and by its rapid rate of reproduction: females have large litters of up to 15 offspring. The young mature in about 8 to 10 months.

The unseen threat

Pollution in Eastern Europe has reached major proportions, has caused untold damage to wildlife, and is likely to cause much more in the future. Acid rain is destroying many forest habitats and waterways. Rivers and lakes are heavily polluted by industrial and agricultural wastes that devastate fish populations and bird life. Existing and planned dams, canals and drainage schemes – especially those affecting the Danube river system – pose a long-term threat to the aquatic and wetland life of the region. The widespread use of pesticides, which enter the food chain, has had a serious impact on many birds, especially birds of prey, destroying their sources of food and causing breeding failure.

SAVING THE BISON

The European bison and the aurochs (a large wild ox) once roamed Europe's great primeval forests. By the 17th century the aurochs was extinct, and the bison was reduced to two isolated populations, one in the Bialowieza forest and the other in the Russian Caucasus. In 1914 there were over 700 in Bialowieza, but by 1918 all were dead. Fortunately a few survived in private collections and in 1929 six were brought together to form a new breeding herd. This has produced more animals for reserves in other parts of their former range.

The Caucasian race is now extinct, but hybrids of the two survive, preserving some of the genetic content of the Caucasian race. In the past, European bison interbred with the North American bison, to which it is closely related, and some bison released into reserves are descended from these hybrids. If what remains of the distinct genetic identity of the two races is to be preserved, further reintroductions and the movement of bison in the wild need to be carefully controlled.

The European bison, which can measure up to 1.8 m (6 ft) in height, is slightly larger than the American bison.

Rodents

The steppes of Eastern Europe provide a rich source of food for small mammals: seeds, fruits, berries, grass, insects and other invertebrates, as well as plant roots, tubers and worms underground. With their many adaptations for a diet of tough plant material, rodents are ideally equipped to exploit these resources: indeed some 23 rodent species out of a European total of about 60 are found here.

The incisor teeth, used for cutting and gnawing, grow continuously through life and are self-sharpening. The cheek teeth grind coarse grass and other plant material into extremely fine fragments, but the key to feeding on herbage is the cecum – a side branch of the intestine – which houses bacteria that break down indigestible cellulose into its constituent sugars (as in rabbits and hares). To ensure full absorption in the gut the semi-digested food is passed out in moist feces. These are eaten again by the rodents until only dry, hard droppings of totally indigestible remains are left.

The supply of plant food is often highly seasonal and patchily distributed and many steppe rodents, especially dormice and hamsters, also eat considerable quantities of insects and other invertebrates. Rodents are great hoarders. Hamsters and dormice have cheek pouches in which they can carry large quantities of seeds back to their nests so that they spend a minimum of time foraging in the open. They store food in their burrows for the winter: a single Common hamster burrow has been found to contain a massive 90 kg (198 lb) of plant material. Species such as hamsters, marmots, sousliks (ground squirrels) and dormice hibernate in the winter, avoiding the harshest weather and the period of greatest food scarcity. Their main aerial predators are eagles and buzzards. Terrestrial predators include

The Common hamster (*right*) was once widespread in steppe and farmland throughout central Europe and the Soviet Union. With the spread of modern agricultural practices, however, its numbers have declined dramatically.

Garden dormouse
Eliomys quercinus

Hazel dormouse
Muscardinus avellanarius

Forest dormouse
Dryomys nitedula

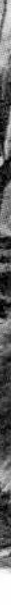

Northern birch mouse
Sicista betulina

Southern birch mouse
Sicista subtilis

Species of rodents The Common or Hazel dormouse hibernates from October to April, rolled up in a tight ball. The nocturnal Forest dormouse lives mainly in deciduous woodlands, and has a mixed diet of insect larvae, berries and seeds. The Garden or Oak dormouse, a woodland species often found in large gardens, is also an omnivore. The Northern and Southern birch mice both hibernate in winter. The southern species is more nocturnal than the northern.

foxes, weasels and polecats. Sousliks post lookouts at the entrances to their burrows; a piercing whistle as a predator approaches soon sends other members of the group scampering for cover. Many steppe rodents are nocturnal, or at least crepuscular, emerging at dusk and dawn to avoid the eyes of daytime predators. However, they are subject to attack from owls and other night hunters.

Burrows and nests

Many rodents are social mammals. Sousliks and Common rats live in complex communal underground tunnel systems, with sleeping rooms, storage rooms and escape tunnels. Although a solitary animal, the hamster also lives underground; it sets aside a blind tunnel for defecation. The round-headed Lesser molerats are the most expert tunnelers of all. They use their strong incisors to loosen the soil, then their powerful forefeet and hind feet push the soil out of the way, throwing up small heaps of earth. Molerats seldom need to emerge onto the surface; they feed mainly on underground roots, bulbs and tubers.

Some rodents live above the ground. Dormice are tree dwelling, and have cushioned paws for gripping branches. The Harvest mouse is an expert climber, using its prehensile tail as an extra limb for gripping grass stalks and for balancing. It has adapted to the changes brought by farming, and builds its woven nest on stalks in haymeadow and wheat fields. Many other steppe rodents have adapted to the activities of humans by taking advantage of new food supplies.

Most rodents breed extremely rapidly, so that populations can expand very fast when food is plentiful. Among the most prolific breeders is the Common vole – females are sexually mature only 8 days after birth and ready to mate at 13 days. Voles disperse to new areas if the population becomes too dense, exploring new sources of food. House mice, on the other hand, deal with the problem in another way. If conditions become too crowded, the younger females become temporarily infertile. Voles and hamsters can reach plague proportions in good years, when they devour vast quantities of cereals.

Tunneling rats The Lesser mole rat is blind and lacks external ears. It lives in a tunnel system punctuated by resting nests and storage chambers for food. The aquatic muskrat nests in grassy banks and digs tunnels, often with underwater entrances.

ANIMALS OF TUNDRA AND TAIGA

LEGACY OF THE ICE AGE · ADAPTING TO EXTREMES · PUSHING NATURE TO THE LIMIT

The Soviet Union contains within its vast boundaries animals as diverse as Polar bears, African wildcats, Snowy owls, flamingoes, walruses and cheetahs. Atlantic sturgeon inhabit its western waters, while the Japanese crane and Oriental cuckoo are found in the east of the region. Much of the diversity comes from areas on the margins of the Soviet Union, such as the high peaks of the south and the mixed forests that span the European frontier. The richest areas are those that were left relatively unscathed by the last period of glaciation: the forests around the Ussuri on the Pacific coast, and the zone between the Black and Caspian Seas. Animals that took refuge here from the advancing cold of the ice age later mingled with others that invaded from the warmer regions to the south as the ice receded.

COUNTRIES IN THE REGION

Mongolia , Union of Soviet Socialist Republics

ENDEMISM AND DIVERSITY

Diversity Low to medium
Endemism Low to medium to high (for example Lake Baikal)

SPECIES

	Total	Threatened	Extinct†
Mammals	275	23	1
Birds	730*	38	1
Others	unknown	34	0

† *species extinct since 1600 - Steller's sea cow* (Hydrodamalis gigas), *Spectacled cormorant* (Phalacrocorax perspicillatus)
* *breeding and regular non-breeding species*

NOTABLE THREATENED ENDEMIC SPECIES

Mammals Manzbier's marmot *(Marmota menzbieri)*, Russian desman *(Desmana moschata)*
Birds none
Others Amur sturgeon *(Acipenser schrencki)*, Balkhash perch *(Perca schrenki)*, Caucasian relict ant *(Aulacopone relicta)*

NOTABLE THREATENED NON-ENDEMIC SPECIES

Mammals tiger *(Panthera tigris)*, leopard *(Panthera pardus)*, European bison *(Bison bonasus)*, Asiatic wild ass *(Equus hemionus)*, Przewalski's horse *(Eqyus przewalskii)*
Birds Crested shelduck *(Tadorna cristata)*, Steller's sea eagle *(Haliaeetus pelagicus)*, Siberian crane *(Grus leucogeranus)*, Relict gull *(Larus relictus)*, Great bustard *(Otis tarda)*
Others Caucasian viper *(Vipera kaznakovi)*, Large copper butterfly *(Lycaena dispar)*, Freshwater pearl mussel *(Margaritifera margaritifera)*

DOMESTICATED ANIMALS (originating in region)

horse *(Equus 'caballus')*

LEGACY OF THE ICE AGE

In the dark, silent boreal coniferous forest or taiga, which covers about a third of the land area of the Soviet Union, the animals are not abundant, and are generally hidden from view; the number of species is low in most parts of the forest. North of the taiga, the tundra (treeless plains where the subsoil is permanently frozen) also has relatively few species. However, in summer the tundra bustles with activity, for the species that are there are represented in huge numbers. Many are migratory birds that come to exploit the abundant insect food of the Arctic summer and feed their chicks; they then fly south at the end of the brief breeding season to escape the Arctic winter.

Twenty thousand years ago tundra would have covered most of the Soviet Union. Gradually, as the Earth became warmer and the tundra decreased in area, animals such as the reindeer, Arctic fox and Snowy owl returned to the most northerly areas. During the ice age so much water was locked up in the glaciers and ice caps that there was a worldwide fall in the sea level. In the far north this created a land bridge between Asia and America, across what is now the Bering Strait. Animals hardy enough to survive so far north were able to move freely between the two land masses; as a result, many of the species resident in the Arctic tundra of the Soviet Union are also found in the tundra of North America.

As the land grew warmer and the sea level rose, about 10,000 years ago, the animals of the two continents were separated. Despite their isolation, most are still technically single species: if brought together they can interbreed. In time they will undoubtedly become different species, but for now they remain linked by their shared past. Similarly some of the taiga animals have counterparts in the boreal forest of the New World: for example, the elk (moose), the Brown bear and the wolverine.

The legacy of the period of glaciation can also be seen in the southern mountains of the Soviet Union. Here, tundra and taiga animals, such as the Mountain hare, the Hawk owl and the Red crossbill, live in the same mountain ranges as invaders from the subtropical south, for example, babblers, Paradise flycatchers and Asian wild dogs. The cold-adapted species normally associated with the north would have been widespread in the surrounding lowlands during the ice age. As the climate gradually warmed up they moved up into the mountains to escape the unfamiliar heat of the plains, and settled as isolated populations surrounded by hot desert.

The warm, dry south

The animals of the region's deserts have a less complicated history than those of the mountains. Most of the species are the descendants of southern immigrants from the deserts of central Asia and the Middle East, which provided a link with the Sahara. This corridor of desert and semidesert enabled animals such as the African wildcat, the cheetah and the ratel or Honey badger to migrate from Africa

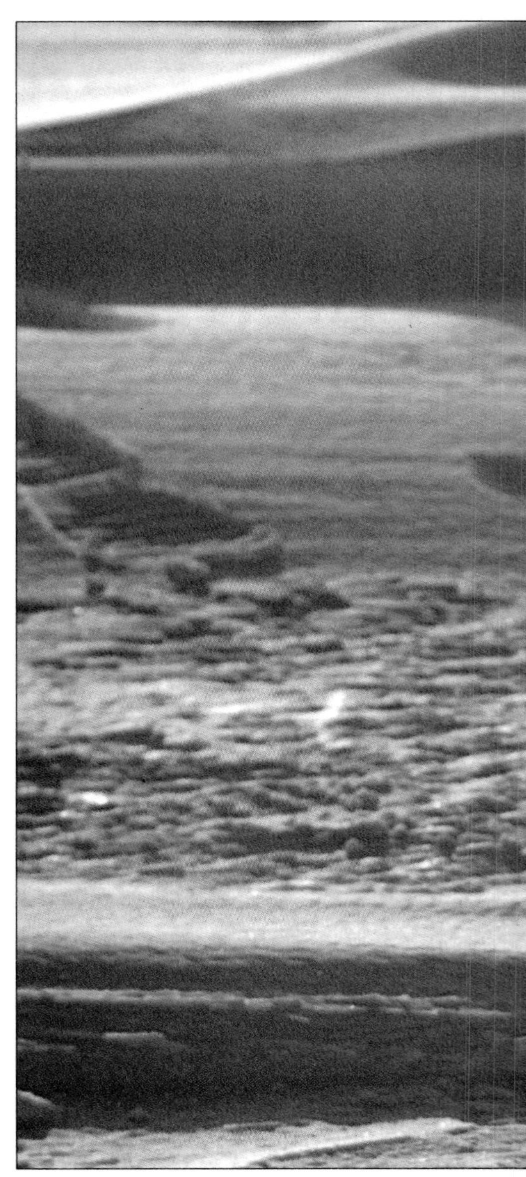

Hunter of the night *(above)* The Striated scops owl is the most widespread nocturnal predator of the steppes. Compared with most owls, it has very long wings, and is capable of gliding flight. It migrates in the fall to the warmer climates of the south.

Camouflaged for survival *(below)* An Arctic fox in its white winter coat merges with the snowy landscape. In summer its coat turns brownish-gray. Arctic foxes are often observed following Polar bears in order to scavenge from their kills.

into Asia. The cheetah may now be locally extinct due to hunting, but there are plans to reintroduce it into a nature reserve. The deserts of the Soviet Union also support many reptiles, including tortoises, skinks, geckos and many species of snake.

Many desert animals also inhabit the steppe grasslands, especially in the drier areas, and some animals have moved in from the mixed woodlands to the northwest. These outsiders share their domain with true steppe mammals, such as hamsters, sousliks (ground squirrels) and Saiga antelopes, along with birds such as larks and bustards.

Specialists and exotics

Endemic species are relatively rare in the Soviet Union, possibly because so few areas were unaffected by the ice age. Clusters of endemics are found only in isolated and enduring habitats such as Lake Baikal in southeastern Siberia. The phenomenal depth of the lake prevented it from freezing solid during the ice age. Lake Baikal supports about 1,200 animal species, of which 60 to 80 percent are endemic. Among them is the Baikal seal, a distant relative of the Ringed seal whose ancestors must have swum inland about 10 million years ago. The peaks of the southern mountains also support some endemic species: for example, each mountain range has its own species of snowcock – specialized mountain birds whose ancestors probably originated in the Himalayas. The Caucasus, the area of land between the Black and Caspian seas, is also rich in endemic species having once been isolated by water.

ADAPTING TO EXTREMES

Only two areas in the Soviet Union have a consistently benign regime: the mild, damp coastal area between the Pacific coast and the Ussuri river in the far southeast of the region, and the subtropical lowlands between the Black Sea and the Caspian. Throughout much of the Soviet Union climatic conditions are extreme, and many animals find survival difficult in summer and winter.

The Arctic ground squirrel has adapted to the difficult conditions by entering a state of suspended animation not just once but twice a year. It hibernates in winter to escape the severe weather; during the driest part of the summer, when the vegetation withers, it repeats the process to avoid starvation.

The effects of the cold mass of air that hangs immovably over the continent in winter are felt even in the deserts of the southern Soviet Union. Although hot and dry in summer, temperatures can fall to below freezing in winter. The jerboas – desert rodents commonly found throughout northern Africa and the Middle East – do not usually hibernate, but in these deserts they do so for as much as seven months of the year in order to avoid the cold weather.

Another group of desert rodents, the gerbils (or jirds) have evolved into a specialized genus, *Meriones*, within these inhospitable central Asian lands. They live communally, unlike other gerbils, and may store food for use in winter. Digging a deep communal burrow and huddling together helps them to conserve their body heat.

Birds of plains and deserts The majestic Imperial eagle breeds throughout central Asia from Iran to China, but its numbers are not known; the European population has been reduced to barely 100 pairs. A bird of desert and steppe, Pallas's sandgrouse flies long distances every day in search of water. Its breast feathers are specially adapted to soak up water, which it carries back to its thirsty chicks.

A scarcity of food

Cold is only one aspect of the problems that these animals face. Lack of food is also common, especially in the taiga. Conifers offer relatively little in the way of food, with their tough, waxy, indigestible leaves. Resins – which give conifers their aroma – are there to defend the wood and needles from attack, and they prove noxious to most animals. With the exception of a few insects, the most successful needle-feeders are the large, sturdy game birds known as capercaillies; these vegetarian birds specialize in feeding on pine needles and must have some mechanism for detoxifying and digesting their food.

Broadleaf trees, notably hardy birches, and undergrowth shrubs such as cranberry provide an alternative food source for many herbivores. Where the canopy is closed, however, little undergrowth can spring up so this potential source of food is also lost. More open areas and swampy ground – of which there are plenty – are the feeding grounds for many animals. The elk, for example, browses on succulent undergrowth and wades into the water to feed on aquatic plants.

The only form of rich and trouble-free food that conifers provide is their seeds, packaged inside woody cones. The cones have evolved to protect the seeds from predators, and are far from easy to open. Mammals invariably gnaw them but this is slow and uses up energy. Animals that can tackle the cones, such as crossbills (members of the finch family) and squirrels, are an important link in the forest food chain, providing sustenance for birds of prey, martens and other predators. The unusual beaks of the crossbills are specialized cone-opening tools – the abnormal jaw joints allow the lower half of the beak to move from side to side, not just up and down.

Unfortunately for the seed-eaters, conifers are inconsistent in the amount of seed they produce. A poor cone-bearing year, especially if it follows a good year, may result in an "irruption" of crossbills scattering from their Siberian homelands in search of other food; they often fly as far as Spain and Portugal.

Many animals store food as a means of ensuring that they have a regular supply throughout the long winter. For example, the Siberian chipmunk, a small ground-dwelling squirrel, has enormous cheek pouches in which it carries seeds to its burrow. A single animal may cache up to 6 kg (13.5 lb) of food before winter sets in.

A unique meeting place

In the unspoiled wildernesses around the Ussuri, river valleys provide north–south migration corridors between the mountains. Here animals from Siberia have mixed with others from China, Japan and

A NOSE FOR THE STEPPES

The saiga is an unusual animal; although often referred to as an antelope, it has in fact many anatomical features that are more like those of a goat. It is a true steppe animal, but in winter it migrates southward to the warmer lowlands around the Caspian Sea. In the 1920s overhunting in the Soviet Union brought the saiga close to extinction; since then strict protection has allowed its numbers to increase, and there are now more than a million.

The saiga's bulbous nose contains enlarged nasal passages that are lined with unusually luxuriant hairs and mucus-producing glands. The nose helps to warm the cold air that the animal inhales, and to filter out the dust that the hooves of the herd churn up as it runs. Why the nostrils should point downward is something of a mystery. During the mating season, the male's nose enlarges, which suggests that a large nose may be intimidating to other males or attractive to females. If this is so, then sexual selection may have played a part in the evolution of this unusual feature.

southeast Asia. A wealth of different habitats favors this abundant and diversified wildlife.

Among the species from the south are giant silk moths, numerous exotically colored butterflies and beetles, praying mantids, and the only softshelled turtle in the Soviet Union. Mandarin ducks, whose center of distribution is in China, nest in tree holes close to forest streams, while the rare Blakiston's fish owl snatches fish from the surface of the larger rivers and lakes. Other exotics include a type of roller (a large, colorful insect-eating bird), a Paradise flycatcher, orioles and drongos – all birds that originated in subtropical forests. Leopards, Asian black bears and Sika deer live here too. Yet this mountainous area also boasts Mountain hares on some of its peaks, and Siberian flying squirrels in its coniferous forests.

Mammals of steppe, desert and forest The saiga antelope of the central Asian steppes has evolved excellent vision and the ability to run very fast to escape predators. Pallas's cat is a rare predator of Soviet deserts, slopes and mountains, where it preys almost exclusively on rodents. The Mongolian jird is adapted to life in arid areas; it is well camouflaged, has long hind legs for jumping and running, and well-developed claws for burrowing to avoid summer heat and winter cold. It breeds rapidly to compensate for high predation. The sable is perhaps the most prized of all Soviet animals. Its thick pelt, which evolved to withstand the rigors of the Siberian winter, is highly valued in the international fur trade. As a result, it has been hunted almost to extinction, but is now protected in several reserves.

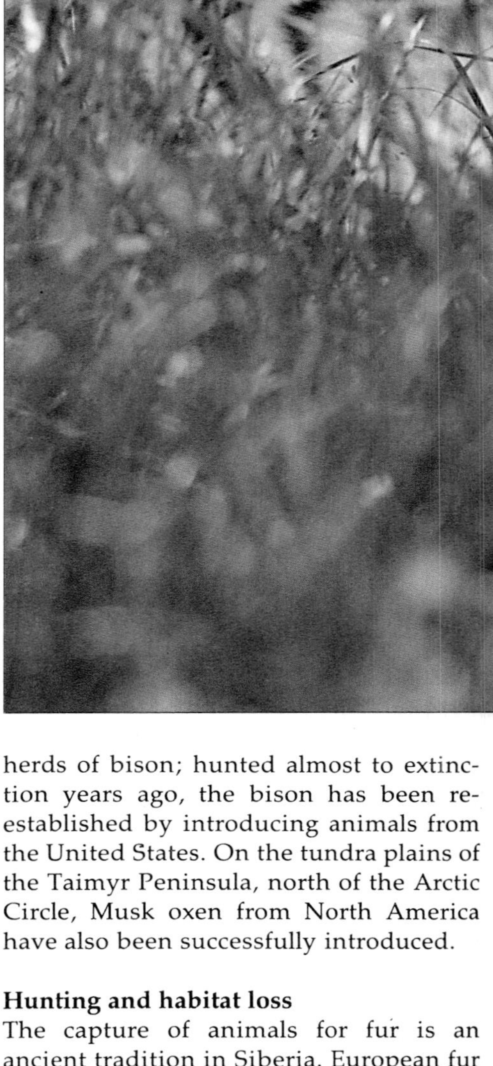

A seal rarity *(above)* The freshwater Baikal seal is believed to be related to the northern Ringed seal. The females give birth in early spring in solitary snow lairs over breathing holes in the ice.

Northern tiger *(right)* The endangered Siberian tiger – the largest subspecies of tiger – has a thick coat to protect it from the Siberian winters. Only about 200 individuals survive today, for the tiger needs a very large area of wilderness to catch sufficient quantities of prey.

PUSHING NATURE TO THE LIMIT

It was once an article of faith in the Soviet Union that nature was boundless and inexhaustible. Such a view is understandable, given the vastness of the land, but it can no longer be justified. Many animals face a bleak future, and if the present range of wildlife is to survive, the scale of the human impact on the environment must be reduced.

Threats to steppe- and forest-dwellers

The animals that have suffered most from human activity are the steppe-dwellers, as their native habitats are particularly well suited to cereal growing. As in North America, this has resulted in the disappearance of millions of hectares of native grassland in favor of wheat fields. Some animals can adapt and thrive in the new environment, but many cannot, and the heavy use of chemical pesticides makes life yet more difficult. The Great bustard is now endangered by both changes to its habitat and overhunting; it is now part of a captive breeding program.

All the surviving areas of steppe are protected as nature reserves. Their total area, however, is no more than 12,000 ha (30,000 acres), and most reserves are very small. The wildlife of these tiny islands of protection tends to become more and more impoverished as time goes by.

To the south, the rich subtropical forests of Transcaucasia have also given way to cultivation – mostly to extensive tea plantations. The varied wildlife of the area has suffered greatly as a result; for instance, the Anatolian leopard is now an endangered species.

The mixed forests of the Baltic republics and European Russia are intensively exploited for timber, with inevitable consequences for animals such as the beaver, Wild boar and Russian desman (a small aquatic mammal). The latter has declined greatly as a result of water pollution and competition from introduced coypus and muskrats. Reserves in these forests may enable some of the animals to survive. A few of the reserves now support small herds of bison; hunted almost to extinction years ago, the bison has been reestablished by introducing animals from the United States. On the tundra plains of the Taimyr Peninsula, north of the Arctic Circle, Musk oxen from North America have also been successfully introduced.

Hunting and habitat loss

The capture of animals for fur is an ancient tradition in Siberia. European fur traders moved into the area at the beginning of the 15th century, and from then on huge numbers of furs were taken. The

CAPTIVE BREEDING – EFFECTIVE RESCUE

The Soviet Union has had considerable success in increasing the numbers of some of its rarest animals by captive breeding: for example, peregrines, lammergeiers (Bearded vultures), waterfowl and desert wildlife such as the Great bustard and Goitered gazelle. The aim is to provide captive populations from which animals can be reintroduced to their former habitats.

Operation Siberian Crane is a joint program between the Soviet Union and the International Crane Foundation based in Wisconsin in the United States. The female crane lays two eggs, but in the wild only one chick usually survives. In this program the second egg is removed and hand-reared, or placed with foster parents of other species of crane. The program is being extended to include the rare Oriental stork. A further measure has been the establishment of a new population at the Oka River Nature Reserve to the southeast of Moscow. The hope is that this population will settle and overwinter here, thus avoiding the threat from hunters while on migration.

Captive breeding of the rare Snow leopard has also been successful – though under zoo conditions the young have no opportunity to learn how to hunt from their mother. The leopards are released into reserves so that they can return to base to be fed until they gradually become independent.

At Askaniya Nova in the southern Ukraine animals such as Przewalski's horse and the Wild ass or kulan have been saved from extinction and reintroduced to the deserts of central Asia. The African eland is being domesticated here, and ostriches, rheas and emus are bred in order to supply zoos. Consequently the need to capture more wild animals is avoided.

Russian desman was once a staple of the fur trade, with tens of thousands of skins being exported annually to meet the demands of western Europe. Stoats, too, have suffered, with as many as 12,000 animals being taken every year from the Kamchatka peninsula alone in the far northeast of the region. The species that have suffered most from the trade are big cats such as the Amur leopard and the Turanian tiger; both are now endangered.

The sable has required legal protection since the 17th century; one of the first forest nature reserves was set up by Peter the Great nearly three centuries ago to protect the sable's habitat. Today sable are being reintroduced, with some success. Hunting has also taken a toll on birds. It has been estimated that 100 million game birds are shot every year. Migrants such as the Siberian crane and Japanese crane are especially vulnerable.

Pollution and destruction of habitats are ultimately a far greater threat to wildlife than hunting. Many birds have been killed by poisoned grain scattered by aircraft to kill rodent pests in agricultural areas: in 1970 50 cranes were killed in one incident and 200 Great bustards died in another.

Large areas of wetlands around the Caspian Sea have been lost because of land reclamation and the use of water for irrigation. This has destroyed most of the wintering grounds once used by wildfowl. Farther east the water level of the Aral Sea is also falling and this, combined with widespread pollution from pesticides used in cotton growing, threatens many thousands of species, while exploration for oil and minerals threatens parts of the Arctic tundra.

Migrant birds of the tundra

In summer the vast treeless Arctic tundra is alive with birds. The myriad pools of snowmelt are the breeding grounds of millions of insects, providing a rich diet for warblers, wagtails and wheatears. The thin layer of thawed soil contains succulent grubs and other small creatures favored by waders, there are plenty of tender young shoots for the geese, and the Arctic seas abound with fish and crustaceans in summer.

Gains and losses

Many millions of migrant birds take full advantage of the abundant food and long summer days to raise their young in the tundra, so the temporary glut of food is put to good use. Warblers and wheatears, make astonishingly long journeys from southern Africa, Southeast Asia, Australia and New Zealand. One exceptional bird, the Pectoral sandpiper, makes the journey back via Alaska to spend the winter in South America.

Such long migration routes represent a vast expenditure of energy and food resources, and the birds face enormous risks on the way from exhaustion, bad weather and predators. The breeding success of these birds has to compensate for the loss to their numbers that occurs during migration. Many of the birds, especially the warblers and other small species, switch from a diet of seeds to one of insects to feed their young. Were they to remain in the south to breed, they would have to compete with the resident birds for a food supply that, in many parts of the tropics and subtropics, does not produce a seasonal glut.

Frontiers of exploration

The flight paths of these migratory birds tell us a great deal about the recent past of the region. Twenty thousand years ago, with the world in the grip of the last ice age, tundra covered most of the Soviet Union: Siberia very largely escaped the glaciers that scoured North America, Scandinavia and Britain but, with the soil permanently frozen even close to the surface, conditions would have been arid, profoundly cold, and inhospitable to most forms of life. As the Earth gradually grew warmer migratory birds pushed back the frontiers of exploration a little more each year. The track of the wheatear shows that it has slowly progressed northward and then eastward and westward from its African wintering grounds. The Pectoral sandpiper worked its way north into Alaska with successive summers, then edged westward to reach Siberia.

Most birds are remarkably conservative in their choice of migration routes, faithfully following the paths by which their first ancestors reinvaded the north. The wheatear still spends the northern winter in Africa, despite the fact that similar conditions may be found in South America and Southeast Asia. Greenland wheatears fly across the Atlantic to Britain before they head south, and birds from eastern Canada traverse the Soviet Union and central Asia.

By contrast, other species, such as the Lesser golden plover, have two distinct populations: one in North America, the other in the Soviet Union. Each has its own migration route – the American population flying to South America, the other migrating to Southeast Asia, the Pacific islands, Australia and New Zealand. Whether these routes have diverged with time, or whether they represent two distinct centers of recolonization of the north, is not known.

The northward migration of birds is not confined to the tundra. Little by little, many birds of the taiga and the temperate forests are also extending their range northward; the progress of the Collared dove across Europe during the present century is just one example. These birds are still extending their breeding grounds – a sure indication that gradual changes in the climate are taking place. The final acts of the great ice age drama are still being played out.

Living in perpetual summer (*above*) The Arctic tern migrates from the coasts of northern Europe, North America and the Soviet Union to the Antarctic and back each year, a round trip of some 40,000 km (25,000 mi), taking advantage of both polar summers for hunting.

The Little stint (*left*) is the smallest breeding wader of the tundra, with a body little bigger than a sparrow's. Despite its small size, it makes the long migration from the northern tundra to Africa, Southeast Asia, Australia or Tasmania in the fall.

Invading the Arctic (*right*) The northern Soviet Union is part of a great belt of boreal forest, tundra and Arctic coast that extends to Greenland and North America. Since the ice retreated following the last glaciation, birds have been reinvading the region from the south. They still migrate south in winter.

Migration routes

Wheatear The wheatear has recently expanded its range from Eurasia to Greenland and eastern North America. Migrating birds still follow the routes by which they colonized the region in the first place, traveling great distances back to Africa.

Brent goose Many of the Brent geese that breed in the Arctic winter in the British Isles and northwest Europe, but one population heads for the United States' east coast, and another flies south from Alaska down the west coast.

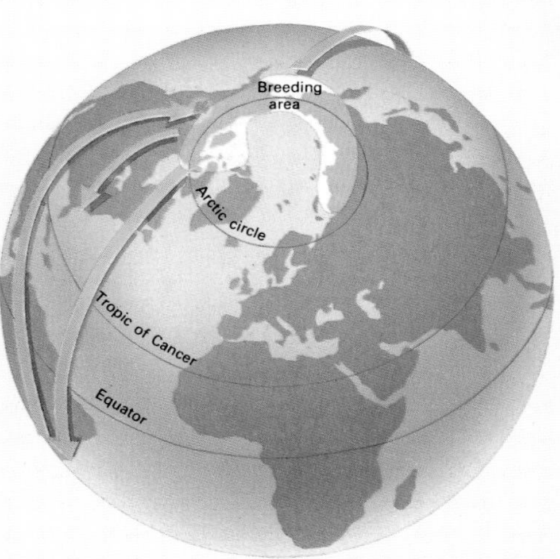

American golden plover Birds following one migration route cross over the Atlantic Ocean to South America and return by a different route across Central and North America. Another route crosses the Pacific Ocean to New Zealand.

A CROSSROADS FOR MIGRATION

A UNIQUE MEETING GROUND · ADAPTING TO ARIDITY · RETREATING WILDLIFE

Although the usual image of the Middle East is one of unrelieved aridity, in reality it contains a surprisingly rich variety of habitats for wildlife. There are bears and Red deer in the deciduous forests of Turkey; porcupines and more than a dozen species of warbler in the woodlands of Syria and Lebanon; and Snow leopards and wild sheep on the alpine pastures of northern Afghanistan. Millions of migrant birds overwinter on the huge lakes and river swamps of Iran and Iraq. The deserts are inhabited by a large number of animals that have successfully adapted to living with scarce supplies of water, including gazelles and oryx; there are no less than 400 species of false ground beetle. The coral reefs of the Red Sea and the Gulf are among the finest in the world; rare turtles and Sea cows (dugongs) still survive here.

COUNTRIES IN THE REGION

Afghanistan, Bahrain, Iran, Iraq, Israel, Jordan, Kuwait, Lebanon, Oman, Qatar, Saudi Arabia, Syria, Turkey, United Arab Emirates, Yemen

ENDEMISM AND DIVERSITY

Diversity Low to medium
Endemism Low to medium

SPECIES

	Total	Threatened	Extinct†
Mammals	193	12	0
Birds	650*	15	0
Others	unknown*	23	1

† species extinct since 1600
** breeding and regular non-breeding species*

NOTABLE THREATENED ENDEMIC SPECIES

Mammals Mountain gazelle *(Gazella gazella)*, Arabian oryx *(Oryx leucoryx)*, Arabian tahr *(Hemitragus jayakari)*
Birds Yemen thrush *(Turdus menachensis)*
Others Latifi's viper *(Vipera latifi)*, cicek fish *(Acanthorutilus handlirschi)*

NOTABLE THREATENED NON-ENDEMIC SPECIES

Mammals Dugong *(Dugong dugon)*, Asiatic wild ass *(Equus hemionus)*, Dorcas gazelle *(Gazella dorcas)*
Birds waldrapp *(Geronticus eremita)*, White-eyed gull *(Larus leucophthalmus)*, Dalmatian pelican *(Pelecanus crispus)*, Great bustard *(Otis tetrax)*, Houbara bustard *(Chlamydotis undulata)*, Imperial eagle *(Aquila heliaca)*
Others Caucasian viper *(Vipera kaznakovi)*, Egyptian tortoise *(Testudo kleinmanni)*, Green turtle *(Chelonia mydas)*

DOMESTICATED ANIMALS (originating in region)

dog *(Canis 'familiaris')*, donkey *(Equus 'asinus')*, sheep *(Ovis 'aries')*, goat *(Capra 'hircus')*, cattle *(Bos 'taurus')*, pig *(Sus 'domesticus')*, dromedary camel *(Camelus 'dromedarius')*, pigeon *(Columba livia)*

A UNIQUE MEETING GROUND

Only a few million years ago Eastern Arabia formed a unique meeting place for the animal species of Eurasia, tropical Africa and the Orient. They were able to move from one continent to another over the land bridges that crossed the Bab al Mandab (the strait at the southern end of the Red Sea) and the Strait of Hormuz at the mouth of the Gulf.

The action of continental drift then caused the Arabian Peninsula to move farther east, away from Africa. These land bridges disappeared, leaving the Sinai isthmus as the only corridor for migration, and the Indian Ocean flooded in to form the Red Sea and the Gulf. It was then the turn of the Indo-Pacific marine animals to colonize the Middle East – which they did in profusion.

The evolution of species

Once the populations of land animals from Africa and Asia were isolated in Arabia, some gradually evolved to become distinctive endemic races and species. The Arabian oryx, for example, is closely related to the African species of oryx, while the Arabian tahr is clearly descended from Asian goats. Hundreds of other species of insects, reptiles, birds and mammals that are found in western Arabia – from the Lime swallowtail butterfly to the porcupine, and from the Gray hornbill to the Hamadryas baboon – are at the eastern extent of their range; the center of their distribution is in Africa south of the Sahara.

Many species from northern Eurasia were driven south into the region during the last period of glaciation; some of these animals later returned to the north when the climate warmed up, while others ascended the high mountain ranges of southwestern Arabia, where conditions were cooler and more moist and the vegetation more lush. Isolated on these mountain refuges by the intervening arid plains, many animals developed into separate races or even species.

urine and thus save water, and a keen sense of smell to detect fresh vegetation and water.

There are at least 60 species of mammal in the Middle East, including nine of the world's 35 cat species, such as the caracal, Pallas' cat and the Sand cat. The region also boasts more than 180 breeding bird species, 140 species of reptile and more than 400 of the world's 1,000 or so species of false ground beetle.

Among the most characteristic animals found here are the true gazelles (those of the genus *Gazella*) whose range stretches from Africa to Asia. The Mountain gazelle is confined to Arabia and has no fewer than five subspecies, all of them rare. The Dorcas gazelle originally had two Middle East subspecies, but it appears that the Saudi Arabian one has become extinct. The third of the region's gazelles is the Goitered gazelle, which has one Middle East subspecies.

Ships of the desert (*left*) A caravan of camels bearing salt rests in the desert before continuing its journey. The one-humped Arabian camel, or dromedary, has been domesticated for thousands of years and is superbly adapted for survival in arid lands.

Preparing for ambush (*below*) A desert viper has buried most of its body deep in the sand as it waits to strike at unwary passers-by. The snake's hard, scaly skin protects it from water loss in the parched conditions of the desert.

A wealth of wildlife

For thousands of years nomadic people have grazed their animals throughout the Middle East, gradually pushing back the wildlife from the more populous areas into the empty places, especially the deserts. Even here, however, animals are hunted on a large scale. Wild oryx and ostriches have become extinct, and wild gazelles and Houbara bustards are greatly reduced in numbers; but despite the hunting, there is still a great wealth of desert wildlife in the region.

The deserts are by no means uniform in character. Their features include sparse acacia groves, *Artemisia* and saltbush shrub steppe, vast and seemingly barren boulder-strewn areas called *harrats*, stony expanses known as *regs* and rolling sand dunes or *ergs*, honed to razor-edged crests by the wind. Far from being empty, these areas support a large number of animals that have successfully adjusted to the arid conditions. Adaptations include nocturnal activity patterns, the development of highly efficient kidneys to concentrate

ADAPTING TO ARIDITY

The most characteristic adaptations of the animals of the Middle East result from the need to survive in very arid conditions. For many land animals the main source of water is dew, which collects on the surface of plants and sand grains. Many false ground beetle species have developed a technique for capturing this moisture. They construct small mounds and trenches in the sand at the crests of dunes, where dew gathers. Water condenses in these traps during the chilly hours before dawn, and the beetles "bask" in them, extracting the water. Shortly after sunrise the sand mounds dry up and crumble, and the beetles dig their way into the sand to avoid the heat of the day.

Other animals make use of even rarer and more transient sources of water. Within hours of pools forming after a rain shower, the water churns with the frantic activity of tadpole shrimps – small freshwater crustaceans. Before the water evaporates, the females lay their eggs. Each egg immediately turns into a hard

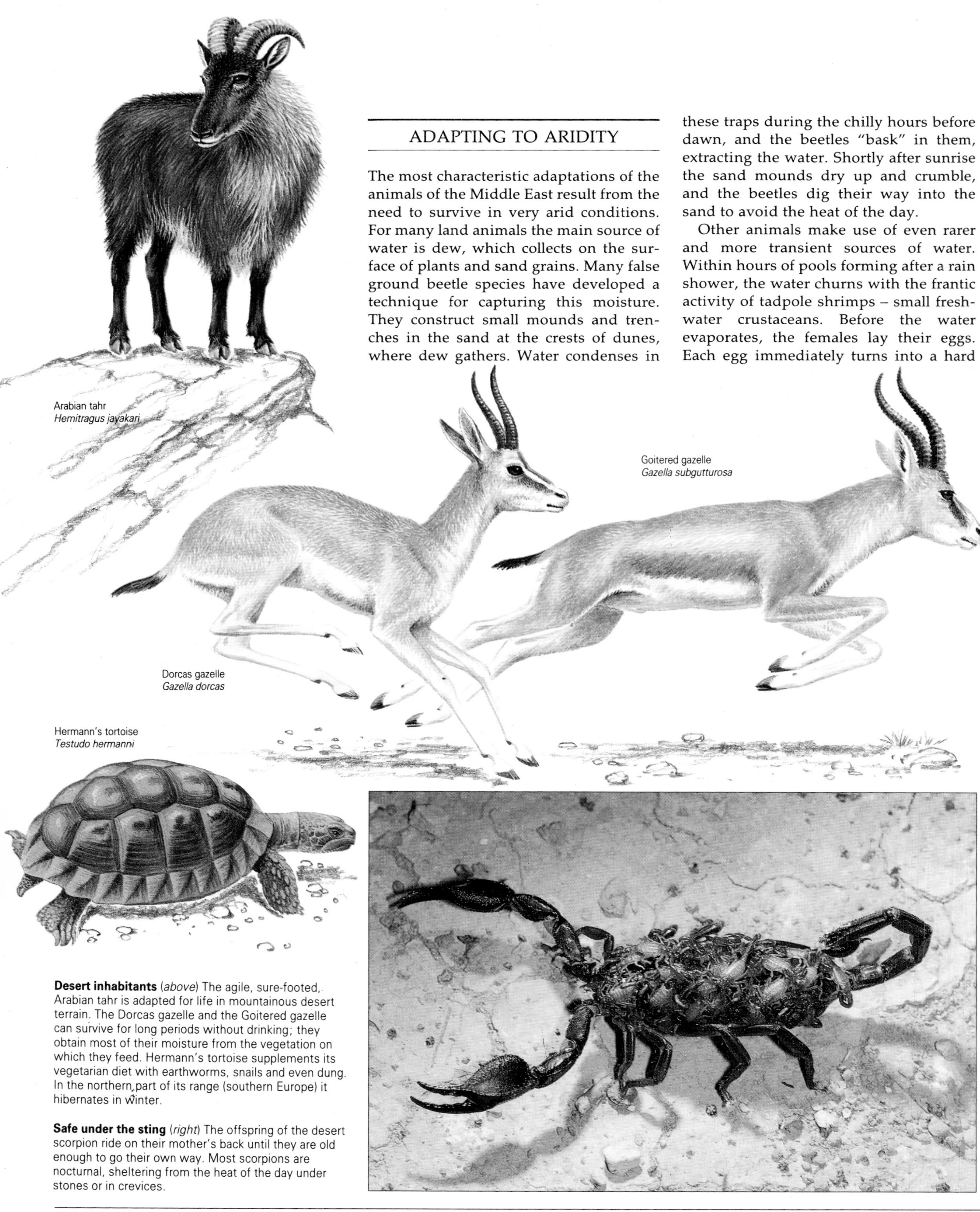

Desert inhabitants (*above*) The agile, sure-footed, Arabian tahr is adapted for life in mountainous desert terrain. The Dorcas gazelle and the Goitered gazelle can survive for long periods without drinking; they obtain most of their moisture from the vegetation on which they feed. Hermann's tortoise supplements its vegetarian diet with earthworms, snails and even dung. In the northern part of its range (southern Europe) it hibernates in winter.

Safe under the sting (*right*) The offspring of the desert scorpion ride on their mother's back until they are old enough to go their own way. Most scorpions are nocturnal, sheltering from the heat of the day under stones or in crevices.

cyst that sinks down into the sand grains, where it remains dormant for seven years or more. As soon as rain falls again the cyst embryo hatches; the shrimp grows, mates, lays eggs – if female – and dies, all within the few days that the water lasts.

Flourishing filter feeders

Permanent open water does exist in parts of the Middle East: set in the vast plains of Iran, Iraq and Turkey are a number of large, shallow and rather salty lakes, such as Lake Tuz in central Turkey. Here, thousands of Greater flamingoes take full advantage of these productive conditions, which provide them with an abundant food supply in the form of tiny algae, crustaceans and other invertebrates.

The flamingoes' long legs enable them to wade in mud and water up to 80 cm (30 in) deep, stirring up the mud to find food organisms which they collect in their large distinctive scythelike bills. Using its tongue, the flamingo sucks in the water and food particles, and then pumps out the water; the food is strained out against rows of plates on the sides of the bill. Young flamingoes, which have straight bills, are fed by their parents with a secretion from their crops until their bills become curved.

Masters of the desert

The one-humped dromedary or Arabian camel originally evolved in the New World, and spread to north Africa, Arabia and Eurasia when the landmasses were still connected several million years ago. The dromedary has been domesticated since about 4000 BC, and though it is extinct in the wild today, it is widespread in its domesticated form.

Camels are extremely well adapted to cope with the harsh conditions of the desert. They have been known to survive for up to 10 months without drinking, storing moisture in their convoluted gut lining, and conserving it by producing small amounts of concentrated urine and dry feces. A camel can sustain a loss of up to 27 percent of its body weight without harm; yet it can drink enough water – more than 100 liters (22 gallons) – to restore its body weight in just 10 minutes.

The large hump stores fat, which is a rich source of energy, and some water in extreme conditions; because this fat is stored in one place, and is not distributed all over the body, it allows the rest of the body to act as a radiator for cooling purposes. The body temperature of a camel drops rapidly at night, and warms very slowly during the day so that water loss through sweating is kept to a minimum. The camel's dense woolly coat acts as an insulator, reducing heat absorption during the day and heat loss at night. Other features that enable camels to survive in the desert are broad, flat pads on their feet that prevent them from sinking into soft sand, forward-pointing lower incisors that enable them to crop thorny vegetation, long eyelashes to protect their eyes during sandstorms, and specially adapted nostrils that can be closed.

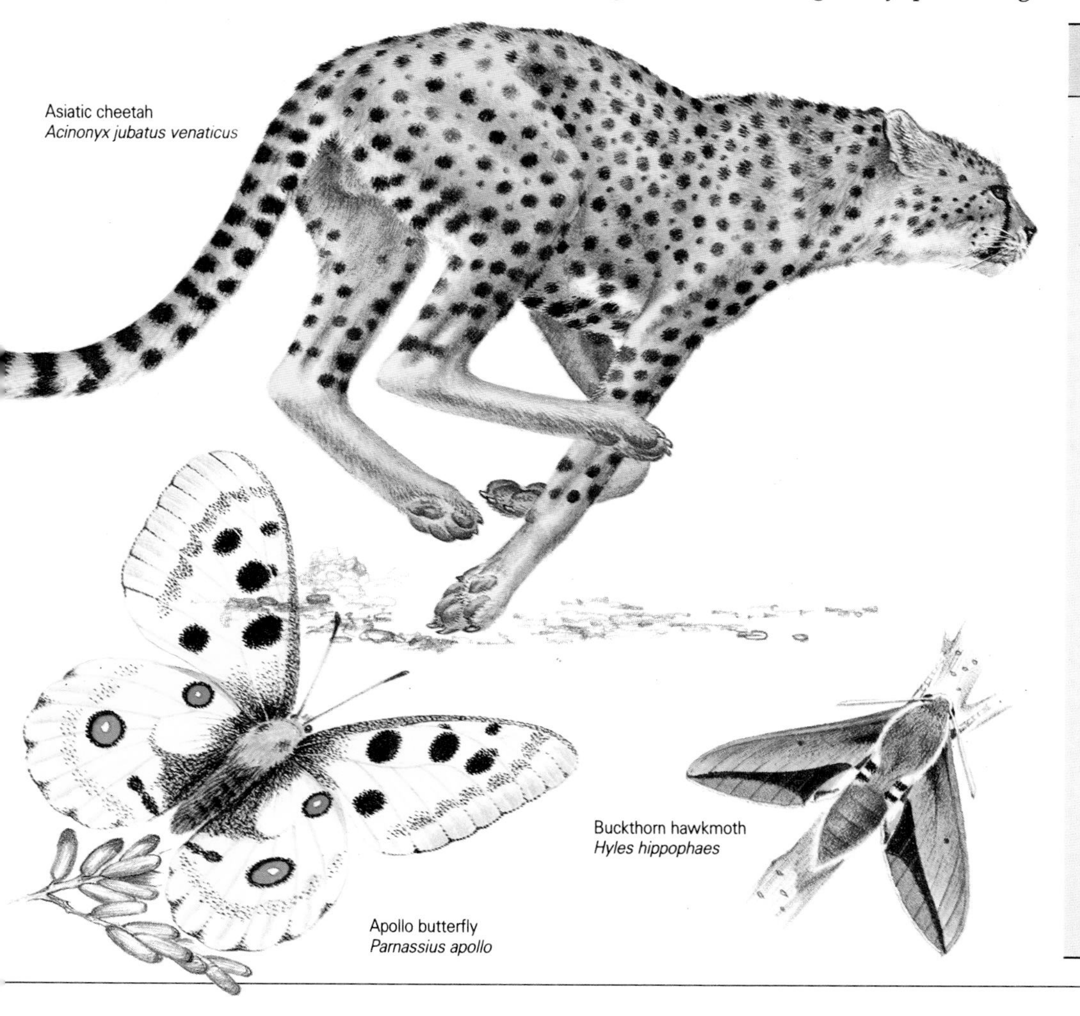

Desert beauties The Asiatic race of the cheetah is in grave danger of becoming extinct, its numbers reduced by hunting and trapping. The attractive butterflies of the Apollo group can survive at high altitudes. The meager vegetation also supports several species of hawk moths.

CHEETAH IN PERIL

The cheetah is superbly adapted to a life on the open steppes and savanna. The fastest land mammal, it can reach speeds of 96 kph (60 mph) when chasing gazelles and other small animals. It is built for speed – with long legs, a slender body, flat ears for improved streamlining and a long tail for balance; its spine is remarkably flexible, giving the cheetah a very long stride.

The cheetah's range extends from Africa and the Middle East to the southern Soviet Union. Numbers are declining as a result of the fur trade, which encourages widespread poaching. The main threat, however, comes from the loss and fragmentation of habitats caused by the expanding human population. The cheetah needs a large home range in which to hunt if it is to find enough prey all year round; as it can only maintain its high speed for short distances, it can only hunt effectively in open country with good cover for stalking its prey.

If the hunting grounds are reduced by human encroachment, the young become more vulnerable to predators, and many die of starvation; even in a large territory, it is rare for more than a third of the cubs in a litter to survive into adulthood.

RETREATING WILDLIFE

Throughout the Middle East human pressure from agricultural and urban expansion and intensive hunting is to blame for the reduction in numbers and local extinction of many species. According to the World Conservation Monitoring Centre (in Cambridge, UK), no fewer than 50 species of globally threatened animals are to be found in the Middle East: 7 invertebrate, 4 fish, 12 reptile and amphibian, 15 bird and 12 mammal species. A large number of other species, such as the Arabian bustard, the cheetah, the leopard and the Wild ass, have either already become extinct or are on the verge of disappearing from the Middle Eastern part of their ranges.

Gazelles, for example, are now rare everywhere and are unlikely ever to recover their numbers. The herds migrate long distances every year in search of fresh vegetation, often using traditional routes, but their predictable behavior pattern puts them at great risk from hunters. Hunting has become an increasingly popular activity among people from the growing industrial and urban centers, who can afford to equip themselves with modern weapons. The construction of new roads between towns has allowed easy access to the migrating gazelles. Hunting has seriously affected several other large species including the Arabian tahr (a goat antelope), the Arabian oryx (which is now extinct in the wild) and the Arabian ostrich (totally extinct). Most of the destruction took place during the remarkably brief period between 1940 and 1975.

Overgrazing

The native wildlife is also threatened by pastoralism, since grazing livestock destroy the vegetation. Sheep and goats were first domesticated in the region more than 5,000 years ago, and the wealth of a farmer is still measured by the size of his sheep flock or goat herd. The numbers of sheep and goats have increased so much that pastoralism is now causing significant damage to the environment, as well as reducing the amount of grazing land available for the native animals. The combined sheep and goat population of Iran, Saudi Arabia and Turkey is more than 126 million head, 30 percent more than the total human population of these countries. With so many animals grazing the land, the vegetation is prevented from regenerating and the soil is compacted underfoot, soon leading to desertification. A vicious circle then ensues; as more range is degraded or converted to cropland, more animals are reared in order to increase or even just maintain the already low level of meat and milk production. Concentrated in ever smaller areas, these animals inevitably accelerate the rate of range destruction.

Birds in peril

Falconry, the traditional hunting sport of the Arab aristocracy, has depleted the populations of many of the finest birds of prey in the Middle East. The damage is not only confined to the region: rare birds of prey and their live eggs are illegally

Saved from extinction (*right*) Arabian oryx now survive only in protected areas. Their pale coloration is typical of many desert mammals; pale colors reflect heat, helping to keep the animal cool. Like the camel, the oryx has splayed, shovel-like hooves that prevent it sinking into soft sand.

Ancestor of the domestic cat (*below*) The African wild cat may be spotted, striped or almost plain; its color ranges from yellowish-gray or sandy in desert cats to brown or reddish-brown in cats from areas with a wetter climate.

smuggled out of European countries such as Great Britain to supply Middle Eastern dealers.

The rarest bird in the Middle East is the waldrapp or Hermit ibis. Once found across most of Northern Africa, central Europe and the Middle East, this rather ungainly cliff-nesting species is now restricted to two separate populations of about 90 pairs each in Morocco and Algeria and, until 1988, one or two pairs near the village of Birecik on the upper reaches of the Euphrates river in east central Turkey. Even these failed to nest after 1988 and only juveniles appear to remain in the wild.

The explanation for the ibis's demise seems to be the gradual drying out of the moist alluvial plains where the bird feeds. This is the result of a combination of climatic change, the conversion of the land for agriculture, and a reduction in the abundance of the bird's insect prey, which is made worse through the widespread use of insecticides.

It seems inevitable that the bird's Middle East population will eventually become extinct in the wild. However, as it breeds freely in captivity (there are more than 400 birds in 33 collections around the world), with care the Hermit ibis should not disappear altogether.

As in other countries that fringe the Mediterranean, the coastal areas of the Middle East are undergoing rapid urbanization and increased pollution as a result of tourism. This destroys coral reef, dune and saltmarsh habitats, and threatens the breeding beaches of rare sea turtles, especially in Turkey, and the rare Mediterranean monk seal.

There are hopeful indications that countries throughout the Middle East are becoming more aware of the need to conserve their wildlife. Afghanistan, Iran, Israel, Jordan, Oman and Turkey all have long traditions of setting aside protected areas that comply with international standards of management and protection. In Saudi Arabia, a national commission was established in 1986 to conserve and develop the country's remaining wildlife stocks. Lebanon set up its first nature reserve at Benta'el in 1987, and Yemen has also begun to take a keen interest in wildlife conservation.

THE ARABIAN ORYX

The Arabian oryx has the distinction of being the first animal to have become extinct in the wild, and later to have been successfully reintroduced to its natural habitat after being bred in captivity. It once roamed throughout the desert plains of the Middle East, but numbers began to decline from the middle of the 19th century onward. In October 1972 the last few animals were shot or captured by an Arab hunting party, marking what seemed to be the end of the wild oryx.

Fortunately, breeding programs that were based on the few animals held in captivity proved successful, and more than 500 oryx were reared around the world. Between 1982 and 1984, 21 animals were released in Oman and were diligently guarded by local tribesmen. The animals soon took to the wild and began to roam over a wide area. However, the ultimate success of this restoration work will hang in the balance until the oryx are able to survive without human help.

The hunting of the bustard

For many centuries aristocratic hunters in Arabia and Iran have practiced the sport of falconry. The falcon generally flown is the saker, which breeds in eastern Europe and central Asia, migrating to winter mainly in eastern Africa. As the sakers fly across Arabia, skilled trappers capture them using baited cages, nets and even live pigeons adorned with snares. The favorite quarry is the Houbara bustard – a large bird whose coloring acts as camouflage in the dry, open landscapes of the Middle East. In winter large numbers of bustards migrate south into Arabia from breeding grounds in the Kazakh steppes of the central Soviet Union.

Changes in falconry

Traditionally a typical falconry party comprised a small group of men on camels, with perhaps six sakers. A good bag would be two or three bustards after a day or two of hunting; the birds would be boiled and served with rice. At the end of the winter season, the falcons would be released back into the wild, apparently none the worse for having been used as hunting birds for a few months. A rich folklore of hunting prowess grew up around the falcon and the bustard. However, the traditional values have now given way to keen competition – partly due to the great oil-derived wealth of the hunters, and to the development of fast modern desert transport. Instead of catching and training a few sakers each winter, the falconers began to purchase them in large numbers. The camel was replaced by four-wheel drive vehicles; semipermanent base camps were set up in the hunting grounds; and local guides, equipped with radios, were employed to find the wintering flocks of bustards.

By the mid-1960s the Houbara bustard population in Arabia had been all but destroyed. From 1968 falconers began to extend their hunting grounds into Iran, Iraq, Pakistan and Northern Africa. A hunting expedition in winter could cost upward of $2 million, or about $5,000 per bustard caught, making the bird a valuable economic resource. The host countries were only too pleased to welcome the falconers.

By 1980 the Houbara bustard had predictably disappeared from many parts of its remaining wintering range in the Middle East. The Arabs responded to calls for conservation by trying to breed the birds in captivity so that they could be released into the wild. They invested millions of dollars in captive breeding schemes, but the birds failed to breed in sufficiently large numbers.

Only in the late 1980s did attention turn to habitat protection, the single viable long-term means of ensuring the future of both Houbara bustards and falconry itself. Reserves have been established in Oman (where falconry is not practiced) and in Saudi Arabia; these countries have some of the last remaining Houbara bustard breeding grounds in the Middle East. Other measures that must be introduced include a reduction in the length of the hunting season, banning the use of weapons and limiting the size of hunting areas. Some control is also needed over the capture of sakers, which are now becoming scarce.

A prized trophy (*above*) The Houbara bustard is very well camouflaged against the desert soil, but is all too easily flushed out by hunting parties equipped with falcons. Hunting has exterminated this slow-breeding species from many parts of the Middle East.

A Bedouin hunt (*right*) High speed is not necessary when hunting bustards, so the Bedouins set out on camels. Bustards are large, heavy birds that prefer the ground to the air; their name derives from the Latin meaning "tardy".

Agents of death (*below*) Falcons have been used for centuries to hunt large game birds and are highly prized by their owners. In the wild pairs work together, one flushing out the quarry for the other to kill.

Cooperation in the Red Sea

Coral reefs grow well in the warm, clear, shallow waters of the Red Sea. Colonies of thousands of tiny organisms – the corals – secrete skeletons of calcium carbonate around themselves that gradually build up to form the reef. The corals are connected by their skeletons and a thin membrane of living tissue; the tiny, vulnerable individuals gain security from being part of a massive structure. The delicate ecosystem of the coral reef depends on a complex series of interrelationships. Algae living in the corals' tissues provide them with necessary food and calcium; and the algae benefit from the shelter provided by the coral. The dazzling array of fish found near coral reefs also participate in this system of mutual self-help. Large fish receive health care from smaller fish such as gobies and wrasses, who remove damaged skin and parasites from the head and gill covers – and sometimes even the mouths – of their "clients". The large fish benefit from the grooming, while the cleaners obtain food.

Reef-dwelling forms of anenomes – which are related to corals – can reach 1 m (3 ft) across. They feed on invertebrates and small fish using stinging cells on outstretched tentacles to immobilize their prey. The clownfish that live among the anenome's tentacles are unusual in being able to tolerate their poison. The fish snap up particles of food and receive protection, while chasing off larger predators.

A colorful spectacle With their warm, constant climate and huge variety of microhabitats, coral reefs are the marine equivalent of tropical rainforests, supporting a remarkable diversity of animal life.

DESERT SURVIVAL

DOMINATED BY THE DESERT · SURVIVAL AGAINST THE ODDS · THE SAHARA'S CHANGING FACE

In Northern Africa animals from Eurasia and Africa mingle, compete and coexist. As the climate and vegetation have fluctuated in recent geological time, species have moved into and away from the area. During periods of drought, desert species from the Middle East, such as wheatears and Desert larks, spread westward. In wetter times, animals moved south from Europe and the Mediterranean lands, and north from the African tropics. Many of these species have now vanished, but some have survived in moist refuges such as desert oases and cool mountains, and in the lush lowlands of the basin of the river Nile, the Sudd swamps of southern Sudan and the Mediterranean coastlands. A few, such as the Egyptian vulture and the African wildcat, have adapted to the harsh conditions of the desert.

COUNTRIES IN THE REGION

Algeria, Chad, Djibouti, Egypt, Ethiopia, Libya, Mali, Mauritania, Morocco, Niger, Somalia, Sudan, Tunisia

ENDEMISM AND DIVERSITY

Diversity Low (Sahara desert) to high (some groups in Ethiopia and Somalia)
Endemism Low (north Africa and Sahara) to high (Somalia and Ethiopia)

SPECIES

	Total	Threatened	Extinct†
Mammals	400	41	1
Birds	1,100*	28	0
Others	unknown	10	1

† *species extinct since 1600 - Red gazelle* (Gazella rufina)
* *breeding and regular non-breeding species*

NOTABLE THREATENED ENDEMIC SPECIES

Mammals Barbary macaque *(Macaca sylvanus)*, Simien jackal *(Canis simensis)*, Cuvier's gazelle *(Gazella cuvieri)*, beira *(Dorcatragus megalotis)*, addax *(Addax nasomaculatus)*
Birds Prince Ruspoli's turaco *(Tauraco ruspoli)*, Djibouti francolin *(Francolinus ochropectus)*, Algerian nuthatch *(Sitta ledanti)*
Others none known

NOTABLE THREATENED NON-ENDEMIC SPECIES

Mammals Mediterranean monk seal *(Monachus monachus)*, Grevy's zebra *(Equus grevyi)*, Dorcas gazelle *(Gazella dorcas)*
Birds waldrapp *(Geronticus eremita)*, White-headed duck *(Oxyura leucocephala)*, Houbara bustard *(Chlamydotis undulata)*, Audouin's gull *(Larus audouinii)*
Others Nile crocodile *(Crocodylus niloticus)*, Egyptian tortoise *(Testudo kleinmanni)*, Orange-spotted emerald butterfly *(Oxygastria curtisii)*

DOMESTICATED ANIMALS (originating in region)

cat *(Felis 'catus')*

DOMINATED BY THE DESERT

A fascinating mix of animal species inhabit Northern Africa, often in a harsh environment. In the oases of the Sahara desert nesting birds include the Blue tit, typical of European climates, and the Fulvous babbler, a species that probably originated in the deserts of the Middle East. Among the desert massifs Pygmy sunbirds feed on flowers, far from their relatives in tropical Africa, while Eagle owls, more familiar in the coniferous forests of Scandinavia, hunt among the crags. In the Ethiopian Highlands, there are brightly-colored butterflies such as the swallowtail.

Wildlife of the Sahara

The wildlife of the region is dominated by the Sahara. The desert is a relatively new phenomenon, the product of just a few thousand years of climatic change, accelerated by human exploitation. There has been little time for new species to evolve within the Sahara itself. Many of the species that are unique to the region – mostly insects and spiders, together with a few rodents, lizards and snakes – live on the desert fringes along the northern margin, or on the cool, damp sand dunes of the Atlantic coast. This suggests that they evolved before the desert formed. Other Saharan species are descended from ancestors that migrated into the region as the climate became more arid, from dry areas to the north or south and from the ancient deserts of the Horn of Africa in the southeast of the region.

On the gentle green slopes of the Atlas Mountains in Morocco and Algeria, to the north of the Sahara, the last remaining populations of Cuvier's gazelle can be found. They were once widespread, but their range has been reduced by the arid conditions of the true Sahara that now prevail; only the more resilient species such as the Dorcas and the Slender-horned gazelle can survive there.

North of the Atlas Mountains, along the Mediterranean coast, remnant populations of the Spur-thighed tortoise (often kept as a pet in Europe) are found. The well-watered lowlands along the Nile, to the east of the Sahara, offer a refuge to flamingoes, Egyptian geese, crocodiles and Nile monitor lizards.

On the southern fringes of the desert in the Sahel, where the frontier of vegetation

is still retreating as the desert spreads southward, many species maintain a precarious foothold. The Sahel is home to the White-throated bee-eater, and the Beisa oryx and other antelopes, as well as predators such as the Sand fox, the Banded or African striped weasel and the caracal. These last two species also live north of the Sahara, though not in the desert itself. Like many animals they have been forced to retreat from the heart of Northern Africa by the steady and relentless process of desertification.

Oasis and mountain species

Desert oases are isolated refuges for animals that once roamed freely over Northern Africa. Here, toads can live and breed in permanent water. Many birds, including coots, blackbirds and babblers – songbirds with a rich variety of calls – are also permanent residents. Migrant birds stop off at the oases for welcome food and water on their long, hazardous journey across the Sahara.

Easy prey (*above*) A Marsh harrier feeds on a Graylag goose killed by a hunter on Lake Ichkeul in Tunisia. Thousands of birds stop over here when migrating between Europe and southern Africa.

The Fennec fox (*right*), the smallest member of the dog family, is well adapted to its desert habitat. The many blood vessels in its large ears, which can measure up to 15 cm (6 in), help to keep the animal cool, and its pale fur serves as camouflage.

Mountain ranges with their cooler conditions and permanent springs and pools, are also important refuges. Earthworms are found here, as well as frogs, toads, baboons, long-limbed Patas monkeys, and even moorhens. The agile aoudad, or Barbary sheep, lives on the rocky massifs; its superb climbing ability enables it to reach pools of standing water high up in the mountains.

Animals of the Horn

The deserts of the Horn of Africa have far more endemic species than the Sahara, because of their greater age. Here, too, are some animals of Eurasian origin, whose

ancestors probably migrated southward during the last ice age. Among them are the Golden-winged grosbeak (closely related to the European hawfinches) and the Somali linnet. The latter is an unusual species with a massive seed-crushing beak. Other species in the Horn of Africa – for example, the tiny Dikdik antelopes, the Bat-eared fox and the spindly termite-eating aardwolf – show remarkable similarities with inhabitants of the Kalahari Desert of southwest Africa. The probable explanation for this is that in more arid epochs a corridor of desert stretched diagonally across the African continent, enabling the species to migrate from one region to the other.

The bleak, eroded pinnacles of the Ethiopian Highlands represent a unique habitat: a cold upland island in the midst of tropical Africa. Here a similar mixture of European, Asian and African species is found. These volcanic rocks, scoured by glaciers during the last ice age, are still relatively poor in species. This is a consequence of their turbulent geological history, the harshness of their dry, cold climate and their isolation from the rest of Africa. Many forest and mountain species do not occur here, simply because their route is blocked by the mighty Nile or by the semidesert zone of northern Kenya. In their place are species that originated in Eurasia – such as the ibex – some of which moved south during the ice age. Living alongside them are species that evolved here, such as the long-haired Gelada baboon, an unusual grass-eating monkey, and the Mountain nyala, a powerfully built antelope.

SURVIVAL AGAINST THE ODDS

The Ethiopian Highlands are a difficult habitat for animals to exploit, and the same is true of much of Northern Africa. Lack of water is usually the main problem faced by desert animals, but they must also contend with intense heat during the day, bitter cold at night, a shortage of food, and violent winds, often laden with dust and sand.

The Gelada baboon has adapted to life on the bleak Ethiopian Highlands by accepting a diet of grass, which it finds in plenty on the mountain slopes. This is a poor diet for a monkey, however, since grass is abrasive, indigestible and low in energy value. To save energy and conserve its body heat in the cool climate, the Gelada feeds in a squatting position and shuffles from one clump of grass to another without rising.

Coping with the heat

A great many desert animals cope with unfavorable conditions simply by avoiding them. A variety of species, from spiders to Eagle owls, take shelter underground and emerge only at night or during the cool of the evening and early morning. At about 1 m (3 ft) below ground the temperature and moisture content of the air remain more or less constant, both day and night, so a burrow at this depth makes life relatively easy.

Burrowing is strenuous, however, and not all animals are suitably built for such work. Taking over the burrow of another species may be the only option for some

Birds of the North African plains The plumage of the Cream-colored courser is an effective camouflage in the desert. When threatened, the courser prefers to run away rather than fly. The Crab plover has a massive bill for dealing with its tough prey: here it is catching crabs, its young looking on. Two female ostriches display to attract a male; after mating, they will leave their eggs to be cared for by the male. The Egyptian plover is a close relative of the courser and not a true plover. It is an insect-eater that prefers living near water. A bird of cultivated land, the White stork stands on its nest while performing one of several displays which are characteristic of this genus.

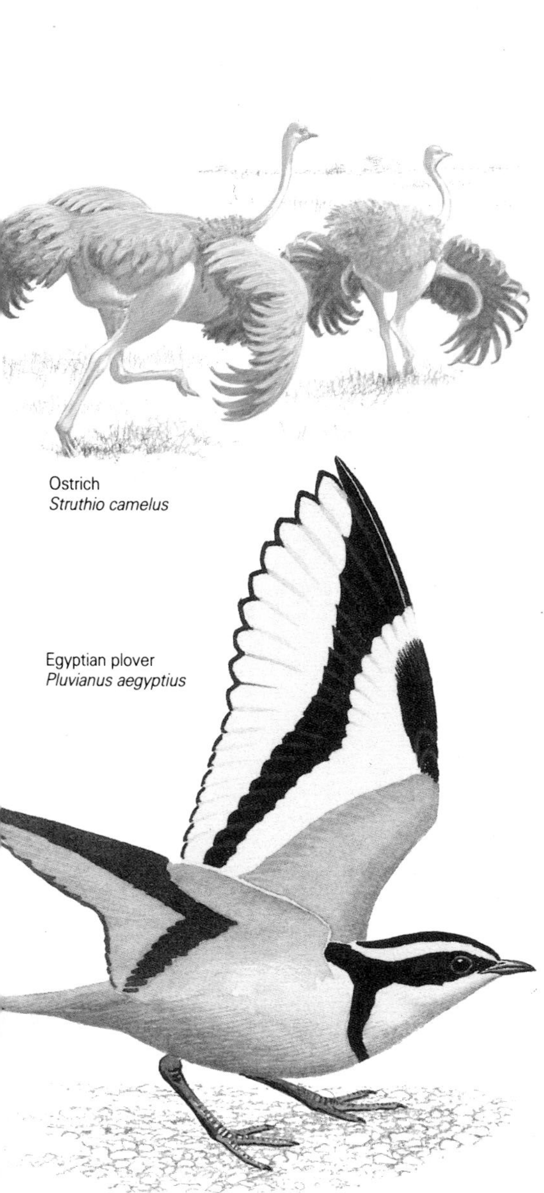

desert dwellers. Gazelles often take refuge in the disused burrows of aardvarks, and may share them with other animals that are hiding from the sun, including lizards and owls. A species of wood louse builds communal burrows. As many as 100 wood lice cooperate to excavate vertical shafts, taking them down 30 cm (1 ft) below the surface in an unusual example of cooperation among crustaceans. Termites also burrow, sometimes so deep that they reach the water table and benefit from its moisture.

Desert species that can brave the heat of the day have various adaptations to high temperatures and dryness. Among the best adapted are the scorpions: even in the driest conditions they lose only one part per 10,000 of their body moisture, the lowest recorded for any animal. A sturdy cover of wax, impervious even when heated to 65°C (149°F), is responsible for this remarkable waterproofing. When they do dehydrate, scorpions take a long time to suffer any ill effects as they can survive losing as much as 40 percent of their body fluid.

The ancient Nile crocodile *(above)* This formidable predator has a splendid array of sharp teeth for tearing and holding its prey: fish and birds, and mammals that come to drink at the water's edge.

A short and frantic life

Perhaps the most intriguing of all the desert animals are those that exploit the intermittent rainstorms. They are dependent on heavy downpours that leave puddles and lakes dotted over the surface of the desert. These temporary bodies of water, which can last from one to three months, are soon transformed into cauldrons of life seething with animals that are engaged in a frantic race for survival.

To stand any chance of passing on their genes, these struggling life forms must mature and lay eggs before the pool dries up. The eggs they lay must also be tough enough to resist cooking under the Saharan sun for months or even years, until the next rainstorm triggers off the process once again. Among the animals that survive in this way are a variety of small crustaceans, various midges and other tiny microscopic invertebrates, as well as a few species of toad. The rains also bring a host of brightly colored plants into flower, and these provide food for bees, hoverflies, butterflies and moths. Like the animals of the temporary ponds, they sprint through their life cycles in a remarkably short time, and then enter a long period of suspended animation as the desert dries out once more.

GUARDIANS OF THE NILE RIVERBANK

Before the spread of the Sahara across Northern Africa, when a string of lakes and rivers linked the Nile to the Niger, the Nile crocodile was scattered throughout the region; and as recently as the beginning of the 20th century some survived in the Saharan zone, mostly in the isolated waters of the mountain ranges. Today, however, the crocodile is confined to the river Nile, where its numbers have been much reduced by hunting for its skin and by destruction of its habitat.

The female crocodile lays up to 80 eggs in a hollow that she excavates near the river. She then covers them and guards them until they hatch, a process that may take up to three months. Other animals take advantage of the deterrent effect of the crocodile's presence: turtles creep past on their way to lay their eggs in the sand. The Gray-headed kingfisher uses an old crocodile nest to start its own nest, sometimes enlarging the hole to a depth of 60 cm (2 ft).

When the crocodile eggs hatch, the mother carries her young to the water. The turtle eggs also hatch at about the same time and sometimes a female crocodile, attracted by the movement of the young turtles, will pick them up in her mouth and carry them carefully to the water, as if they were baby crocodiles. More than half the crocodile's young and eggs – and those of the turtles – are taken by predators such as Monitor lizards, mongooses, baboons and birds of prey. Smaller birds, such as wagtails and starlings, eat the shells to gain calcium in order to make their own eggshells.

THE SAHARA'S CHANGING FACE

Ten thousand years ago the Sahara was a green and fertile land. Giraffes, elephants and rhinoceroses ranged over a savanna landscape and were hunted by tribes that have now vanished. These people left paintings of the animals they hunted in rock shelters among craggy massifs – now surrounded by endless gravel plains and sand dunes, but then set amid grass and trees. Crocodiles and hippopotamuses wallowed in the rivers and marshes, fish were plentiful, and lions hunted the herds of grazing antelopes.

Today large areas of the Sahara appear to be barren and devoid of all plants. However, if the thin gravelly surface is scraped away, termites, which feed on dead wood or leaves, can be found in the dry soil beneath. Termites survive in this apparently unvegetated wasteland by feeding on the buried trunks of large tamarisks and other trees, which last put forth leaves more than 5,000 years ago. When the trunks of these trees fell, they were covered in alluvial silt brought down by rivers in full spate, and they have lain there ever since, while the valley above them slowly dried out to its present parched and lifeless state.

Human pressures

The climate began to grow drier about 4,000 years ago, with a sudden deterioration 2,000 years later. The indigenous tribes, who had already become pastoralists, overgrazed the remaining vegetation with their cattle, steadily reducing the sparse plant cover and leaving the soil exposed to the forces of erosion. This process still continues on the edges of the Sahara; here the grazing goats and camels of present-day pastoralists whittle away the desert's green edges. The removal of the last trees and shrubs for fuelwood exposes the bare soil to the blazing sun and swirling wind.

Hunting has also contributed to the loss of wildlife. Lions, cheetahs and ostriches were exterminated from much of the region relatively recently by big-game hunters of the colonial era. The last lion to inhabit the Air mountains in north central Niger was shot in 1932.

Desert antelope (*left*) Addax are said to be able to survive without water for a whole year. Pressure of hunting has made the addax almost extinct over much of its former desert range, and it is now found mainly in reserves. Properly managed, addax could be a valuable source of meat in arid regions.

Journey's end (*right*) Throughout the Mediterranean, millions of migrating songbirds are trapped every year for food, to sing in cages, and often simply for sport. This tradition is beginning to have serious effects on the numbers surviving to breed farther north. Bird protection societies are pressing for legislation against this practice.

The sacred scarab (*below*) Dung beetles accelerate decomposition by dragging balls of dung undergound to feed their larvae. This helps to recycle organic matter and dispose of disease-breeding waste. The dung beetle, or scarab, was often featured in ancient Egyptian, Greek and Roman art, perhaps because of its association with death and renewal.

There is a long tradition of hunting in Arab countries, and recent improvements in firearms, and the advent of four-wheel-drive vehicles that can career across deserts, have led to increased slaughter. Even fleet-footed desert animals such as the gazelle are now vulnerable. Even though there is greater awareness of conservation, at least in some Northern African countries, it is impossible to police vast tracts of inhospitable desert.

The wetlands under threat

Wetland areas in Northern Africa are seriously threatened by human demand for water. Ambitious irrigation schemes have been undertaken, using the latest advances in engineering skills. These will result in the destruction of large areas of important wildlife habitats. In southern Sudan, the proposed Jonglei Canal, already partly constructed, threatens to drain the great Sudd marshes, home to some of the richest wildlife in northern Africa. This project has been abandoned because of the civil strife in Sudan, but may be resumed in the future.

Some countries are now realizing the potential of their wetlands for wildlife tourism. Tunisia is host to many visitors who come to watch the birds on Lake Ichkeul, but if the region's wetlands and the wildlife they support are to survive, conservation must be a priority.

THE MEDITERRANEAN MONK SEAL

The Mediterranean monk seal is one of three species of monk seal, all of which are adapted to life in warm waters, and are endangered by human encroachment on their habitat. The Caribbean monk seal may now be extinct, the Hawaiian monk seal is endangered, and the fate of the Mediterranean species remains uncertain: there are probably fewer than 500 individuals. The largest populations are in the Aegean Sea and along the Turkish coast; scattered breeding colonies can also be found along the Mediterranean coast of northern Africa, with a few outlying groups along the Atlantic coast of the western Sahara.

Mediterranean monk seals mate in the sea but haul out on to land to give birth. Once they used rocky shores all along the north coast of Africa – as in other parts of the Mediterranean – but their fear of humans has driven them to use sea caves and other secluded spots. These are often less safe, and there have been many breeding failures. Disturbance by boats and people can cause the females to abort or to abandon their suckling pups. Other dangers include the animals becoming entangled in fishermen's nets or being attacked by the fishermen themselves, who regard the seals as pests. Their main chance of survival probably depends on special reserves being established in Greece and Turkey.

Gundis

Small balls of soft fur scampering up vertical rock faces, trilling and chirping like flocks of birds: these are the gundis, a curious family of rock-dwelling rodents that are found only in northern Africa. Four species of gundi live in the Sahara, one in the Horn of Africa.

Although the five species are all approximately the same size and have similar behavior patterns, they are different enough anatomically to be placed in four distinct genera. Gundis provide a striking illustration of the way new species are formed when populations of a parent species become isolated by geographical barriers. Gundi colonies are restricted to rocky outcrops and mountain ranges. In their isolated refuges, separated by the encroaching desert, the five species have evolved in different ways. The only two species that are similar enough to be put into the same genus – the Northern gundi and the Desert gundi – are also the only two with overlapping ranges; this suggests that they evolved from a parent species relatively recently.

Adapted to desert life

The gundis' way of life depends on fissured rocks to shelter them from the desert heat and the eyes of predators such as snakes and hawks. They can flatten their rib cage to squeeze into narrow cracks, and their large eyes enable them to see in the semidarkness within. Although their ears are very well developed and can detect the slightest sounds made by approaching predators, the external earflap is generally very small; when squeezing through tight spaces in a rocky habitat an external ear could easily be torn. In the Northern gundi of Algeria, Libya, Morocco and Tunisia the flap is little more than a circlet of fur; in the Mzab gundi, a species found in southern Algeria, Chad and Niger, it is nonexistent.

A talkative gundi (*left*) Speke's gundi has a particularly wide range of vocalizations, which include chirps, chuckles and whistles. Low-pitched calls carry well across the rocky desert; short, sharp calls warn of predatory birds; longer calls denote the presence of terrestrial enemies. Gundis also thump the ground with their hind feet to alert the colony to approaching dangers.

On the alert for danger (*right*) The Mzab gundi's ears are flat and immovable. Like many desert rodents, all gundis have acute hearing that is capable of detecting the weak low-frequency sounds not only of snakes gliding over the ground, but also of hawks hovering overhead. Young gundis, left in a rocky shelter, make a continuous chirruping noise that helps their mother to find them again.

The gundis' diet is mainly composed of the leaves, shoots and seeds of desert plants, but they will also eat stems and small twigs when food is scarce. They derive all the moisture they need from their food, and so can survive without water. As a special adaptation to this parched diet, they have highly efficient kidneys that produce a very concentrated urine, so preventing water loss. Female gundis with young conserve moisture by producing very little milk. The young are fed on leaves, which the mother has softened by chewing, within hours of their birth. They are remarkably well developed for rodents, being able to run about even as newborns.

Food can be scarce in the desert, and gundis cope by conserving energy as much as possible. One of the methods they use is to avoid wasting energy on generating body heat, an activity that requires a great many calories in most mammals. In the early morning gundis can be seen basking in the sun outside the crevices where they have spent the night: it is reasonable to suppose that they allow their body temperature to fall during the night, as northern animals do during hibernation. Once they have warmed up sufficiently, the gundis begin to forage for food, but they retreat to the cool interior of a rock to escape the fierce heat of the noonday sun.

Species and their calls (*below*) Each species of gundi has a different repertoire of calls for family and colony communication. The North African gundi makes distinctive chirping noises that help members of this species to recognize each other in the habitat they share with the Desert gundi, a species that whistles. The Mzab gundi is the least communicative gundi, with only an infrequent chirp, while Speke's gundi has a rich vocabulary of calls. The Felou gundi makes a harsh "chee-chee" sound when in danger.

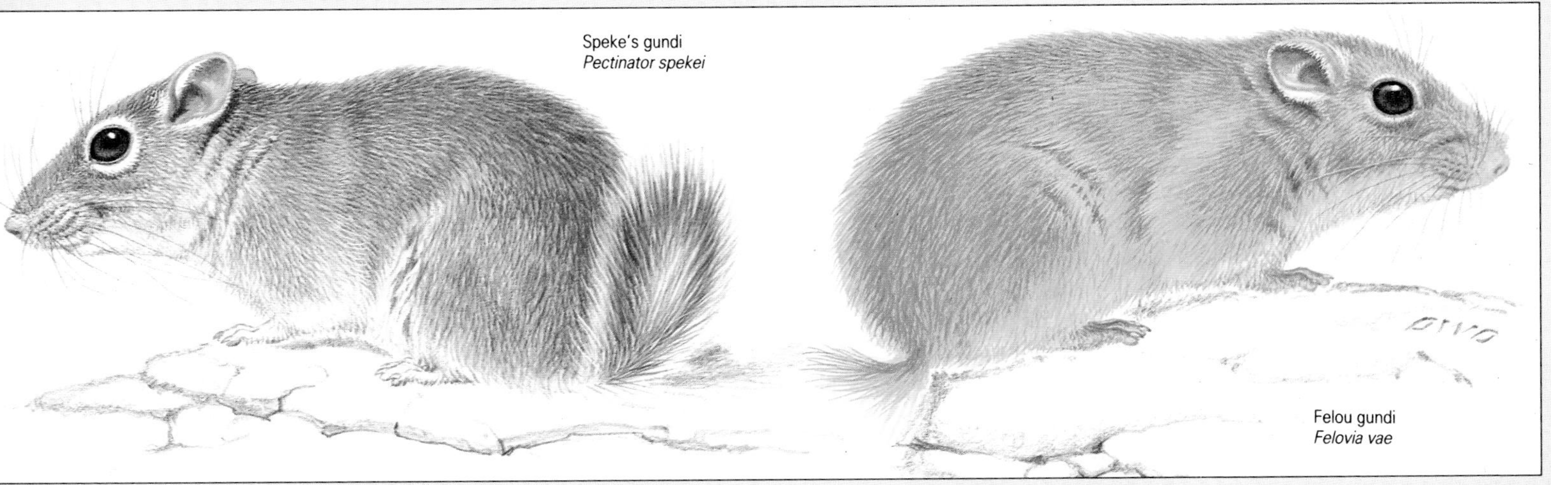

Army on the march

The migratory locusts of Africa are not so much adapted to desert life as to making the most of intermittent rainstorms. When not moving in great swarms across the desert, these insects live a solitary life in wetter coastal areas, along the Nile valley, in the mountains or in the scrubby vegetation alongside seasonal streams. They have evolved a nomadic lifestyle that exploits the periodic flushes of vegetation that follow sporadic rainfall. By migrating huge distances with the prevailing winds, which carry them to the point where conflicting air-masses meet, locust swarms arrive at the same time as the rainstorms. Locust swarms can devastate large areas: a single swarm may contain 50 thousand million locusts and cover 1,000 sq km (386 sq mi).

The locusts make the most of the ephemeral vegetation that follows rain by reproducing explosively when conditions are right. They gear their life cycle to the food supply by responding to the aromatic substances that some desert shrubs produce as they grow new leaves. These scents stimulate the migratory locust to mature sexually and reproduce. By the time the vegetation withers again, a new generation of locusts will have hatched and be ready to move on.

Voracious hordes A swarm of migratory locust hoppers marches across the desert, devouring every living plant in its path.

ANIMALS OF RAINFOREST AND SAVANNA

RAINFOREST AND GRASSLAND · STRATEGIES FOR SURVIVAL · A CONFLICT OF INTERESTS

Cooling, life-giving water, or indeed the absence of it, is the most powerful influence on animal life in tropical Central Africa. It divides the region sharply into two zones, each with its own characteristic wildlife. The year-round moisture in the equatorial rainforests produces lush plant growth and a generous supply of food for many animals. The dry habitats of the savanna, thorn forest and steppe demand a more flexible approach to life. Here migration is the key to survival for many species, producing some of the greatest wildlife spectacles on Earth: the mass movements of vast herds of zebras, wildebeeste and other grazing animals. The abundant water of Central Africa's great winding rivers, huge lakes, marshes and mangrove swamps is exploited by hippopotamuses, crocodiles and other amphibious species.

COUNTRIES IN THE REGION

Benin, Burkina, Burundi, Cameroon, Cape Verde, Central African Republic, Congo, Equatorial Guinea, Gabon, Gambia, Ghana, Guinea, Guinea-Bissau, Ivory Coast, Kenya, Liberia, Nigeria, Rwanda, São Tomé and Príncipe, Senegal, Seychelles, Sierra Leone, Tanzania, Togo, Uganda, Zaire

ENDEMISM AND DIVERSITY

Diversity High to very high
Endemism High to very high

SPECIES

	Total	Threatened	Extinct†
Mammals	550	85	0
Birds	1,400*	97	1
Others	unknown	288	1

† *species extinct since 1600 - Seychelles parrot* (Psittacula wardi)
* *breeding and regular non-breeding species*

NOTABLE THREATENED ENDEMIC SPECIES

Mammals Nimba otter-shrew *(Micropotamogale lamottei)*, Mountain gorilla *(Gorilla gorilla beringei)*, chimpanzee (*Pan troglodytes*), bonobo (*Pan paniscus*), drill *(Mandrillus leucophaeus)*, Pygmy hippopotamus *(Choeropsis liberiensis)*, Ader's duiker *(Cephalophus adersi)*
Birds White-breasted guinea fowl *(Agelastes meleagrides)*, Bannerman's turaco *(Tauraco bannermani)*, Sokoke scops owl *(Otus ireneae)*, Seychelles magpie robin *(Copsychus sechellarum)*
Others Goliath frog *(Conraua goliath)*, Lake Victoria cichlid fish (250 species), African blind barbfish *(Caecobarbus geertsi)*,

NOTABLE THREATENED NON-ENDEMIC SPECIES

Mammals African wild dog *(Lycaon pictus)*, Black rhinoceros *(Diceros bicornis)*, African elephant *(Loxodonta africana)*
Birds Wattled crane *(Bugeranus carunculatus)*, White-winged flufftail *(Sarothrura ayresi)*, corncrake *(Crex crex)*
Others Nile crocodile *(Crocodylus niloticus)*, Green turtle *(Chelonia mydas)*, Coconut crab *(Birgus latro)*

DOMESTICATED ANIMALS (originating in region)

guinea fowl *(Numida meleagris)*

RAINFOREST AND GRASSLAND

The animals of Central Africa reflect the region's sharp division of habitats. Forest and grassland each have their distinctive species, and relatively few species range from one to the other. The often remarkably sharp dividing line in tropical Africa between rainforest and savanna is due to the pronounced change in rainfall patterns at the edge of the forest zone and the scourge of grassland fires. These fires cannot penetrate the forest, but they exclude forest trees from the edges of the savanna. In other parts of the world, one habitat and its wildlife often merges almost imperceptibly into the next.

Luxuriant forests

The animals of the rainforest enjoy an abundance of moisture, warmth and food in which the environment itself offers few checks or setbacks. But such luxuriance creates its own problems, and the seething life of the rainforest is fiercely competitive – the "struggle for survival" between species reaches an unparalleled intensity. There is strong reason to suppose that this factor contributes to the extraordinary richness of species found in tropical rainforests.

Although they contain a remarkable number of species, particularly insects, Central African rainforests are less diverse than their counterparts in Amazonia and Southeast Asia. This has been attributed to the presence of large herbivores such as forest elephants, buffaloes and okapi (a relative of the giraffe), which keep the vegetation of the forest relatively open as they push their way through the undergrowth. The physical impact of these large animals on the lowest storey of the forest, and on germinating seedlings on the floor, may reduce the number of microhabitats available to other animals.

Endemic and immigrant species

The history of the African continent may also have contributed to the reduction of species' diversity. In Southeast Asia and South and Central America the native animals evolved during a long period of isolation; then, as a result of continental drift between 15 and 3 million years ago, these regions were invaded by animals from Australasia and North America respectively. This process of isolation and unification greatly enriched the wildlife. Africa also experienced eons of total or partial isolation, during which some of its unique species evolved: elephant-shrews, hyraxes, Old World monkeys, cane rats (a primitive type of porcupine) and mouse-birds – so-called because of their habit of crawling through the thick foliage. Other animals, such as the hippopotamuses,

Learning from mother *(above)* Lion cubs remain with their mothers for up to two years learning how to hunt. Lionesses cooperate to hunt large prey. Male lions are not skilled hunters, and may die of starvation if expelled from a pride.

Nomads of the savanna *(left)* Grant's gazelles are well-adapted to long-distance travel in search of grazing. With their small hooves and long legs, they are able to build up a good speed and, twisting and turning, can outrun lions and even cheetahs.

survived in the isolation of Africa while dying out elsewhere in the world. The northward movement of the continent and the falling sea level subsequently reunited Africa with Eurasia, and there was an influx of new animals from the north. But this occurred much earlier than parallel events in Southeast Asia and the Americas, as much as 25 million years ago. The passage of time since then has probably allowed the mix of species to settle down; some animals are thought to have ousted others in the process, thus reducing the total number of species.

STRATEGIES FOR SURVIVAL

The animals that dominate the Central African savanna – the many antelopes and zebras – evolved to cope with a diet of tough, abrasive grass. The complex structure of the cheek teeth results in their remaining sharp as they wear down. Grazing animals are conspicuous against the open plains so, rather than relying on stealth to outwit predators, they became either fast running or large, belligerent and well defended. The majority followed the former course, evolving long legs and a light build, while a few, such as the Cape buffalo, pursued the more defensive strategy. Faster-running plant-eaters led in turn to speedier predators, and the two groups evolved together, developing their speed to the limits of their physiology – best exemplified by the cheetah, which can reach a speed of more than 90 kph (50 mph) while chasing its prey.

Birds also adapted to conditions on the savanna. The Secretary bird, which feeds on snakes, evolved very long legs for stalking through the grasses; and the ostrich can reach speeds of 50 kph (31 mph) to escape from predators.

Living in a dry country

Some animals survive the arid conditions of the grasslands by remaining underground. The Naked molerat of northern Kenya feeds on the large underground tubers produced by herbaceous plants. It never emerges on the surface, but lives in large burrowing colonies. The totally subterranean lifestyle of these animals has allowed them to dispense with fur, vision and even the ability to regulate their body temperature.

The lungfish is also adapted to survive long periods underground. When the swamps in which it lives dry out it burrows into the mud and cocoons itself in a layer of mucus until the rains return.

FOREST GIANTS

Hollywood has given the gorilla the reputation of being an aggressive, dangerous creature. This is totally undeserved – very few animals are as peaceable. Unlike its close relatives, humans, chimpanzees and orangutans, the gorilla eats no meat under normal circumstances. It sustains its great bulk on leaves, tender young shoots and the pith of plants. The lowland subspecies also includes fruit in its diet.

Leaves are not a high-energy food, and a band of gorillas – usually about a dozen animals – requires prodigious quantities each day. Their feeding technique is almost a form of agriculture, because it encourages more of their preferred food to grow. By feeding in what seems a destructive way, knocking down small trees and stripping whole branches of their foliage, gorillas maintain a relatively open habitat with plenty of undergrowth and vines. When they have finished feeding, the immediate vicinity may be almost bare of leaves; but when they return to the spot some months later, a flush of new leaves will have grown to provide them with food.

The rarest gorilla The Mountain gorilla's survival is threatened by loss of its forest habitat, being trapped in snares set for other animals, and human disturbance. There are now fewer than 400 Mountain gorillas left in the wild, and none in captivity.

Central African monkeys and baboons A Yellow baboon of the lowlands sleeping in a tree. The Gray-cheeked mangabey, with its light build and long tail, spends most of its time in the treetops. The Mustached monkey bobs its head from side to side when threatening, to "flash" its mustache. Allen's swamp monkey includes shrimp, fish and snails in its diet. The Talapoin is the smallest African monkey, measuring only 34 cm (13 in) from its head to the tip of its tail. The Patas monkey is the swiftest of all primates, and can attain a speed of 55 kph (34 mph).

Other animals survive the season of shortage and aridity by migrating. As the grass withers to a lifeless sea of brown, waterholes dry up, and the baked earth cracks in the riverbeds, the herds of wildebeeste, zebras and Thomson's gazelles trek off, following the great seasonal cycle of the rains.

In the arid season elephants feed on the dry leaves and bark of the savanna trees and dig for water in the stream beds. They, too, may travel great distances in search of food and water, relying on the memory of their ancient matriarch. Chimpanzees and other animals living in the seasonal thorn forests of Tanzania's western mountains move to the forests that line the rivers. These fingers of tropical rainforest extend into the grassland and scrub, thanks to the permanent water around their roots, providing a welcome food source in the dry season.

The butterfly *Catopsilia florella* flies long distances in search of flowers on which to feed. Traveling in their thousands, the insects' white and yellow wings create clouds as much as 20 km (12 mi) long and 5 km (3 mi) wide. The locust is another seasonal insect migrant; huge swarms of Red locusts, for example, fly northward every June and August in pursuit of the rains. They delay mating and egg-laying until they reach an area of plentiful recent rainfall where there will be ample green vegetation for their offspring to feed on.

Birds have less difficulty reaching supplies of water. Sandgrouse will fly up to 80 km (50 mi) each way every two or three days to reach a waterhole. Between the breast feathers and the skin of the male sandgrouse lies a mass of microscopic filaments, produced by the inner surfaces of the feathers. These filaments form a spongelike structure that can soak up water, and the sandgrouse uses them to bring drinking water to its chicks. While drinking, the male steeps his breast feathers in the water, and on his return, the chicks cluster around him to sip the moisture retained there.

The unchanging forest

In the well-watered zones of the rainforest, there are virtually no seasonal differences. From a plant-eater's point of view, this means that supplies of leaves, flowers, nectar, fruit, nuts, sap and other food are available all year round in great abundance. In the course of rearing a brood of young a male Silvery-cheeked hornbill may carry some 24,000 fruits to its mate and chicks. Hornbills take fruit from a wide range of trees, but other species in the forest are specialist feeders. The Palm-nut vulture, for example, has abandoned a carnivorous diet to feed exclusively on the greasy fruits of the oil palm and the raffia palm.

Insects are another abundant source of food in the Central African rainforest. But in response to the high level of predation they have evolved a great many defenses to ward off insectivores. Some are well camouflaged, others are quick to escape, while a third group rely on toxic or foul-tasting chemicals in their bodies to repel predators.

One group of mammals, the pottos and lorises – small nocturnal primates related to bush babies – have specialized in feeding on this last type of prey. With apparent impunity, they eat caterpillars covered with irritant hairs, poisonous millipedes, stinging ants and beetles that exude noxious fluids when alarmed. Close observation shows that they are not oblivious to the distasteful nature of their food: they will take less noxious insects if offered a choice. The irritant hairs are carefully rubbed from caterpillars before they are eaten, and the potto wipes its lips and nose clean afterwards. No doubt there are enzymes in the animal's body that can quickly break down the toxins in its diet. Because such insects make no effort to conceal themselves, relying on their toxic defenses, the pottos and lorises have secured for themselves a reliable, abundant and easily caught food supply – even if it does leave an unpleasant taste in the mouth.

Trees in the rainforest are just as well defended as insects, and their leaves are loaded with toxins. Leaf-eating animals such as colobus monkeys are remarkably resistant to these poisons – some of this resistance undoubtedly stemming from their own abilities to break down poisons, although the bacteria living in their stomachs also help. As a result, a Black-and-white colobus can make a meal of leaves that would induce immediate and violent vomiting if eaten by humans,

Talapoin
Miopithecus talapoin

Patas monkey
Erythrocebus patas

or eat handfuls of seeds so poisonous that only two or three will kill most mammals.

Treetop travel

Many animals have evolved adaptations to help them travel easily in search of food in the dense canopy of branches of the tall rainforest trees. There are gliding animals that swoop from tree to tree on extended "flight membranes". They are not true fliers: they cannot flap these membranes, and can become airborne only from an aerial launch pad, never taking off from the ground or ascending. The African rainforests are home to a unique group of gliding mammals: the Scaly-tailed flying squirrels. Despite their name, they are a group of rodents with no other close relatives; they are probably another example of an animal that evolved during Africa's long period of isolation. Their tails are scaly near the base, rather than furred. Like most gliding animals, Scaly-tailed flying squirrels are active at night. In the daytime they would be much more vulnerable to predators.

Taking to water

The emergence of two distinctive types of habitat in tropical Africa is the explanation for the two species of hippopotamus that are found here. The hippopotamus of the savanna is the better known of the two; an enormous, lumbering, mud-loving creature with bulging froglike eyes that protrude from the water when the rest of the animal's bulk is submerged. Yet this species is a relatively new arrival to the African scene. The compact Pygmy hippopotamus of the rainforests, standing only 90 cm (3 ft) tall and averaging less than a tenth of its cousin's weight, is the more ancient of the two. Hippo ancestors resembling today's Pygmy hippopotamuses once inhabited the dense forests that covered all of Africa millions of years ago. As the forests gradually retreated and the grasslands advanced, these animals found it difficult to adapt to their new environment because their skin was lacking in waterproofing layers, making them vulnerable to moisture loss.

Eventually the larger species evolved a set of characteristics to cope with the new conditions. It became amphibious and nocturnal, spending the day submerged in lakes and rivers, and grazing only after sunset. Its greater size made it less susceptible to temperature changes and moisture loss, and better able to defend itself against the evolving predators of the savanna. Its head slowly changed, the eyes moving to their present periscopic site on the top of the skull, while the ears and nostrils acquired closable flaps for use underwater.

Other animals have adapted to life in lowland swamps and marshes. The sitatunga is one of the world's few antelopes specialized for a wetland existence. Its long, splayed hooves distribute weight, allowing the animal to move swiftly over the marshy ground without sinking in, and it is a fair swimmer. When alarmed it will submerge almost totally, so that only its nostrils are visible above the water. It inhabits riverine swamps, dense reedbeds and marshes.

A CONFLICT OF INTERESTS

The burgeoning human population of Central Africa puts great pressure on wildlife and its habitats. Forests are steadily felled for firewood and to make way for agriculture, while the savanna is being taken over as grazing for herds of cattle and goats. Formerly nomadic peoples, such as the Masai of Kenya and Tanzania, who used to migrate with their herds are becoming settled pastoralists, erecting fences to protect their grazing land from wild herbivores. In many parts of the region such cattle fences bar the migration routes of large mammals; this causes many to die from starvation and

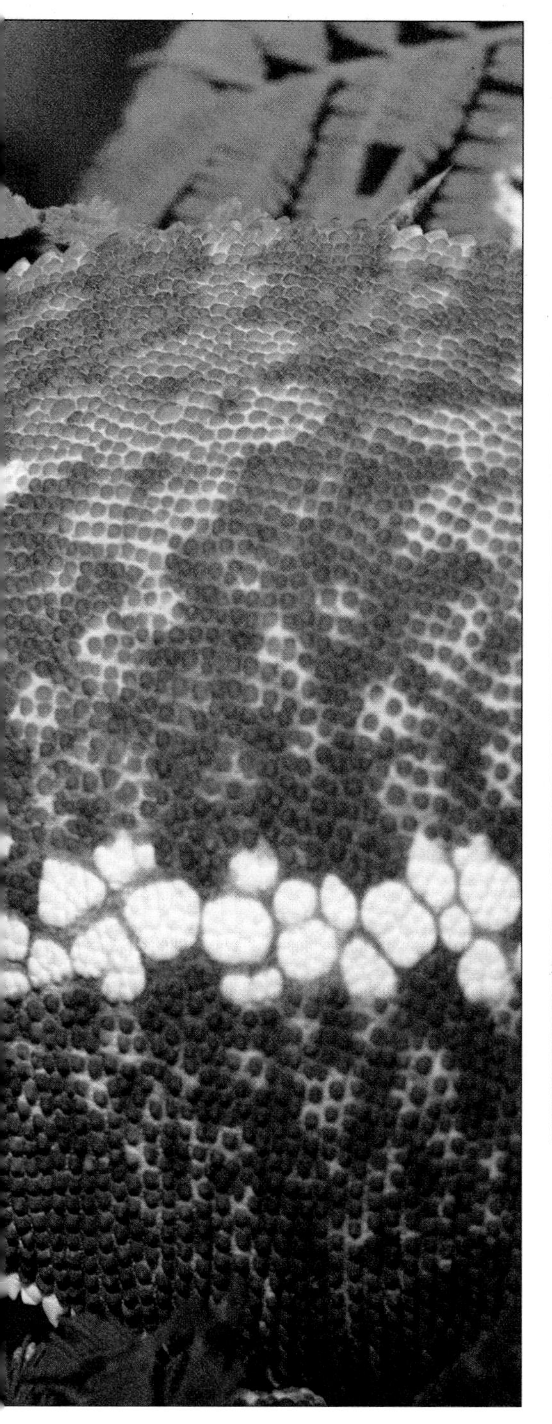

Africa's largest venomous snake (*above*) A 3.4 m (11.5 ft) Black mamba slithers up a tree in search of birds and small tree mammals. Once it has injected its venom, it retains its grip to prevent the prey from falling from the tree.

A fearsome display (*left*) The Flapnecked chameleon swells its dewlap (throat pouch), displaying bright orange lines of skin, and opens its mouth to reveal the brightly colored lining, in a spectacular gesture of threat and defiance.

thirst, and forces the wild animals into cultivated areas, where they come into conflict with people and their livestock.

Migration routes are further interrupted by the loss of corridors of natural habitat that provided grazing and water en route. Still more natural habitat is lost as overgrazing by domesticated livestock and the felling of trees for fuel causes desertification. Huge areas of natural habitat, and the animals that inhabit them, are likely to disappear if plans to log extensive tracts of tropical forest in Cameroon, Nigeria and Zaire go ahead.

Poaching and protection

Poaching is a major threat, especially to large mammals such as elephants and rhinoceroses. "Bush meat" has been a traditional mainstay of local people for centuries, and the increased demand from the rapidly growing population now threatens to outstrip the supply. A new and far more serious threat, however, comes from the poaching of ivory on a commercial scale. The traditional ivory poacher was usually a poor peasant, for whom the sale of a single elephant tusk represented more than a year's income from other endeavors. Today's poachers belong to large gangs, typified by the Somali raiders who wreak havoc in Kenya's game parks; they are often armed with automatic rifles and even rockets left over from the region's guerrilla wars. Proceeds from the sale of ivory and rhino horn do not improve the life of local people but enrich the dealers and traders. Defending the wildlife, even in national parks, can be a matter of life and death for the park rangers.

Yet Central Africa has some great conservation success stories too. The many national parks of the region protect large numbers of animals despite the strong economic pressures its countries face. For many, especially in eastern Africa, wildlife tourism is a major source of foreign currency. Educating the public has been remarkably successful. The Mountain gorilla project launched in 1978 in Rwanda is a good example. Carefully managed wildlife tourism is allowed in this forest reserve; visitors are taken to watch certain habituated gorilla groups, while leaving the rest in peace. This profits local communities, and a concerted effort with videos and lectures has helped to change the attitudes of schoolchildren and adults to their local wildlife.

"FEATHERED LOCUSTS"

The Red-billed quelea is reckoned by some to be the most numerous bird in the world. Its total population may be as great as 10 billion. Far from being disturbed by the spread of agriculture across former tracts of Central African savanna and forest, it has profited from the increased supply of grain to the extent of becoming an agricultural pest. Flocks containing thousands or even millions of these birds descend on fields of maize, millet or sorghum, stripping them completely bare.

The quelea is nomadic, following the shifting rain belts. In areas where the rains have just fallen the seeds have germinated, so in the absence of this food source the queleas turn to insects. They then migrate in the opposite direction to the rains, flying back to where it rained weeks or months earlier. Here the grasses and grain crops have set new seed, and the queleas take the chance to breed. Nestlings are fed on insects for the first week but on seeds thereafter. The young fledge quickly and fly off with the flock less than 4 weeks after the eggs were laid. The migration continues until the flock reaches an arid zone; then it turns and flies back again. By following the rains and breeding opportunistically, queleas raise three or four clutches a year, with about three young in each.

Elephants under threat

A matter of trust (*above*) An orphaned elephant calf takes food from an attendant at Nairobi National Park, Kenya. As poachers have already killed most of the large bulls, females are now being shot. Their orphaned calves will die unless rescued by rangers.

Dusted with gold Elephants in Tsavo National Park, Kenya, are coated with dried mud after rolling in a mud bath. Mud-bathing helps to protect the elephants' skin from parasites and the heat. The large surface area of their ears also helps to keep them cool.

African elephants are the largest land mammals on Earth today. A fully grown bull can stand 3.8 m (12 ft 6 in) tall and weigh over 10 tonnes. This giant mammal is strictly vegetarian, eating leaves, buds, twigs, bark, roots and occasionally fruit. It can consume up to 200 kg (440 lb) of food a day. Thanks to its size, the elephant has little to fear from any predator, apart from humans. The calves are vulnerable, of course, but the intelligent cooperation of the elephant herd and the ferocity of the females in defense of their young will defeat most predators. While an elephant's tusks are used mainly for digging up roots or prising bark from trees, they are also formidable weapons.

The terrible slaughter

Loss of suitable habitat affects elephants in Africa as it does wild animals all over the world. The biggest threat to the elephant, however, is poaching for ivory. It is estimated that 50,000 elephants are killed every year out of a total population of between 400,000 and 600,000 – a degree of loss that no slow-breeding mammal can sustain for long. Female elephants reproduce only once every two or three years, and elephant numbers are now down to just one-third of 1979 levels.

Most worrying is the switch from killing mature bulls, which have the largest tusks, to young bulls and females. It is a change that has been made out of necessity as virtually all the older males have been slaughtered. To sustain the same yield of ivory, far more elephants are now killed each year; and when females are taken, the orphaned calves are unlikely to survive. The overall damage to the herd is consequently very much worse now than it was in the past.

Matriarchal society (*above*) Families of female elephants and calves live in a clan led by an older female, the matriarch. Adult males live apart from the cows. A sire bull usually has one or two young males for company, but very old bulls are often solitary.

Controls and conservation

Past restrictions on ivory trading agreed on by the Convention on International Trade in Endangered Species (CITES) probably helped to inflate the price of ivory by making it more unobtainable. At the same time, the certification process for "legal" ivory was routinely flouted by international dealers, and over 90 percent of ivory sold came from poachers rather than legitimate sources.

Many African governments have made sincere attempts to stop the poaching, though corruption often thwarts their efforts. In Zimbabwe, in southern Africa, however, an innovative and successful scheme allows villagers to manage their own elephant populations, culling surplus animals and marketing the meat and ivory to improve the community's standard of living and to pay for elephant protection. This has almost eliminated poaching by local people.

In 1989 it was decided that the only way to save the African elephant was to prohibit the ivory trade entirely. President arap Moi of Kenya had already demonstrated his country's commitment to saving the elephant by publicly burning 12 tonnes of confiscated ivory. The countries of southern Africa protested that their successful methods of protection depended on the sale of elephant products; nevertheless 100 countries agreed to ban the trade worldwide under CITES. Almost immediately, this step was followed by a noticeable reduction in largescale poaching. A temporary exemption was given to ivory dealers in Hong Kong to allow them to clear their stock, reputed to be worth close to $150 million. There is now evidence that this action undermined the effectiveness of the ban, enabling dealers to channel illegal ivory through Hong Kong.

The great migration

One of the greatest wildlife spectacles in Africa is the annual migration of big game animals. This has evolved so that the animals can exploit the seasonal resources of the savanna grasslands. In May and early June they set out on a northward journey of 200 km (125 mi) or more to their dry-season grazing grounds, where there is permanent water and trees for shade. When the first December storms begin, the herds move southward to the central plains, where they can find fresh grazing and escape the worst excesses of the tsetse flies. During the wet season, from January to March, the herds wander in a roughly anticlockwise direction to wherever they can see, smell or hear rain falling.

The herds are dominated by wildebeest, zebras and gazelles (though in areas where rainfall is more uniformly distributed the wildebeest remain all year round). These species have different food requirements, and do not compete with each other. The zebras travel ahead of the main column, grazing and trampling the taller grasses, exposing the lower-growing plants for the wildebeest and gazelles. The migrating herds have to counter attacks from lions, hyenas, cheetahs and other predators through whose territories they pass; they also have to cope with more persistent camp followers such as young hyenas, who travel with them.

A hazardous crossing Migrating wildebeest cross a river in the Masai Mara National Park in Kenya. Many animals drown in the crush to get across, while others are killed by the predators that haunt these popular sites for ambush.

BOUNTIFUL LANDS

FROM THE SAVANNA TO THE SEA · SAFETY IN NUMBERS · DISTURBING NATURE'S BALANCE

The savanna grasslands of Southern Africa support the world's largest herds of grazing animals. Flamingoes and hippopotamuses thrive in the wetlands; the deserts of the Kalahari and Namib have their own highly specialized animals such as gazelles, ostriches, foxes and rodents. The cool nutrient-rich waters of the Benguela Current, which flows northward along Africa's southeast and southwest coasts, provide food for vast shoals of fish; these in turn support large colonies of fur seals, gannets, cormorants and penguins. Madagascar is the greatest natural treasure house of southern Africa. Its separation from the mainland in the early days of mammal evolution allowed many primitive mammals, such as the lemurs and the shrewlike tenrecs, to survive and evolve into a unique wildlife.

COUNTRIES IN THE REGION

Angola, Botswana, Comoros, Lesotho, Madagascar, Malawi, Mauritius, Mozambique, Namibia, South Africa, Swaziland, Zambia, Zimbabwe

ENDEMISM AND DIVERSITY

Diversity Low (for example mammals on Mauritius and Réunion) to very high (for example some birds on Madagascar)
Endemism High to very high

SPECIES

	Total	Threatened	Extinct†
Mammals	450	83	3
Birds	1,200*	58	32
Others	unknown	256	22

† *species extinct since 1600, for example quagga* (Equus quagga), *bluebuck* (Hippotragus leucophaeus), *Lesser Mascarene flying fox* (Pteropus subniger)
* *breeding and regular non-breeding species*

NOTABLE THREATENED ENDEMIC SPECIES

Mammals Juliana's golden mole *(Amblysomus julianae)*, Golden bamboo lemur *(Hapalemur aureus)*, indri *(Indri indri)*, Brown hyena *(Hyaena brunnea)*, Mauritian flying fox *(Pteropus niger)*, Riverine rabbit *(Bunolagus monticularis)*, Mountain zebra *(Equus zebra)*
Birds Madagascar serpent eagle *(Eutriorchis astur)*, Cape vulture *(Gyps coprotheres)*, Pink pigeon *(Nesoenas mayeri)*
Others Angonoka tortoise *(Geochelone yniphora)*, Cape platana or clawed toad *(Xenopus gilli)*, Fiery redfin *(Pseudobarbatus phlegethon)*

NOTABLE THREATENED NON-ENDEMIC SPECIES

Mammals African wild dog *(Lycaon pictus)*, Black rhinoceros *(Diceros bicornis)*, African elephant *(Loxodonta africana)*
Birds Wattled crane *(Bugeranus carunculatus)*, corncrake *(Crex crex)*, White-winged flufftail *(Sarothrura ayersi)*
Others Nile crocodile *(Crocodylus niloticus)*, Hawksbill turtle *(Eretmochelys imbricata)*

DOMESTICATED ANIMALS (originating in region)

none

FROM THE SAVANNA TO THE SEA

Huge populations of buffaloes, zebras, antelopes, elephants and giraffes inhabit the moist tropical savannas of southern Africa all year round. Among them live predators such as lions and leopards, and the scavenging hyenas and vultures. In some places on the savannas the dry season lasts for many months, and the land is arid and thorny; here the vast grazing herds are continually on the move from June to December in search of adequate grazing.

Wetlands and drylands

A large area of today's savanna was once covered by a vast lake, but in time this became filled in with windblown sand. The resulting salt pans and swamps – including the Etosha Pan of Namibia, the Makgadikgadi Pan, which stretches across the wide Kalahari Desert, and the Okavango Delta of Botswana – are rich in wetland wildlife, especially birds.

For most of the year the Etosha Pan is a parched wasteland. The seasonal floods bring the place to life, providing a breeding ground for thousands of flamingoes. The flamingoes' breeding is finely synchronized to the rains and water levels of the Etosha Pan: the female will lay her single egg only if the conditions offer a chance of success; sometimes the flamingoes go no further than nest building. The shimmering lagoon is shallow, bitter with dissolved salts, and so rich in algae that it resembles pea soup. The algae are food for the filter-feeding flamingo and its chick, but it is only a temporary feast. Within days of the departure of the flamingo and its fledgling, the pan dries out to cracked clay.

The Okavango river forms a vast inland delta in northern Botswana. More than 27,000 million liters (6,000 million gallons) of water pour into the marsh each day, maintaining its freshness and supporting an immense array of wildlife. Hippopotamuses are important residents because they feed on the water lilies, reeds and papyrus that would otherwise completely block the channels between the islands. Animals generally associated with dry savanna – lions, leopards, cheetahs and elephants – can all be found here, but it is the birds that benefit most from the bounty of the marsh. Saddlebill storks and Goliath herons stalk the shallows, and African skimmers fish the open water. Catfish and tilapia (a bony freshwater fish) throng the waterways, preyed on by a large population of African fish eagles whose calls ring out across the marsh. At night the fish fall prey to Pel's fishing owl, the only fish-eating owl in Africa. Similar groups of animals grace all of southern Africa's large rivers.

Who goes there? Gray meerkats or suricates have an almost human stance as they keep watch at the entrance to their burrow, scanning the skies for enemies such as hawks or eagles. Meerkats live in large packs, often combining to drive off predators.

The animals of the Kalahari Desert are typical of desert species throughout Africa. They include nomadic gazelles and ostriches – grazers adapted for long-distance travel and speedy escape from predators – and vultures, which can cover vast areas in search of scarce food. Here, too, are the desert rodents such as gerbils, which shelter by day in moist burrows, emerging by night to feed on seeds, and nocturnal hunters such as scorpions and desert foxes. The Namib Desert, with its dew-forming fogs and windblown dead organic matter, has its own species that have specialized for living amid moving sand dunes.

Seashore and island life

The Benguela Current, a continual upwelling of plankton-rich Antarctic waters just off the southeast and southwest coasts of South Africa, provides food for vast shoals of pilchards, maasbankers and anchovies. Twenty-three breeding colonies of South African fur seals still survive, though the species was heavily exploited – along with whales – in the 19th century, but unlike other fur seal species, these seals have no need to migrate in search of food and so remain on the breeding beaches for most of the year. Gannet, cormorant and Jackass penguin colonies depend entirely on the bounty of the Benguela Current; they are threatened by a decline in fish stocks due to overfishing by humans. Oil pollution is also a danger, threatening all wildlife along the coasts.

Madagascar, lying to the east of the African continent, offers a complete contrast to mainland Africa. The island broke away from the ancient supercontinent Gondwanaland at about the time when the first mammals were evolving. Over millions of years, its cargo of ancestral plants and animals evolved into a unique range of species, remarkable for its richness and diversity. There are five climatic areas in Madagascar, ranging from desert to tropical rainforest. More than 40 species of lemur evolved to fill the different ecological niches, and, in spite of extensive deforestation, 30 still survive today. The island boasts many other animals that are found nowhere else, such as the shrewlike insectivorous tenrecs and the fossa, a large and powerful relative of the civet.

The coelacanth is a deepwater fish that has survived almost unchanged for more than 100 million years. Its lobed fins, triple tail, hinged skull and rough, heavy scales resemble those of its ancestors, but its swimbladder is now nonfunctional. Until 1938, the fish was thought to have become extinct about the same time as the dinosaurs, 70 million years ago; but then a living specimen was trawled from the deep sea off South Africa's east coast. More coelacanths were subsequently discovered in the waters around the Comoros in the Mozambique Channel. Why the fish, whose fossilized remains are distributed worldwide, is confined today just to this area at depths of between 70 and 400 m (230 and 1,300 ft) is a mystery that will probably never be unraveled.

Walking tall The giraffe's long legs and neck enable it to browse on branches above the reach of other mammals. The blotchy pattern of the giraffe's coat breaks up its outline, making it almost invisible from a distance.

SAFETY IN NUMBERS

Much of southern Africa is a harsh, arid land. It is able to support such a wonderful diversity and quantity of wildlife because the hordes of grazing and browsing animals share the resources. Giraffes feed mainly by browsing on the low branches of trees, so they do not compete with animals feeding at ground level. Wildebeeste eat the upright stems and topi (another species of grazing antelope) only graze the low-growing mats of grass. Buffaloes are the first to move to new areas. They feed on tall grasses and open up the plains to zebras, which eat the coarse parts of grasses, exposing the tender parts for other grazers such as Thomson's gazelles; these also feed on the broadleaf plants growing beneath.

Migration and cooperation

Migration is an essential part of this delicate system of cooperation; without it the animals would destroy the frail habitat on which they depend. However, migration also makes the old and sick, as well as young animals, very vulnerable to attack. The unfit animals from the herd fall prey to the predatory lions, leopards and hyenas that follow in the herd's wake, waiting for the unwary. For this reason large numbers are an advantage: bunched together in herds, individual animals are safer from attack and secure in the knowledge that a warning will be given if a predator approaches.

Ostriches also live in social groups. Six or more females lay their eggs in a single scrape in the ground, where they are incubated by the dominant male and female. The probable reason for this unusual behavior is to protect the eggs from predators. Although it is the largest bird's egg, the ostrich's egg is small in relation to the bird's size, and one bird can cover a great many of them. This leaves fewer nests to be attacked. Brown hyenas also share the care of their infants by creating a communal "nursery den" where the cubs of several females are raised; they are brought solid food by any related adult or subadult, leaving their mothers free to join the hunt.

Molerats take cooperation much further. Colonies of 30 to 40 animals live underground in an extensive network of tunnels that can cover 12 ha (30 acres) and run for very considerable distances. The molerats are specialist feeders, eating the tubers and other underground storage organs of desert plants. Most species almost never need to come above ground, where they would be exposed to the fierce heat of the sun, the drying wind and the eyes of hungry predators.

Molerats have evolved a distinctive social system in which only one female in the colony breeds at a time. Successive litters form castes of workers that perform the various tasks associated with maintaining the colony. Small, young molerats dig the foraging tunnels and bring plant roots and tubers to the community nest; larger workers guard the colony. The very largest males do not work: their task is to mate with the reproductive female. A few large females act as "aunts" to the newborn pups. Should the colony's mother die, one of them will replace her.

Threatened with extinction The numbers of both the White rhino and the Black rhino have been drastically reduced by poaching in recent decades. The White rhino is a savanna-dwelling grazer, while the Black rhino lives in forest and thorn scrub, tearing off leaves with its long, muscular upper lip.

Fortresses of food

Towering colonies of termites are a conspicuous feature of the African landscape. Decaying vegetation imported into the colony forms the basis of these "food fortresses". Mud-built termite colonies – either domed, conical, mushroom-shaped or with pinnacles – sometimes reach 6 m (20 ft) in height and can house several million termites. At the heart of the mound lives the white, bloated queen; she can neither move nor feed herself, but only lay eggs.

The massive termite colonies provide an abundant food supply for any predator able to overcome the fierce defenses of the soldier termites. A range of specialized insectivores take advantage of the termite colonies and of ants' nests; they include the heavily armored pangolins, as well as the aardvark and aardwolf (a species of hyena), both of which are unique to Africa. All have a long slender muzzle for probing the colonies, strong claws for breaking them open, and a long, sticky tongue for lapping up the insects.

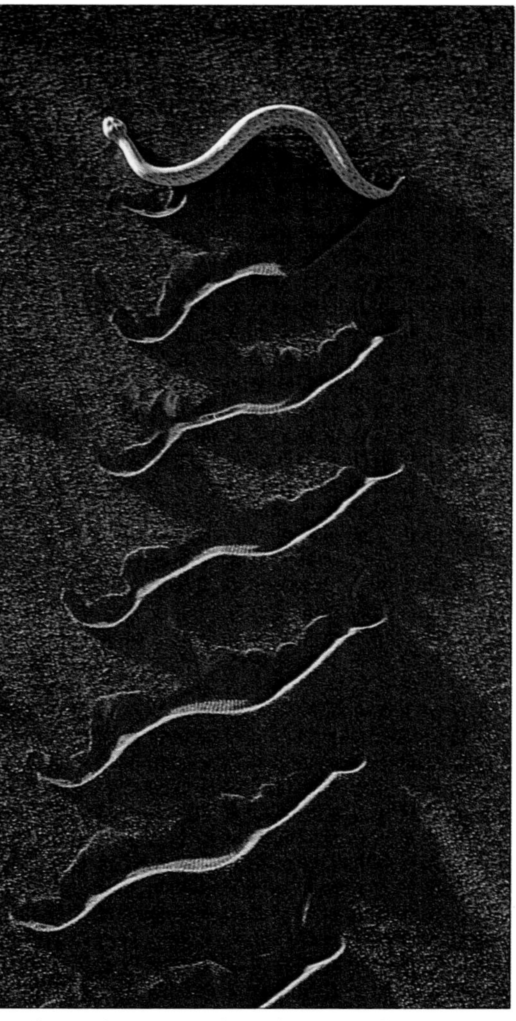

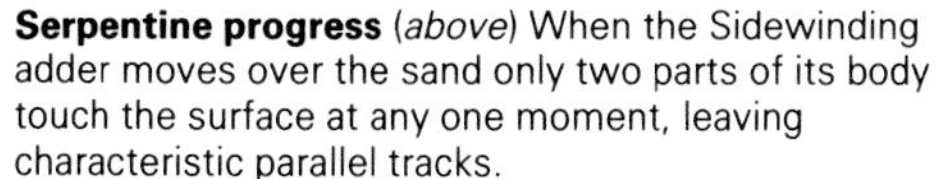

Serpentine progress (*above*) When the Sidewinding adder moves over the sand only two parts of its body touch the surface at any one moment, leaving characteristic parallel tracks.

Antelopes and gazelles (*right*) Coke's hartebeest showing the submissive posture of a yearling when facing an adult. A territorial male impala roaring during the rutting season. A male bontebok initiating butting by dropping to his knees. A male Common duiker "duiking" (diving into dense bush when disturbed). A Yellow-backed duiker showing the erect yellow rump patch that develops as the animal matures. An oribi scent marking a grass stem with its ear gland. A steenbuck scent marking with the pre-orbital gland.

LIVING IN SAND

The Namib Desert is a unique region of sand dunes moistened by coastal fogs. The ever-changing slip-faces of the dunes are an important habitat. By day the surface of the sand becomes extremely hot; but beneath the ground, at depths of 30–50 cm (12–20 in), the temperature will vary by only a few degrees over the year.

The animals of these dunes lead relatively solitary underground lives. The Golden mole almost never emerges, swimming through the loose sand in search of insects that have taken refuge in its cool depths – lacking eyesight, its supersensitive hearing enables it to catch its prey. The Dune gecko, a lizard that emerges to hunt in the cool of night, has curiously webbed feet that enable it to dig burrows in the loose sand. The Sand-diving lizard hunts by day; it cools its feet by leaning on its tail and raising diagonally opposite limbs in the air in what looks like a dance.

The Sidewinding adder moves swiftly across the surface of the soft sand. Its method of locomotion, which creates the maximum traction with the minimum of body contact, prevents it from sinking. When not on the move, it lies concealed in the sand with only its eyes showing. In an example of convergent evolution, unrelated species of sidewinding snakes have evolved in sand deserts in Asia, North America and the Sahara as well as the Namib.

Cichlids of Lake Malawi
In Lake Malawi, at the southernmost end of the Great Rift Valley, there are vast congregations of a very different kind. About four hundred species of cichlid, fish of brilliant colors and many fascinating forms, inhabit the lake, which was created by millions of years of volcanic activity along the Rift valley. The ancestors of Lake Malawi's cichlids were river fish. Faced with the problem of an uncertain habitat – always liable to dry out or change course – the female cichlids evolved the strategy of picking their eggs up after fertilization and carrying them in their mouth until they hatched.

The courtship also changed. The males developed "dummy egg" markings on the anal fin. As the female sucks up her own eggs after spawning, she also tries to collect the dummies, sucking up milt (sperm) emitted by the male, so fertilizing all the eggs in her mouth.

This reproductive strategy has been highly effective, leading to the colonization of Lake Malawi by a kaleidoscope of brilliant cichlid species. While the ancestors may have been dull-colored fish, their progeny developed beautiful coloration and elaborate courtship dances.

DISTURBING NATURE'S BALANCE

The savannas and deserts of southern Africa are among the last wild places left on Earth large enough to support great herds. The landmasses of Europe, India and North and South America all once boasted great migratory herds of plant-eating animals. These have largely been replaced by just one species: the cow. In Africa, however, cattle ranching is possible only in areas free from the tsetse fly, which transmits diseases that are lethal to animals and humans. Rinderpest kills the cattle and sleeping sickness kills humans. However, the development of pesticides to control tsetse flies has enabled cattle ranching to spread into formerly wild habitats. The new cattle fences that have been erected interrupt the migration routes of elephants and antelopes, and can cause thousands of animals to die from starvation and thirst.

In some areas of the region game reserves have been established, at first protected for sport but now preserved for conservation and tourism. The protection of southern Africa's animals has always been a problem; hunting is integral to the way of life of the indigenous people, and ivory, rhinoceros horn and other wildlife products command very high prices on the international market. Poachers have been able to earn the equivalent of a year's income for their families in a few days. Over the years poaching has become big business, with ivory being exported by a chain of dealers to Arab Gulf states, eastern Asia, western Europe and the United States. The flood of weapons from guerrilla wars in Angola, Mozambique and other parts of southern Africa has increased the power of the poachers to kill and maim.

Poaching rhinoceroses
Four out of five of the world's species of rhinoceros will probably be extinct by the end of the 20th century. The likely survivor – the White rhino – has been saved by determined conservation management in South Africa; but the future for the Black rhinoceros of southern Africa is not so good. Rhinoceros poaching is now a very high-risk business – with the horns fetching up to $30,000 each. This has had fatal repercussions for humans as well as for the animals.

Rich harvest *(above)* A fisherman of the ba Yei people clears his nets of the catfish and tilapia caught in the Okavango river, Botswana. Crocodiles and birds of prey such as African skimmers and fish eagles also profit from the river's abundant supply of fish.

Forty-three poachers from Zambia were shot in the first three months of 1987 crossing the Zambezi river to hunt Black rhinoceroses in northern Zimbabwe, and others have died since. A total ban on trade in rhinoceros horn imposed in 1976 has done nothing to help the species; numbers fell from 50,000 then to about 3,500 in 1990. Conservationists now deliberately cut off the rhinoceroses' horns so as to save the animals. Another approach is to move them from highly vulnerable areas to the safety of reserves protected by high electrified fences.

Long-suffering elephant

The elephant has suffered for more than 1,000 years from the human desire for ivory. In spite of legal protection, its numbers are falling dangerously. Some of the countries of southern Africa – Botswana, Zimbabwe, Malawi and South Africa – have succeeded in maintaining stable populations in protected national parks. In the wild, elephants would normally migrate to find new grazing. Because that is now impossible, it has become necessary to cull the elephants in order to prevent them destroying the habitat that has been set aside for them. Elephants are also killed when they persistently invade cultivated land. By selling the ivory, hides and meat, the hard-pressed national parks have gained revenue for management.

In Zimbabwe the scheme "Operation Windfall" encourages game ranching by local communities, who benefit by exploiting animal carcasses and charging big-game hunters. But Zimbabwe and other southern African countries are now suffering from the ban on ivory trading imposed by the 1989 Convention on International Trade in Endangered Species. The ban was agreed to stop the slaughter of elephants elsewhere in Africa and the laundering of poached ivory through legal channels.

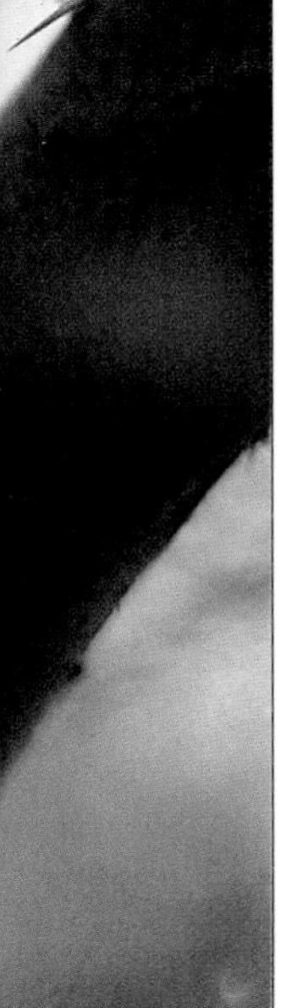

A rare return *(left)* The Mauritius kestrel was once almost extinct, but conservation programs mean that today the birds are limited only by a lack of suitable habitat.

Preserved in the deep *(below)* The coelacanth was thought to have died out some 70 million years ago until one was hauled from the deep sea off southern Africa in 1938. The coelacanth's ancestors gave rise to the first four-footed animals, the amphibians.

THE FATE OF THE ISLETS

Small islands or islets have always been especially vulnerable to harm from human activity. Sailors would break long voyages to stop, pick up water and catch fresh food. This enabled ships' rats to make their way ashore, where they would devastate populations of island birds. In the South Atlantic the Tristan da Cunha flightless rail is now found only on Inaccessible Island, some distance from the main island. When the native wildlife of such islands dwindled, sailors often released goats to assure themselves of food on return journeys. These voracious herbivores have destroyed the natural vegetation of countless islands.

Then came the settlers. The original vegetation of St Helena – more than 1,800 km (1,100 mi) to the west of the southern African coast – was quickly destroyed by deforestation and livestock grazing. Mauritius in the Indian Ocean suffered in the same way. Before European colonization began, forests of ebony covered the mountains of Mauritius. Now over 40 percent of the island is covered in sugarcane fields. The island was the only home of the dodo, a large, bulky bird that had lost the ability to fly as it had no natural predators. Most of the dodos were killed by European sailors in the 17th century; the rest became extinct when pigs and monkeys were brought to the island. The Mauritius kestrel, which was very common in the late 19th century, is one of the world's most endangered birds; its survival depends on a program of captive breeding and on the wise management of its remaining natural habitat.

The lemurs of Madagascar

The lemurs symbolize the unique nature of Madagascar's rich wildlife. They are primates – members of the same order of mammals to which monkeys, apes and humans also belong. The common ancestor of all these species was a lemur, probably sharing similarities with the presentday Ring-tailed lemur. Sometime in their evolutionary history, the modern prosimians, which include the lemurs, became distinct from the evolutionary line that led to the monkeys and apes.

All 30 species (divided into 5 families) of presentday lemurs are found on Madagascar. They retain many primitive characteristics, and in this sense are closer to the ancestral primates than are the anthropoids. It is not clear how they arrived on the island: one theory is that they colonized the newly formed land 50 million years ago by clinging to rafts of vegetation that were swept out to sea from the river deltas of eastern Africa. Once on Madagascar, where they were protected from competition with other primates, they evolved to fill a wide range of ecological niches.

An array of species

At one time there were probably as many as 40 species, including the *Megaladapis,* which had a skull measuring 30 cm (12 in) long and a body as large as that of an orangutan. Giant lemurs like this are now extinct, but there is evidence to show that they were still around when people first arrived on Madagascar about 1,500 years ago. They were subsequently hunted for food by the local inhabitants. Fire and competition with domestic animals may also have accelerated their disappearance.

The smallest of the lemurs, which belong to the family Cheirogaleidae, are the dwarf and mouse lemurs, the adult weight of some being as little as 55 g (10 oz). They are active mainly at night, their large eyes having a reflecting layer at the back that enables them to make the best use of the small amount of light available. Most of them are omnivorous, eating insects as well as small fruits, but some are specialist feeders. The Fork-crowned dwarf lemur, for example, eats the resinous gum that exudes from injured tropical trees: it has developed specially formed teeth to scrape away the gum, which it is able to digest because of the particular bacteria in its gut.

Most of Madagascar's lemurs belong to the family Lemuridae. They are roughly cat-sized and are agile tree climbers, using all four legs to travel and often supporting themselves on thin branches and twigs. They feed mainly on fruit and leaves, occasionally taking small animals. The Gentle lemur is a bamboo-eating specialist, and during the dry season the Mongoose lemur feeds almost exclusively on nectar, which it licks from flowers.

The sifakas and the indri (family Indriidae) are larger than other lemurs and use a different method of moving through the trees: their back legs are long, and they leap from one upright branch or trunk to the next. They feed on fruit and leaves and, unlike other lemurs, are active mainly by day. The gut of the indri has a large sidebranch where the plant food is able to ferment.

The last family of Madagascan lemurs, the Daubentonidae, consists of a single species, the aye-aye, the most unusual lemur of all. It is a nocturnal animal, with large eyes and sensitive ears that are capable of detecting a grub moving

A versatile mover The sifaka is rare among lemurs in being almost as agile on the ground as in the trees. Its long arms and grasping hands are well suited to leaping tremendous distances from tree to tree, but it can also bound rapidly over the ground on two legs.

beneath the bark of a tree. Once it has marked its prey, it gnaws a hole in the bark with its rodentlike canine teeth and inserts its long, wire-thin middle finger. It uses this to mash up the insect, and then extract it and place it in its mouth.

The aye-aye is an endangered species, and time may be running out for all Madagascar's others lemurs. The island's forests are being felled at a rapid rate for shifting agriculture and for fuel and timber, and this has brought many of these primitive mammals close to extinction by destroying their habitat and food sources and disturbing their breeding cycles. They will only be saved by stringently enforced protection and changes in agricultural practice.

A pair of Verreaux's sifaka (*left*) rest in the shade. Groups of Verreaux's sifaka number from 3 to 12 individuals, often with more than one breeding adult of each sex. Females can reproduce at about three years of age. The single newborn infants are almost hairless and blackskinned. They cling to the mother's abdomen at first, before transferring to her back.

Madagascan lemurs (*below*) The 30 species have a variety of characteristics for identifying and being identified. There are often marked sexual differences in coat color as, for example, in the Black lemur and the Mongoose lemur, with the males shown here above the females. Only the male White-fronted lemur lives up to its name: the female has gray fur instead of white. Because they live in dense forests, lemurs communicate by smell, and have scent glands that they use to mark branches or even one another. A Gray gentle lemur marks a branch with the scent glands on its wrist, while a Brown lemur marks its tail. A Ruffled lemur engages in anogenital marking. Unlike most lemurs, which are sociable animals, the Sportive lemur spends most of its time alone.

Gannet city

Gannets are large seabirds that breed from the Arctic circle through the tropics to Antarctica. They are sociable, noisy birds that nest in huge colonies on near-level ground at the edges of cliffs. The most highly prized nest sites are the ledges of the cliffs below, and these sites are soon occupied. The colonies usually contain hundreds of thousands of pairs of gannets. It is thought that the noise of these colonies helps to stimulate the birds to breed and lay their eggs at the same time, thus reducing the constant danger from predators.

Gannets pair for life. In such crowded conditions, pairs need a strong sense of identity; this is reinforced by certain displays, notably the sky-pointing display in which both birds point their bills skyward and raise their wings – a display also used by boobies and albatrosses. When threatened by an intruder, "site-ownership" display is used: the gannet sweeps its head up and down and shakes it from side to side, calling harshly. Gannets lay single eggs. Although the female will retrieve pieces of nest material that have fallen down and return them to their place, she will fail to recognize her egg if it rolls away – even if it is within reach of her bill it will simply be ignored.

Cape gannets – a species adapted to high temperatures – on Malgas Island, South Africa. Gannets are quarrelsome birds; the nests of this crowded colony are spaced so that each is just out of pecking distance of its neighbor.

LAND OF THE TIGER

FROM THE MOUNTAINS TO THE SEA · OF HEIGHTS, HEAT AND DUST · FEARED AND REVERED

The Indian subcontinent has a rich mix of wildlife, with about 500 species of mammal, 1,300 bird species and several hundred species of amphibian and reptile. This diversity is a result of the region's varied topography, altitude and climate. The subcontinent's turbulent geological history has further enriched the wildlife with animals more typical of other regions. Many Indian mammals are rarely seen, as most are nocturnal forest dwellers; but this is compensated for by the birds, which are conspicuously active by day. Wildlife ranges from the Snow leopard of the remote Himalayas to the crocodiles that inhabit the Sundarbans mangrove swamps of the Ganges delta; from the tiger and rhinoceros of the dense jungles to the bustards and gazelles of the Thar Desert of northwestern India and eastern Pakistan.

COUNTRIES IN THE REGION

Bangladesh , Bhutan, India, Maldives, Nepal, Pakistan, Sri Lanka

ENDEMISM AND DIVERSITY

Diversity High
Endemism High

SPECIES

	Total	Threatened	Extinct†
Mammals	400	42	0
Birds	1,400*	81	3
Others	unknown	52	0

† species extinct since 1600
** breeding and regular non-breeding species*

NOTABLE THREATENED ENDEMIC SPECIES

Mammals Lion-tailed macaque *(Macaca silenus)*, Hispid hare *(Caprolagus hispidus)*, Indus river dolphin *(Platanista minor)*, Indian rhinoceros *(Rhinoceros unicornis)*, Pygmy hog *(Sus salvanius)*, Swamp deer *(Cervus duvauceli)*
Birds Lesser florican *(Sypheotides indica)*, Jerdon's courser *(Cursorius bitorquatus)*, Western tragopan *(Tragopan melanocephalus)*, Forest owlet *(Athene blewitii)*, Great Indian bustard *(Choriotis nigriceps)*
Others gharial *(Gavialis gangeticus)*, Malabar tree toad *(Pedostibes kempi)*, Green labeo *(Labeo fisheri)*, Relict Himalayan dragonfly *(Epiophlebia laidlawi)*, Scarce red forester *(Lethe distans)*

NOTABLE THREATENED NON-ENDEMIC SPECIES

Mammals dhole *(Cuon alpinus)*, Snow leopard *(Panthera uncia)*, tiger *(Panthera tigris)*, Asian elephant *(Elephas maximus)*, Musk deer *(Moschus moschiferus)*
Birds Greater adjutant *(Leptoptilos dubius)*, Green peafowl *(Pavo muticus)*, Siberian crane *(Grus leucogeranus)*
Others Desert monitor *(Varanus griseus)*, Kaiser-I-Hind butterfly *(Teinopalpus imperialis)*

DOMESTICATED ANIMALS (originating in region)

mithan *(Bos 'frontalis')*, Water buffalo *(Bubalus 'bubalus')*, yak *(Bos 'grunniens')*, Asiatic elephant *(Elephas maximus)*, chicken *(Gallus gallus)*

FROM THE MOUNTAINS TO THE SEA

The region's wildlife is a mix of species of diverse origins. India was once part of the ancient supercontinent, Gondwanaland, which also included the landmass of present-day Africa. Consequently, some of its animals, such as the Asiatic lion, the bustards and the sandgrouse, have affinities with African animals. After the Indian subcontinent collided with the Eurasian mainland some 40 million years ago, Indochinese species such as the tiny mouse deer or chevrotains – which are intermediate in form between pigs and deer – and lorises (short-tailed relatives of bush babies) have colonized from the east. Some Chinese species, including the small Lesser or Red panda, the goral and the serow (both goat antelopes) and many species of babblers (from a large family of songbirds) have extended their range into the Himalayas. Other animals have advanced from the northwest to colonize new habitats.

Many distinctive species live in the mountains and forests of the Himalayas. The Snow leopard, one of the world's rarest and most elusive animals, still survives in the more remote areas. Several species of wild sheep and goat also live here, including the nayan, the world's largest sheep. The ungulates (hoofed mammals) include the primitive little Himalayan Musk deer; the males are easily distinguished by their protruding canine teeth. The Brown bear inhabits the open ground of the alpine meadows above the treeline, while the Himalayan black bear takes advantage of the protection offered by the forests lower down the slopes. The birds of the Himalayas are very varied; they include gorgeously colored pheasants with their striking iridescent plumage and long tail feathers, and noisy gregarious babblers that flitter through the forest canopy and forage in the undergrowth.

Plains, desert and plateau animals

To the south of the Himalayas lie the great floodplains of the Ganges, Indus and Brahmaputra rivers where some of the most important reserves for the Indian tiger are located. The tigers prey on the numerous deer of the plains, especially the Spotted or Axis deer. The subcontinent's two largest animals, the Asian elephant and the Indian rhinoceros, also survive in reserves, much of their natural habitat having been lost to cultivation by farmers. In western India the dry rolling hill country of the Gir National Park provides the last refuge of the Asiatic lion.

The region's wetlands – in particular the sanctuary at Keoladeo in north central India – provide a wonderful spectacle of bird life. In wet years many thousands of large waterbirds breed in the sanctuary, and tens of thousands of ducks overwinter here, along with the extremely rare Siberian crane.

West of the Gangetic plain lies the Thar or Great Indian Desert where blackbuck, Wild asses and graceful Indian gazelles were once abundant. Now their numbers are much reduced. Notable among the desert birds are the endangered Great Indian bustard and various species of sandgrouse. Monitor lizards – the largest and heaviest lizards in the world – are powerful predators that prey on small desert mammals, birds, other lizards and their eggs, and they also take carrion. They have a reputation for being very ferocious hunters, often swallowing their prey whole.

South of the plains, Asian elephants and gaur – large wild oxen – still roam wild in the less accessible parts of the dry Deccan plateau. The hill forests of the Western Ghats and Nilgiri Hills on India's west coast shelter a number of species that are endemic to the region; these include the endangered Lion-tailed macaque monkey, the Nilgiri tahr (a goat) and the Nilgiri langur (a leaf monkey). Sri Lanka also has many endemic species in its tropical forests, mangrove swamps and floodplains.

Bird of the high mountains (*above*) A lammergeier or Bearded vulture, which in flight resembles a huge falcon, soars in front of Mount Thamserku in Nepal. This vulture covers a vast area in search of carrion; its long, narrow wings enable it to glide for long distances. It will sometimes descend to scavenge near towns and villages.

The King cobra (*left*), the world's longest poisonous snake, grows to a length of 5.5 m (18 ft) or more. When threatened or curious, the snake can rear to a height of over 1.2 m (4 ft). It lives in the swamps and forests of northern India, and is unusual among snakes in making a nest for its eggs.

Marine and island life

The 2,500 or so scattered coral islands and reefs of the Maldives have just two native mammal species – both fruit bats – and only five species of reptile, all of them introduced. Some 120 species of migrant birds pause here on their journeys, and a number of seabirds, such as the graceful White-tailed tropic bird and the Lesser frigatebird, breed on the islands. The Andaman and Nicobar Islands in the Bay of Bengal are still largely covered in lush tropical forests that support a number of reptile species and 14 endemic species of bird. One of these is the Nicobar megapode, a terrestrial bird about the size of a domestic hen. The eggs are incubated under mounds of heat-producing decomposing humus, which are covered over with sand.

The shallow waters of the Indian seas support a rich diversity of marine life. The once abundant Sea cow or dugong is now seen only rarely, its population having been greatly reduced by overfishing and accidental trapping in fishing nets. Numerous sea turtles nest on the beaches, including the Green turtle and the hawksbill. The Maldives and, farther north, Lakshadweep, are famous for their coral reefs and associated rich variety of fish and invertebrates.

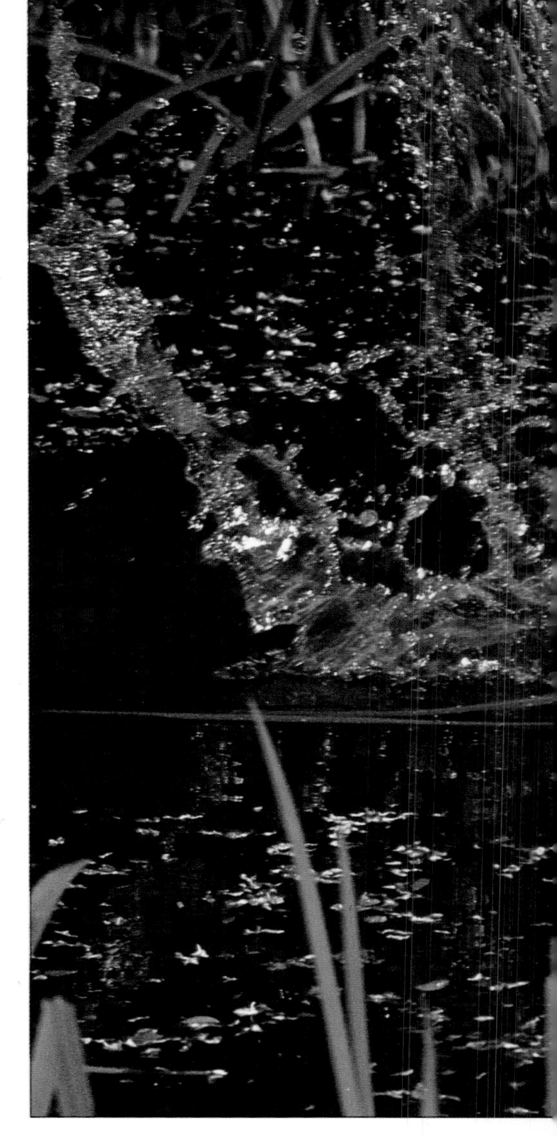

OF HEIGHTS, HEAT AND DUST

The Himalayas are home to one of the world's highest-living mammals, the wild yak. Its long, dense, woolly fur, reaching almost to the ground, provides good insulation. Like some other high-altitude mammals, the yak has developed large lungs that help it obtain sufficient oxygen in the thin mountain air. Another mammal supremely adapted to the peaks is the magnificent Snow leopard; it, too, has a thick coat to shield it from the cold, and fur on the soles of its paws to ensure a firm grip on the ice. Perhaps the most skillful climbers of the high Himalayas are the wild sheep and goats, such as the Blue sheep and the Himalayan tahr. They scale the precipitous slopes with ease in search of the sparse vegetation. Severe competition from domestic stock and overhunting, however, has led to their disappearance from many areas.

Several mountain animals, such as the Snow pigeon, are altitudinal migrants that move to the lower slopes in severe winter weather. Some animals, such as the Himalayan marmot (a member of the squirrel family) survive the winter by hibernating. They excavate deep, snug burrows in which to escape the cold. The large family of adult male and female marmots and their young sleep huddled together in their communal den.

Specialized forms of butterflies, grasshoppers, caddisflies, beetles, springtails, mayflies and stoneflies live at high altitude above the treeline. Some even thrive at a height of 6,990 m (23,000 ft). While many feed on plants such as lichens and mosses, the great majority are scavengers sustained by dead insects and spiders, fungal spores and seeds that have been lifted from the distant Indian plains by hot air currents. On bright summer days numerous beetles and other insects can be seen hunting for such debris on the Himalayan snowfields.

In summer these insects are exposed to dangerously intense ultraviolet radiation, and many have developed dark pigmentation for protection. Strong winds blowing over the high peaks make flight very difficult, so most insects here are flightless. In winter the insects hibernate, relying on the snow cover for insulation against the bitter cold. The snow and ice are crucial to their survival, providing them with all the moisture they need.

Desert specialists

At the other extreme, many animals have adapted to the exposure of hot deserts. The elegant Indian gazelle is tolerant of extreme heat; its summer coat of very glossy hair is thought to reflect the sun's rays. The gazelle can survive for a week without drinking, and is capable of traveling up to 48 km (30 mi) a day to reach water. It grazes on desert plants at first light and late in the evening when dew forms on the leaves, thus deriving the maximum moisture from them. In such an open habitat the gazelle must be constantly vigilant if it is to evade predators; it has acute vision and hearing, and the ability to run at the exceptional speed of 72 kph (45 mph) over a long distance if danger threatens.

Another high-speed desert inhabitant is the Great Indian bustard, a large ground-dwelling bird endemic to the region. The bustard's long, strong legs enable it to run at great speed, but if it senses danger it will squat motionless behind a thorn bush, in an attempt to remain unseen by its enemies. The bird's plumage of a deep sandy buff color irregularly patterned with black acts as effective camouflage in the desert.

Predators of land and water

The Indian subcontinent boasts five species of large cat: the Asiatic lion, the leopard, the Snow leopard, the Clouded leopard and the tiger. Of these, the leopard is the most successful and adaptable. Unlike the tiger, which is mainly nocturnal and restricted to forest habitats, the leopard is equally at home in scrub and open country and has also learned to live near human habitation. However although it ranges throughout the region,

The elusive Snow leopard (*above*) on a rare fishing expedition. This beautiful and solitary creature needs a large territory in the barren mountains of India and Nepal in order to find enough prey.

The smaller bears (*right*) The shaggy Sloth bear uses its long claws and flexible snout to forage for termites and grubs. The Asian black bear is mainly herbivorous, but occasionally takes carrion.

PHEASANTS

Pheasants are among the most brightly colored birds in the world. No fewer than 18 species are found on the Indian subcontinent. The males have the most brilliant plumage with elaborate patterns and long ornate tails; females are generally drab and patterned cryptically for concealment. During the courtship display the male Blue peafowl raises his magnificent long train to form a great arching fan of metallic bronze, green and blue. The male Himalayan monal – the national bird of Nepal – has even more dazzling plumage with nine iridescent colors that glisten in the sunlight. Iridescence also enhances the plumage of the Red jungle fowl – perhaps the most numerous bird in the world – believed to be the ancestor of the domestic chicken.

The majority of pheasants are forest birds. For most of the time such spectacular plumage is a hindrance rather than an advantage; the males tend to skulk in the thickets in order to conceal their brilliant colors from predators. Pheasants have strong feet and legs, often preferring to run away rather than fly. Hunting and forest destruction have reduced their numbers; seven species are considered threatened, including some of the most spectacular, such as the Western tragopan.

Agile goat antelopes (*left*) These species inhabit rocky, steep terrain. The goral rarely moves far from the safety of the rocks. The Himalayan tahr prefers dense forest; the females move up to mountain pastures in the summer. The urial is long-legged and swift, and can survive in desert habitats.

it is very rarely seen. Its tawny coat, which is marked with black rosettes, provides excellent camouflage. Another explanation for its success is its varied diet; the leopard preys on a range of animals that includes deer, monkeys, large rodents and domestic animals. It is an adept treeclimber and can seize prey by jumping from overhead branches or leaping from the ground.

Another powerful predator of the region is the world's largest crocodilian, the rare and endangered gharial. Males of the species sometimes exceed 7 m (23 ft) in length. Gharials live in river pools and are well adapted for aquatic life, having streamlined bodies, webbed hind feet and a powerful tail for propulsion. They feed almost entirely on fish, snatching them in their long narrow jaws, which have more than 100 pointed teeth. The gharial is at serious risk of extinction, owing to loss of its river habitat from dams, human and industrial pollution, disturbance by increasing river traffic and loss of riverbank nesting areas.

FEARED AND REVERED

Animals are intimately associated with culture and religion on the Indian subcontinent. Representations of animals are common in art, folklore and tribal dances, and ancient traditions of animal worship remain strong. Both Hindus and Buddhists regard nature as highly beneficial to humanity; strict adherents of these faiths will not kill animals of any kind. The current abundance of wildlife in Bhutan, India, Nepal and Sri Lanka can be largely attributed to these countries' dominant religions. By contrast, the majority of people in Bangladesh and Pakistan are Muslims, who are keen hunters.

Sacred and honored species

Many animals in the region are widely regarded as sacred; the cow, for example, is sacred to the Hindus. Throughout the subcontinent millions of cows are allowed to consume vast quantities of vegetation; they are even free to help themselves to vegetables from stalls in city streets. Many cows are diseased, but for religious reasons they cannot be culled. Eating beef is regarded as unclean by most Hindus, so this valuable source of protein is wasted in a region where so many people are undernourished. Another sacred animal is the Sarus crane, which pairs for life; the couple's faithfulness to each other is legendary in India.

Large flocks of Oriental white-backed vultures and Long-billed vultures are often seen circling over cities searching for carrion. In some places they eat human corpses: people belonging to the Parsee religious sect leave their dead on

The people's friend (*above*) The Asian elephant holds a special place in Indian culture and history. However, its traditional migration routes are being disrupted by agricultural and urban growth.

Sacred monkey (*right*) The Hanuman langur is a symbol of self-sacrifice for a friend in India. Langurs are often fed by temple worshippers, and are even allowed to steal food.

The Bar-headed goose *(above)* breeds on lake islands where predators cannot reach it. The lakes of China's mountain plateaus are a haven for ducks, geese and other water birds. Like many of China's waterfowl, this goose migrates in winter.

jungle fowl (probably the ancestor of the domestic chicken) and, on the island of Taiwan, the endangered Swinhoe's and Mikado pheasants.

Surviving the ice age

Although China escaped any widespread glaciation during the most recent ice age, the fall in temperature nevertheless killed off many species in the north; they were able to survive only farther south or at lower altitudes, where it was warmer. Toward the end of the ice age, cold-loving species reestablished themselves in the north and on mountainous terrain. It was only because these animals were able to move location as the conditions changed that they survived. Many were isolated for long periods because of the unfavorable climate, and as a result they gradually evolved to form new species; for example, there are moles, shrew-moles and voles that are found only in China (endemic species).

China is also rich in primates. It has several endemic species: three species of the Snub-nosed monkey, and the Tibetan macaque. The Francois' leaf monkey, the Rhesus macaque, three species of gibbon and the Slow loris also live in China.

Animals of the high plateaus

As much as a third of China is covered by mountains, in places rising to more than 8,000 m (26,200 ft) above sea level. Some of the region's greatest gatherings of large mammals are to be found on the upland plains and high plateaus of Qinghai, Xinjiang and Tibet or, as it is called by the Chinese, Xizang. Above 4,000 m (13,000 ft) herds of Tibetan asses and Tibetan antelopes survive, along with Tibetan gazelles, wild yaks and argali sheep – and higher still the bharal or Blue sheep can sometimes be found. Bactrian camels were thought to have become extinct only 50 or so years ago. However, it is now known that there are still some living in the wild in China. Their skulls are occasionally found by travelers in the region, as well as in Mongolia.

All these hoofed animals are well-adapted for life on the arid, poor-quality grassland of the plateaus. They can travel long distances in search of grazing, and their digestive systems are adapted to cope with a poor quality diet. With so many grazing animals to prey on, large carnivores – wolves, Snow leopards and Brown bears – also flourished, though some are now endangered. Smaller predators such as the lynx, Red and Tibetan foxes, and the Steppe cat are also found in these high, dry areas.

The remote lakes of these mountain plateaus are home to thousands of birds during the breeding season, including the Bar-headed goose, Ruddy shelduck, Great cormorant and rare Black-necked crane. Most of the birds, apart from the plant-eating geese and ducks, feed on the carp that abound in the waters of these lakes. Bar-headed geese, Ruddy shelduck and many other birds migrate to the river valleys of Nepal and northeast India.

On the run Wild asses can reach speeds of up to 70 kph (43 mph). They cannot survive for more than a few days without water, and their numbers have fallen as humans have appropriated their water supplies for the irrigation of farmland.

DESERT, RIVER AND FOREST SPECIALISTS

The animals that inhabit the sagebrush scrub and desert of China's high plains and plateaus have to subsist in a seemingly hostile environment. The small mammals include typical desert forms such as the burrowing rodents of the jerboa family and the jird or gerbil subfamily. The jerboas, which are widespread across northern Africa and central Asia, are particularly well represented in China: 7 of the 10 genera and 10 of the 29 species occur in the arid, often cold deserts of the north and west.

Jumping jerboas

The jerboas are nocturnal animals that are particularly well adapted for life in the desert. Their hind limbs are greatly elongated – at least four times longer than the forelegs – enabling them to travel fast and far by jumping and hopping like kangaroos over the sparsely vegetated terrain. This allows the animals to search for food over a very wide area, which is necessary in this unproductive habitat. It also enables them to escape from predators. Jerboas obtain all the moisture they need from their food – insects, bulbs, seeds and roots – because their kidneys produce extremely concentrated urine.

To avoid the extremes of temperature on the surface, jerboas dig burrows in the sand. These are often completely cut off from the outside world – plugged just below the surface and reopened each time the animal emerges. Thus a surface of windblown sand in the western desert of China, marked only by ripples, may hide beneath it a whole family of Feather-footed jerboas, Long-eared jerboas or Pygmy jerboas. Those jerboas that live in sandy soil have tufts of hair under the toes of their hind feet that provide the animals with extra support on loose sand; they also grow bristles in the external ear opening to prevent sand from getting in. In winter jerboas hibernate to avoid both severe weather and food scarcity.

Relics from the past

Among the curiosities of Chinese wildlife are a number of ancient species that live nowhere else. One of these is the Chinese paddlefish which lives in the Chang river. It is an extraordinary filter-feeding fish with a long snout. Its mouth is saclike and gapes open as the fish swims, scooping up crustaceans and other small invertebrates. The purpose of its paddle-shaped upper jaw is not known. Various functions have been suggested: it may be an electrical sensor for detecting swarms of plankton in murky water, a stabilizer to balance the head against the rush of water into its huge mouth, or a shovel for digging in the mud.

Inhabitants of the muddy rivers of northern China include the Chinese river dolphin or baiji, and the Chinese alligator. The baiji is the rarest of the river dolphins and is considered to be a primitive form – its neck vertebrae are not fused, so it can still turn its head. Its eyes are very reduced, and it has a large melon (forehead), thought to be involved in sound transmission; while hunting it emits a series of clicks, finding its fish prey in the muddy water by echolocation. Farther west and south lives another ancient survivor, the Giant salamander, closely related to the Japanese giant salamander. It is the second largest amphibian in the world and weighs up to 45 kg (100 lb).

The best known ancient species is undoubtedly the Giant panda. Pandas feed almost exclusively on bamboo leaves and shoots. They have a modified wrist bone that acts like a sixth digit – an opposable "thumb" – when stripping the leaves from bamboo stems. Because of the generally low energy content of bamboo, pandas have to feed for up to 14 hours a day. They may eat as much as 38 kg (84 lb) of bamboo shoots in one day at certain times of the year – equivalent to 45 percent of the animal's body weight. Much of this food is not digested, and pandas produce copious quantities of feces. They conserve their energy by moving slowly as they roam the high altitude forests in the snow and frost. The closely related northern bears feed on high-energy food such as fruit and meat during the summer, and then sleep through the winter on their accumulated reserves of fat; the panda's diet, by contrast, is so low in energy content that

Species of deer (*right*) Reeve's muntjac or "barking deer" and the Tufted deer are small, primitive deer with tusklike upper canine teeth. When excited, muntjacs bark rather like dogs. The Sika deer has a spotted coat: good camouflage in the dappled light and shade of the forest. Père David's deer has been extinct in the wild for centuries, but captive-bred individuals have recently been reintroduced to China. The Chinese water deer has no antlers, only tusks.

Reeve's muntjac
Muntiacus reevesi

Toad with a secret weapon (*below*) The Fire-bellied toad has an unusual defense when attacked: it arches its back and turns its hands and feet upward, displaying its brilliant red belly. This is a shock tactic and a warning that the toad's skin is toxic.

the animal is unable to put on sufficient fat to lie up during the winter.

Newborn pandas are tiny, weighing only about 100 grams (3 to 4 oz). They grow quickly, however, and reach about 35 kg (77 lb) by the end of their first year. Pandas frequently give birth to twins, but the mother almost always rears only one. As with kangaroos, which also produce very small offspring, the baby panda has not cost the mother much in terms of energy when it is born: species with larger offspring have wasted more energy if one of their young dies.

Goat antelopes *(right)* The argali or Wild sheep is an animal of the high Tibetan plateaus. The takin lives in mountain bamboo forests. The ibex browses in the mountain forests in winter, but moves above the tree line to graze in summer.

THE GIANT CLAM – A UNIQUE PARTNERSHIP

The coral reefs that occur along the coasts of southern China and, farther south, around the islands of the Nansha archipelago, once supported many giant clams. These are unique among bivalved mollusks in having a symbiotic (mutually beneficial) association with a species of green alga that lives in the flesh of the animal. The clam keeps its valves slightly open to expose the algae to the light. The algae photosynthesize and pass a proportion of the carbohydrates they synthesize to the clam. This may perhaps explain why the clams grow so large: up to 137 cm (54 in) in shell length. The three largest species have all been seriously overexploited by fishermen and are on the verge of extinction.

In recent years the farming of giant clams has become popular, especially in Taiwan. Young clams are hatched and grown in tanks. When they are big enough to survive predation they are "planted" out on the reef to grow unhindered until they are harvested.

Largest living mollusk Some species of giant clam are now close to extinction.

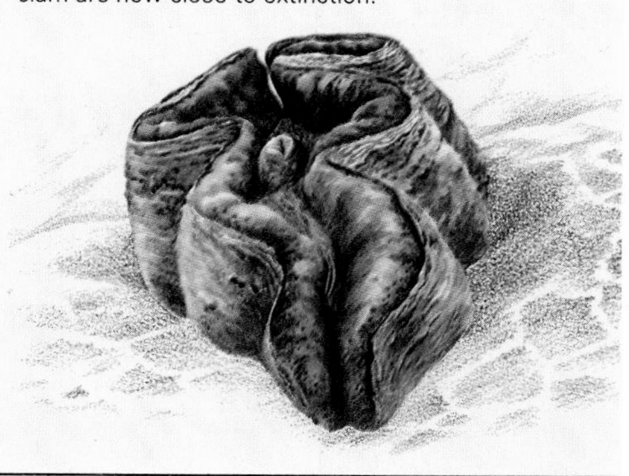

NO ROOM FOR WILDLIFE?

Giant pandas once ranged from Burma through most of southern China to Vietnam. Since they first appeared in the fossil record 3 million years ago they seem to have remained widely distributed until humans arrived on the scene. Although some of their subsequent reduction in range is due to changes in the climate and hence in the vegetation, most of it can be attributed to the destruction of their forest habitat: over the past 15 years there has been a 50 percent reduction in suitable habitat for pandas. This has been due partly to logging for timber, but it is largely the result of population growth.

Large areas of natural habitat, especially forests and wetlands, were destroyed during the Chinese Communists' "Great Leap Forward" of 1958–61, when peasants were encouraged to smelt iron in their backyards. Millions of trees were felled to supply their furnaces, and agriculture was neglected. A terrible famine followed, in which hundreds of thousands of people died. In their desperate hunger, people slaughtered countless wild animals. In 1966 there was a change in land use, with every available patch of land, even woodland and ground unsuitable for crops, being plowed so that farmers could grow grain; they cleared forests higher and higher into the hills to make fields.

Animals suffered serious declines in range as a result: rhinoceroses and wild elephants once thrived in China, but the former is now extinct here, while the elephant survives only in small areas of forest in Yunnan.

Animals for trade and table

Many species have been brought close to extinction, or actually exterminated, as a result of hunting for skin, meat and parts of the body important in oriental medicine. Roughly 3 million birds were reported to have been legally taken for food markets and the pet trade in 1985; many more will have been captured or killed illegally. It is still possible to purchase illegally traded capes made from Golden monkey fur and the skins of officially protected rare and endangered species such as the Clouded leopard and the Snow leopard. A public government campaign to protect the Giant panda – with the threat of the death penalty for trading in skins – has failed to stop poaching, and pandas are also trapped accidentally in snares set for Musk deer.

One of the main victims of hunting has been the tiger. Tigers were once abundant and widespread in southern China. As late as 1950 there were probably more than 4,000 individuals of the South China subspecies, the only one endemic to China. Hunting, encouraged by government bounties for skins, was a major cause of the rapid decline of the South China tiger over the following years. The Siberian tiger, together with the Giant panda and the Golden monkey, were declared protected animals in 1959. But in the same year the South China tiger, leopards, wolves and bears were declared pests, and hunters were called upon to eradicate them as quickly as possible.

Between 1951 and 1979 the mean number of South China tiger skins taken fell from 400 to just 5 per year. The subspecies was declared a protected animal in 1977, but by then it was almost too late. The population had declined to probably no more than 200 individuals living in scattered areas of fragmented forest habitat. The law was not enforced, and poaching has continued. Many people still regard the tiger as a pest, and there is a flourishing trade in the animal's organs – bones, testes and eyeballs – for traditional Chinese medicine. By 1986 there were estimated to be only 50 to 80 South China tigers left.

Resting in a tree *(above)* The Red panda is a shy nocturnal animal of mountain forests and bamboo thickets. Its feet have hairy soles to help it grip wet branches, and its bushy tail may be up to three-quarters the length of its body.

Living at high altitude *(right)* A Tibetan and his yak cross a mountain pass at 6,000 m (19,500 ft). The fermentation of the bacteria in the yak's stomach generates heat and provides a personal central heating system. Its dense fur provides insulation.

Dance of a threatened species *(below)* Fewer than 500 Black-necked cranes survive in the wild. Their ceremonial dancing often precedes mating, and helps to cement the bond between the newly formed pairs. Sometimes a whole flock will dance, it has been suggested simply for pleasure.

Conservation effort

China has a long tradition of conservation: the earliest regulations date from the Zhou dynasty of 1122 to 249 BC. In recent years the government has set up more than 400 nature reserves, and it aims eventually to protect some 2.5 percent of the country.

Attempts have been made to stop hunting and habitat destruction, but continued logging of the forests outside the reserves threatens to isolate them as small

CENTURIES OF CAPTIVITY – PERE DAVID'S DEER

Père David's deer or milu is a large deer with antlers that look as if they have been put on back to front. The species became extinct in the wild some 1,800 years ago and now exists only in captivity or semicaptivity. All the living Père David's deer are descendants of animals that were kept in the imperial hunting park at Nan Haizi, south of Beijing. It was here in 1865 that the French missionary Père Armand David discovered a herd of about 120 deer. By 1900 the animals had disappeared, killed during the civil unrest of the Boxer rebellion; the only survivors were those at Woburn Abbey in England, which had been brought from Nan Haizi several years earlier.

Fossil evidence indicates that milu occurred originally in swampy, lowland areas of north and central China. They are typical swamp deer, with large splayed feet. In their courtship display the males wade into water and emerge with their antlers festooned with water weeds. By 1990 two groups of milu had been reintroduced to China: one to Nan Haizi itself, the other to Dafeng in Jiangsu province (central China) probably nearer to their natural habitat. Breeding has been successful in both herds and the reintroduction of their offspring to further sites within their former range has been planned.

islands of habitat. One approach to combating poaching has been to encourage captive breeding of the species hunted. Both Musk deer and bears have been kept in cages or enclosures to provide a commercial supply of musk and bile respectively. So far such ventures have had little or no impact on poaching, and they are probably a drain on the wild population because of poor animal husbandry.

Other breeding programs are aimed at reintroducing animals to the wild from captive-bred stock. Père David's deer has now been reestablished as a semi-wild breeding herd in an enclosure; a captive herd of Przewalski's horse – which until the 1950s roamed the Mongolian desert – is being raised in Xinjiang (northwestern China); and pandas and many other species have been successfully bred in China's zoos. However, some conservationists feel that it would be better to devote more resources to preserving natural habitat rather than to protecting a small number of spectacular species.

The besieged panda

More than 1,500 Giant pandas still survive in the wild. They are distributed in the narrow band of remaining habitat along the eastern edge of the Plateau of Tibet from southern Sichuan through the Daxiangling, Xiaoxiangling, Qionglai Shan and Min mountains to the isolated Qin Ling mountains of Shaanxi province. The panda was once much more widespread in the lowlands as well, but increasing human population and habitat destruction now restrict it to mountain forest strongholds. Fourteen nature reserves have been established to protect the Giant panda and associated animal species, such as the rare Golden monkey, and the habitat itself, which is particularly rich in plant life. However, encroachment and poaching continue in many areas, despite government efforts to resettle farmers away from their reserves.

Fragmented populations

It is not just the area of habitat lost that has reduced the panda's numbers; perhaps more crucial in the long term is the pattern of habitat loss: the fragmentation of the remaining habitat into more than twenty "islands" of panda population, hemmed in by roads, villages and farmland even within reserves. This makes poaching easier, and can also cause problems during the periodic mass flowering and die-off of the bamboo on which pandas feed: bamboos are curious plants in that each species of bamboo flowers at the same time over wide areas at intervals of up to 150 years.

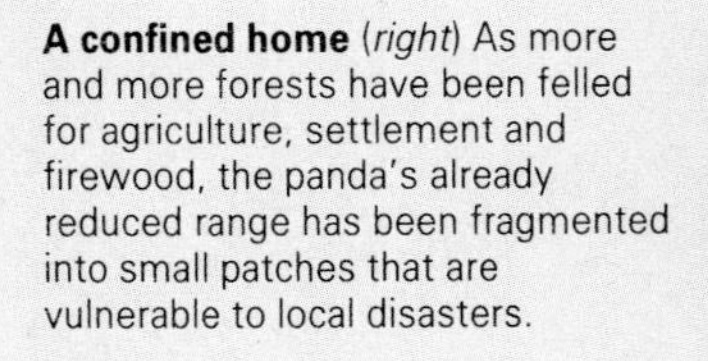

A confined home (*right*) As more and more forests have been felled for agriculture, settlement and firewood, the panda's already reduced range has been fragmented into small patches that are vulnerable to local disasters.

A paintbrush tail (*below*) Giant pandas, especially males, use scent to mark their territories. This is an important form of communication for the secretive pandas, enabling them to avoid aggressive confrontations with each other. The scent is produced by glands under the base of the tail, which acts as a brush to spread the scent. The panda usually marks at prominent points such as crests of hills or on rocks, or it may do a handstand so that it can mark trees up to shoulder height. During the breeding season the scent indicates to others the panda's sex and breeding condition.

A languid life (*above*) The panda's slow pace of life enables it to conserve energy. Because it cannot obtain enough nourishment from its poor diet of bamboo to store up reserves of fat for the winter, it does not hibernate like other bears.

The panda's "thumb" (*below*) The panda has evolved an opposable "thumb", derived from one of its wrist bones, which enables it to hold bamboo stems while stripping the leaves. The panda consumes vast quantities of vegetation, eating almost nonstop both day and night and taking only short rests.

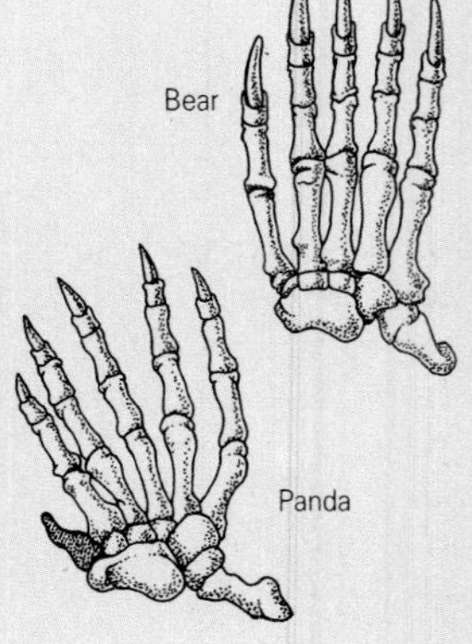

The main species of bamboo eaten by pandas probably flower every 30 to 80 years. For most of the time the bamboos reproduce rapidly and effectively by sprouting shoots from their roots, but after flowering and dropping seeds onto the forest floor the adult plants die off, converting the pandas' feeding grounds into a thicket of dead stems. The seedlings grow so slowly that it is at least 10 or 12 years before they are large enough to provide food for the pandas again. Before the fragmentation of their habitat into isolated areas the pandas would have been able to move down to the valleys or to neighboring areas where other species of bamboo had not flowered: this is no longer possible.

Separation into small, isolated populations also has genetic implications for the pandas. There are dangers of inbreeding as the population size decreases, and this has been shown in some animals to lead to low fertility. Pandas already have a low reproductive rate: only one young can be reared at a time, and females come into season for only about three days every year, so there is little opportunity to replace an infant that has died. In captivity the females are very choosy about mates; they reject "unsuitable" partners even if there are no other males around, thereby missing the chance to breed for another year. Attempts to breed pandas have thus been hampered, though the problem can sometimes be overcome by artificial insemination.

Proposals for rescue

To increase the area of habitat protected, and to facilitate the movements of pandas, 14 new reserves are proposed. This would increase the total protected panda habitat from just under 3,000 sq km (1,150 sq mi) at present to about 5,000 sq km (2,000 sq mi), roughly doubling the number of pandas inside the reserves. Three of the new areas would link existing reserves, and a further three "corridors" are envisaged. These would not have the status of reserves, but they would join populations of pandas at present separated by unsuitable habitat or busy roads. The project is ambitious, involving the replanting of panda habitat. But it is hoped that the flow of animals between presently isolated populations will not only enable the pandas to overcome the problems of bamboo flowering but will also encourage interbreeding between the populations.

China's captive breeding program is the subject of some controversy. A valuable 100 animals remain captive in zoos around the world, a significant proportion of the total population. However, despite extensive research, more pandas die in captivity than are born. With a restricted, low-energy diet and slow reproductive rate the panda needs all the help it can get if it is to survive into the 21st century.

A PARADISE FOR ANIMALS

UNIQUE ANIMAL LIFE · ANCIENT FOREST LIFE · NATURE IN RETREAT

Southeast Asia is home to an extremely diverse collection of animals and an enormous variety of habitats; there are some 150,000 species in the region. The animal life forms part of the Oriental realm, a zoogeographic region which comprises most of Asia. Consequently many of the region's animals also occur in adjacent parts of the Indian subcontinent and southwestern China. New species have evolved on a number of the islands: for example, the Proboscis monkey and the Flying frog of Borneo. On the easternmost islands Asian mainland animals, such as forest pigs and civets, mingle with animals of Australasian origin: for example, marsupials such as the tree kangaroo. The rich Indo-Pacific coral reefs and warm seas support a wealth of marine life, including sea turtles and schools of Sperm whales.

COUNTRIES IN THE REGION

Brunei, Burma, Cambodia, Indonesia, Laos, Malaysia, Philippines, Singapore, Thailand, Vietnam

ENDEMISM AND DIVERSITY

Diversity Very high (second only to South America)
Endemism High to very high

SPECIES

	Total	Threatened	Extinct†
Mammals	650	75	2
Birds	2,000*	200	3
Others	unknown	141	0

† *species extinct since 1600 - Panay giant fruit bat* (Acerodonlucifer), *Schomburg's deer* (Cervus schomburgi), *Javanese wattled lapwing* (Vanellus macropterus), *Caerulean paradise-flycatcher* (Eutrichomyias rowleyi), *Four-colored flowerpecker* (Dicaeum quadricolor)
* *breeding and regular non-breeding species*

NOTABLE THREATENED ENDEMIC SPECIES

Mammals Pileated gibbon *(Hylobates pileatus)*, orangutan *(Pongo pygmaeus)*, Flat-headed cat *(Felis planiceps)*, Malayan tapir *(Tapirus indicus)*, Javan rhinoceros *(Rhinoceros sondaicus)*, kouprey *(Bos sauveli)*, tamaraw *(Bubalus mindorensis)*
Birds Philippine eagle *(Pithecophaga jefferyi)*, Giant ibis *(Pseudibis gigantea)*, Gurney's pitta *(Pitta gurneyi)*, Salmon-crested cockatoo *(Cacatua moluccensis)*
Others River terrapin *(Batagur baska)*, False gharial *(Tomistoma schlegelii)*, Komodo dragon *(Varanus komodoensis)*

NOTABLE THREATENED NON-ENDEMIC SPECIES

Mammals Asian elephant *(Elephas maximus)*, gaur *(Bos gaurus)*
Birds Lesser adjutant *(Leptoptilos javanicus)*, Green peafowl *(Pavo muticus)*, Asian dowitcher *(Limnodromus semipalmatus)*
Others Estuarine crocodile *(Crocodylus porosus)*

DOMESTICATED ANIMALS (originating in region)

Bali cattle *(Bos javanicus)*, Water buffalo *(Bubalus 'bubalis')*, Asian elephant *(Elephas maximus)*, duck *(Anas platyrhynchos)*, chicken *(Gallus gallus)*

UNIQUE ANIMAL LIFE

The rainforests, mangrove swamps and myriad islands of Southeast Asia support a very high number of endemic species. Many of these animals evolved in the tropical rainforests that cover most of the region. Some of the oldest forests in the world are found here, and have provided a stable habitat long enough for new species to evolve. There are in fact whole families and orders of animals that occur nowhere else. These include the tarsiers, tree shrews, flying lemurs, leaf monkeys and gibbons, and the leaf birds. Other birds, such as hornbills and pheasants, are especially characteristic of Southeast Asia. The forests also provide a refuge where many rare large mammals maintain a precarious foothold: tigers, the Indian elephant, rhinoceroses, tapirs, bears, numerous monkeys and the large distinctive orangutan.

Centers of evolution

Southeast Asia's great diversity of animals is attributable in part to the presence of thousands of islands, some of them very large. The western islands of Sumatra, Java and Borneo were once connected to the Asian mainland during past periods of glaciation. Under these conditions land bridges were formed, enabling mammals and birds gradually to colonize the area before the islands had formed. The islands between the Asian mainland and Australia have experienced animal immigration from both east and west. For example, there are marsupials (pouched mammals) in New Guinea, rhinoceroses in Sumatra and Java; orangutans in Sumatra and Borneo and tapirs in southern Sumatra. After the land bridges were submerged and the islands became isolated, some animals continued to colonize by island hopping.

In Indonesia and on the surrounding islands, remote fragments of tropical forests have served as miniature centers of evolution, and are home to many endemic species. These animals have diversified in many cases into individual island species or races that are specially adapted to the local conditions. Wild cattle are good examples of different species having adapted to particular environments: the large gaur lives in upland forests, the wild Water buffaloes in swampy areas, the banteng only in Java and the small anoa in the hills of Sulawesi.

Another species that has diversified is the Sulawesi macaque – related to the Pig-tailed and Crab-eating macaques of nearby Borneo: over time it has evolved not only into a new species but also into several distinct races in different parts of the island.

In 1858 the naturalist Alfred Wallace (1823–1913) proposed a boundary to mark the easternmost extent of the Oriental zoogeographic region. The Wallace line has been recently modified, and now runs along the edge of the Sunda continental shelf, just east of Java, Borneo and the Philippines. A later line was proposed in 1902 by the Dutch zoologist Max Weber to mark the westernmost extent of Australasian wildlife. The recently amended Weber's line runs just west of Australia and New Guinea, following the edge of the Australian continental shelf. In between these two lines lie Sulawesi, the Lesser Sunda Islands and many other islands in which both Australasian and Oriental animals are mixed.

The Rhinoceros hornbill feeds mainly on fruit, which it picks up with the tips of its large mandibles. The large, horny growth above the bill may help the bird to recognize sex, age and species; in the larger species it may also be used when fighting.

Gentle giant The orangutan is superbly adapted for life in the trees, with long arms, hooked hands and handlike feet for gripping. Already decimated by collection for zoos, the orangutan population is now threatened by destruction of its rainforest habitat.

Stump-tailed macaque
Macaca arctoides

Pig-tailed macaque
Macaca nemestrina

ANCIENT FOREST LIFE

Tropical Asia's ancient forests have been the focus of evolution of several groups of animals that are highly adapted to life in the treetops. These include the tree shrews and the tarsiers, both of which are found only in Southeast Asia.

Tree shrews are small, squirrel-like mammals. The ground-dwelling species are larger, with short tails, long snouts and well-developed claws that they use to dig for insects. Tree-dwellers tend to be smaller, with eyes more forward-directed for judging distance, and long tails for balancing. They are thought to resemble closely the early ancestors of the simians (monkeys and apes) and are therefore placed in the prosimian ("forerunners of the simians") animal suborder.

Tarsiers are small, nocturnal mammals confined to Sumatra, Borneo, Sulawesi and a number of small islands in the Philippines. They have huge eyes for night vision which – as in owls – take up so much room in their sockets that they cannot move much; the animal compensates by turning its head through a wide angle. The eyes are directed forward to give good stereoscopic vision, useful to the animal when catching prey with its long, slender fingers, which act like a cage to ensnare swift-moving insects. Tarsiers also use their large, sensitive ears to locate prey. They can easily jump up to 2 m (6.5 ft), landing feet first on neighboring trees, using their enlarged adhesive finger pads for better grip.

The tarsiers' carnivorous diet has led to controversy over their taxonomic position – or classification – in the animal world. Current opinion regards them as intermediate between the prosimians (the suborder that also includes lorises, lemurs and bushbabies) and the primates.

Another prosimian, the Slow loris, is found from Vietnam to Borneo. Like the tarsiers it is nocturnal, and does not compete with other primates. Its fingers are shorter, and it creeps up on its unwary prey, rather like a chameleon. The Slow loris has a remarkable ability to eat prey that other animals find repulsive.

Movement through the canopy

Various primates employ different ways of moving among the trees. Orangutans move slowly, walking along the branches, but will use their heavy weight to bend trees until the gap between one tree and the next has narrowed sufficiently for them to traverse it. The gibbons swing from branch to branch using their long arms and hooked hands alternately, but they walk upright on two feet on broad branches and on the ground.

Animals from many other groups have evolved the habit of gliding from tree to tree. The Gliding gecko has broad flaps of skin along both sides of its abdomen, webbed limbs and toes and a flattened tail that is used as a rudder while gliding. The Flying dragon, another lizard, has broad, often brilliantly colored, winglike membranes. They serve two purposes: gliding, and attraction of the female during courtship. When the lizard alights on a tree trunk, the thin membrane presses close to the bark, concealing the animal's shadow. Flying squirrels also use membranes between their limbs for gliding; this not only helps them to travel easily through the canopy, but also provides a useful means of escape from predators.

Perhaps the most remarkable gliding mammals are the flying lemurs or colugos, which belong to an order of mammals with no known relatives. Their gliding membranes extend from the tips of their fingers and toes to the tip of their long tails, forming an impressive parachute. Other perhaps unexpected gliders are the Flying frog and the Gliding snake. The frog uses widely spread webs between its toes to glide through the air, spreading its limbs and toes wide to present as large a surface area as possible to the air. The Gliding snake controls its fall by curving its undersurface into a concave shape.

Birds also need certain adaptations for flying among trees. The Philippine eagle, the main predator of the Philippines jungle, is capable of attacking monkeys, taking them by surprise from branches. It has short, broad wings and a long tail, enabling it to maneuver swiftly in the upper storey of the forest, its brown-gray plumage providing camouflage.

Moor macaque
Macaca maura

Pastures of the sea

The shallow lagoons that lie behind many coral reefs in this region support extensive meadows of sea grass, a favorite food of sea turtles and the unusual mammal, the Sea cow or dugong. Sea cows and their relatives, the manatees, are the only vegetarian marine mammals. The Sea cow has a fat, streamlined body with plenty of blubber to keep it warm and aid buoyancy, along with paddlelike forelimbs and a flattened tail resembling a dolphin's. Its short snout ends in a fleshy horseshoe-shaped disk armed with stiff bristles around a slitlike mouth. The sea cow has few teeth; it chews using the rough horny pads on its upper and lower palates. It feeds at night, raking up food from the seabed with its muscular disk – earning it the popular name of Sea pig. Its slow movement and relative inactivity enable it to survive on a diet low in nutrients.

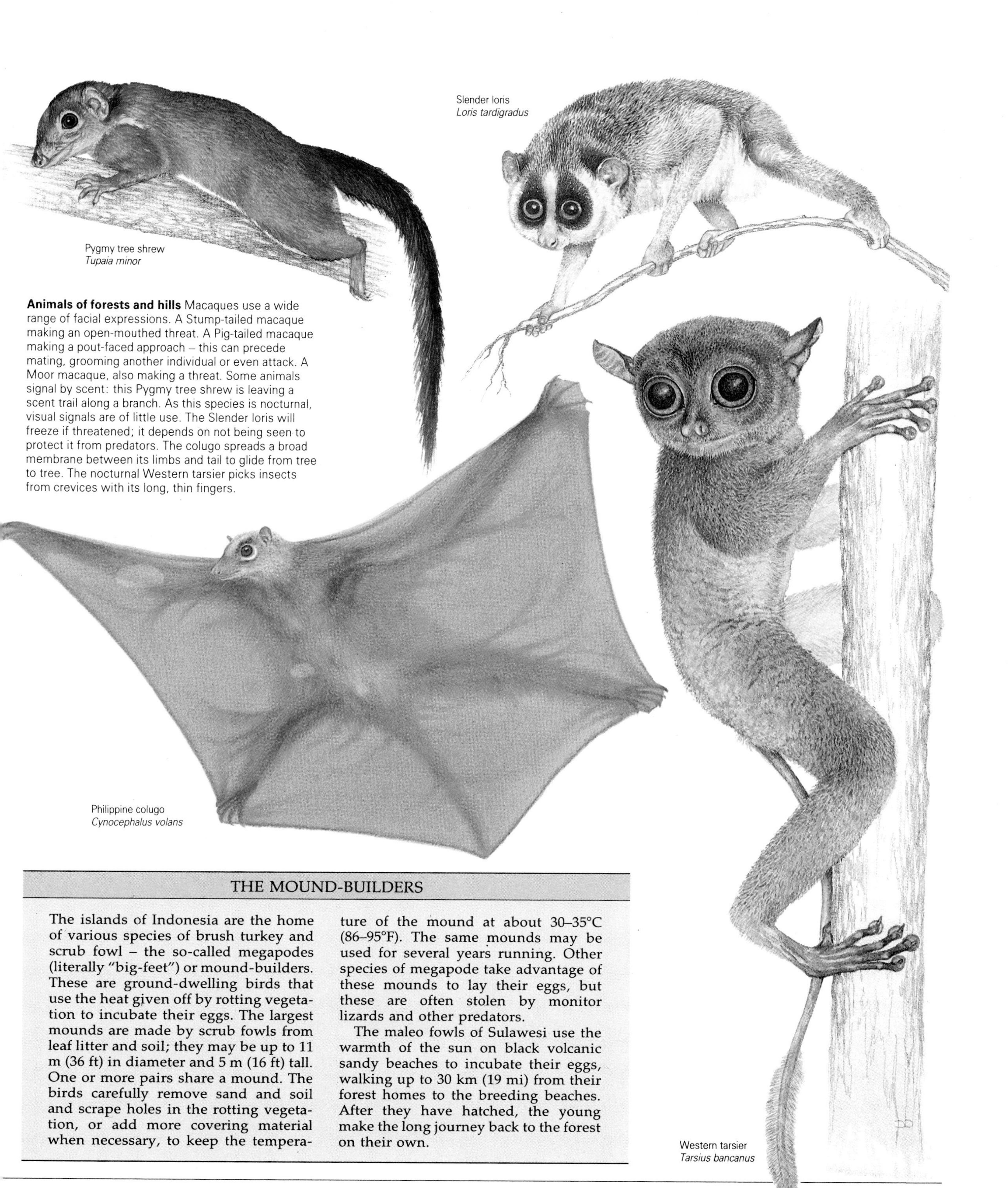

Slender loris
Loris tardigradus

Pygmy tree shrew
Tupaia minor

Animals of forests and hills Macaques use a wide range of facial expressions. A Stump-tailed macaque making an open-mouthed threat. A Pig-tailed macaque making a pout-faced approach – this can precede mating, grooming another individual or even attack. A Moor macaque, also making a threat. Some animals signal by scent: this Pygmy tree shrew is leaving a scent trail along a branch. As this species is nocturnal, visual signals are of little use. The Slender loris will freeze if threatened; it depends on not being seen to protect it from predators. The colugo spreads a broad membrane between its limbs and tail to glide from tree to tree. The nocturnal Western tarsier picks insects from crevices with its long, thin fingers.

Philippine colugo
Cynocephalus volans

Western tarsier
Tarsius bancanus

THE MOUND-BUILDERS

The islands of Indonesia are the home of various species of brush turkey and scrub fowl – the so-called megapodes (literally "big-feet") or mound-builders. These are ground-dwelling birds that use the heat given off by rotting vegetation to incubate their eggs. The largest mounds are made by scrub fowls from leaf litter and soil; they may be up to 11 m (36 ft) in diameter and 5 m (16 ft) tall. One or more pairs share a mound. The birds carefully remove sand and soil and scrape holes in the rotting vegetation, or add more covering material when necessary, to keep the temperature of the mound at about 30–35°C (86–95°F). The same mounds may be used for several years running. Other species of megapode take advantage of these mounds to lay their eggs, but these are often stolen by monitor lizards and other predators.

The maleo fowls of Sulawesi use the warmth of the sun on black volcanic sandy beaches to incubate their eggs, walking up to 30 km (19 mi) from their forest homes to the breeding beaches. After they have hatched, the young make the long journey back to the forest on their own.

NATURE IN RETREAT

It is estimated that some 2 million ha (5 million acres) of Southeast Asia's forests are destroyed each year. Official figures put the deforestation rate of Indonesia alone at a million hectares a year; more than 50 million ha (150 million acres) of rich tropical forest have already disappeared. In the Philippines forest cover is down to 3 percent of its original extent. At the current rate of deforestation it is probable that all lowland forests outside conservation areas will have disappeared by the year 2000.

Such largescale habitat destruction inevitably has an enormous impact on the region's wildlife. The fruit-bearing trees are a haven for fruit-eating birds such as the spectacular hornbills, as well as squirrels and monkeys. Flowers attract nectar-sipping butterflies and tiny, iridescent sunbirds, and the leaves are food for a great many herbivores such as monkeys and small insects. The abundant insect life is in turn preyed on by birds such as flycatchers, babblers and leaf birds.

Coastal reclamation schemes have now cleared more than 1 million ha (2.5 million acres) of primary swamp forest. Up to 600,000 people have been resettled in the environmentally sensitive wetlands of Sumatra, Borneo and New Guinea. Present economic policies favor short-term resource exploitation and undervalue the long-term benefits of conservation.

With the continuing loss of natural habitats, biological diversity is being swiftly reduced. Large mammals such as tigers and rhinoceroses need sizable areas in order to sustain viable populations. Despite full legal protection, the Bali tiger became extinct in the 1960s, and the Javan tiger disappeared in the early 1980s. The future of the Sumatran tiger outside protected areas looks extremely bleak. The Javan rhinoceros is now found only in the Ujung Kulon National Park, where fewer than 50 individuals survive.

Large mammals that exhibit traditional migratory patterns, such as elephants, are the most likely to be affected by habitat fragmentation. It will be impossible to protect the entire range of a population of Sumatran elephants, which could be as large as 1 million ha (2.5 million acres).

Attempts have been made to increase the orangutan population by reintroducing to the wild individuals that have been rescued as orphans or captured for the pet trade. Such programs prove very difficult, as the animals have to be taught how to feed and survive in the wild.

Of the 1,000 or so globally threatened species of bird, more than 200 occur in Southeast Asia. The Gurney's pitta may have just 10 breeding pairs left, at Khao Phra Bang Khran in Thailand. Indonesia harbors more endangered bird species than any other country in the world: a total of 128, of which 91 are endemic. Some species are already in peril. The Bali starling, although numerous in zoos, numbers only 50 in the wild.

The threat of overexploitation

Overhunting is another problem that particularly affects the larger mammals. The Indian cheetah is already extinct, and the tigers and rhinoceroses have been greatly reduced in numbers as a result of overhunting in colonial times; added to this they have suffered from loss of habitat in more recent decades. Another modern trend has been the largescale trapping of monkeys for medical research. The magnificent Philippine eagle has been brought to the brink of extinction by trophy-hunters, and it seems likely that this majestic bird will be finally exterminated by the loss of its remaining fragments of forest habitat.

Crocodiles have been hunted for their skins and eggs and are now scarce in much of their former range. The taking of shells and eggs threatens several species of sea turtle. Collecting has also reduced numbers of the giant Robber crab or Coconut crab, possibly the largest terrestrial arthropod in the world, measuring 1 m (3 ft) across its legs.

Silent hunter *(right)* A Malayan green tarantula waits for insect prey. The rainforests support a wide range of invertebrate predators, such as praying mantids, hunting wasps, carnivorous beetles and robber flies.

Malayan tapir
Tapirus indicus

Lone wanderer *(left)* A Malayan tapir browses in the forest. It uses its muscular prehensile lips to twist and tear off leaves and twigs. Tapirs are skilled swimmers, and often browse on water plants, or submerge themselves in the water to cool off or to rid themselves of parasites.

Wasted wildlife For the poor of Indonesia, capturing wild birds brings in much-needed money. Very few of these birds will survive the journey to their destinations in Europe, the United States and many Asian cities.

THE KOMODO DRAGON

The tiny Indonesian islands of Komodo and Rinca, and Flores, the larger neighboring island, are the home of the world's largest lizard, the Komodo dragon. This enormous monitor lizard can reach up to 3 m (10 ft) long and can weigh as much as 150 kg (330 lb). It is related to the carnivorous dinosaurs: fossils strikingly similar to the Komodo dragon have been unearthed in Australia from chalk deposits dating back 130 million years.

The lizard has no predators on these islands. With its huge powerful body, jaws that can be disarticulated like those of snakes, and sawedged teeth, the lizard preys on deer and Wild boar. Its powerful jaws can also excavate the egg mounds of megapode birds, and it will scavenge carrion. Its stomach juices are so strong that the lizard digests almost every part of its prey, including bones, hair and hooves. It will even devour its own offspring – young Komodo dragons spend most of their first two years hiding in trees, out of sight of the adults.

The world population of Komodo dragons is believed to number less than 2,000. Very few are kept in zoos, although they have been known to breed in captivity. The species is under threat from the loss of its prey to human hunters, and the loss of its natural habitat to urbanization and agriculture.

The world's largest lizard The powerful Komodo dragon is a formidable predator, its only enemy being humans – local people hunt it because it is considered to have palatable flesh.

Overexploitation is nowhere more prevalent than in the seas of Southeast Asia and the Pacific. When coral reefs are mined for building material and lime, the resulting clouding of the water kills the remaining living corals and with them the rich life of the reef. Reefs are frequently plundered so that resorts can be built for the tourists whose very aim is to see the unspoilt beauty of underwater life.

Commercial fishing is wreaking havoc in the Indo-Pacific. Invisible rot-proof gill nets of up to 40 km (25 mi) long drift through the ocean, often abandoned. Every year in the north Pacific alone they kill hundreds of thousands of seabirds, seals, turtles, porpoises and dolphins.

The future outlook

Environmental awareness among the governments of Asian countries is growing. Over the last 15 years Indonesia has set aside 20 million ha (50 million acres) as conservation areas, and aims to protect 20 percent of its land. In 1989 Thailand banned logging in an attempt to halt forest degradation. However, conservation must be linked to economic development among the poor of the region if a real solution to habitat destruction and species loss is to be found.

Dolphins

Dolphins are really small whales. They have elongated jaws forming the so-called beak and a high, rounded forehead, which is called the melon. Their jaws are lined with conical pointed teeth that are often curved for a better grip on their slippery prey. Most dolphins feed on fish, but a few species feed on squid or crustaceans; these species tend to have more rounded foreheads, blunter beaks and, in some cases, fewer teeth.

Marine specialists

Dolphins have managed to exploit virtually all types of marine, estuarine and riverine habitats except for the deepest parts of the ocean. They are found in the open ocean, along muddy shores and in clear rivers. Unlike most mammals, they are well adapted to a marine existence. Their bodies are streamlined, and well endowed with fat for insulation against the coldness of the water. Their forelimbs have been modified to form paddles, and they have a horizontal finlike tail for propulsion. Some species are able to dive to depths of 300 m (1,000 ft), remaining submerged for eight minutes or more. During these prolonged dives, the heart rate and peripheral blood flow are considerably reduced. In dolphins it has been estimated that 80 to 90 percent of the air in the lungs is replaced during each inhalation – a far more efficient process than human breathing. Unlike seals, dolphins give birth in the water. The calves, which are born tail first, are generally large in relation to the size of the mother – this increased surface area to volume ratio minimizes heat loss to the water.

Dolphins have either reduced or absent taste and smell organs, which are of little use in water, but they have very acute hearing, and some species use highly sensitive echolocation to find their prey. The various river dolphins rely on this method to guide them to their prey in the murky water; indeed, the Indus and Ganges dolphins have no eye lenses at all. Species that hunt in clear waters, however, probably rely on vision.

In Southeast Asia only four species live in estuaries and rivers, but no fewer than 21 species of oceanic dolphin migrate through the offshore waters of the tropical Pacific and Indian Oceans. Dolphin migration is usually a response to seasonal changes in water temperature, caused by the monsoons, which also affect the availability of food.

Purse-net danger

In some dolphin species individuals are solitary, while in others they are highly gregarious. The Spotted dolphin and other ocean dolphins may make up enormous groups of 1,000 to 2,000 individuals. They cooperate to concentrate a shoal of fish, and this has proved their undoing when hunting tuna. Every year tens of thousands of oceanic dolphins drown in the purse-seine nets of tuna fishermen, particularly from the South Korean and Japanese industrial fleets, a catastrophe that threatens to exterminate local dolphin populations. Research has revealed that at least one dolphin dies – mostly Spotted and Long-snouted spinner dolphins – for every nine tuna caught. In the eastern Pacific fishermen use the dolphins to lead them to the tuna.

The United States and some other countries now insist that precautions are taken by their tuna fishermen: these may involve using divers to free trapped dolphins, or modifying the methods of pulling in the nets. It is likely that in the early 1990s strict regulations will come into force, such as a ban on industrial tuna fishing in the Pacific within the majority of the 322 km (200 mi) Exclusive Economic Zones that surround every coastline. However, these measures will not do anything to save the tens of thousands of dolphins and porpoises that are drowned each year by becoming entangled in fishermen's vast drifting gill nets that are suspended vertically in the water.

Common dolphin
Delphinus delphis

Dolphins of Southeast Asia (*left*) A Bottle-nosed dolphin scanning the world above the water. This is the species most commonly seen in dolphinaria. A coastal species, it likes to ride the bow waves of boats. The Rough-toothed dolphin is named for the wrinkled enamel on its teeth. Risso's dolphin, or grampus, lives in herds with stable group relationships, and is found in the open ocean throughout the world. The skin of males becomes very scarred with age, due to fights with other males. The Common dolphin often forms herds of several hundred individuals. They give a spectacular display when swimming close to the surface, sometimes leaping clear of the water.

Spinner dolphin (*above*) Bubbles streaming from its blowhole, a dolphin rises to the surface. Spinners are named for their acrobatic displays, usually performed in the evening, when they leap out of the water and spin round before dropping back. Many thousands are killed every year by tuna fishermen.

An engaging smile (*left*) The smile of the Bottle-nosed dolphin is a feature of its anatomy rather than of its temperament, but it makes this the most popular of dolphins. Lone individuals sometimes befriend swimmers and sailors, playing games with them and allowing themselves to be stroked. The Bottle-nosed dolphin is large; its only natural predators are Killer whales and sharks.

Conservation or collection?

As the human population has increased, the threat to wildlife from collectors, who take specimens either for pets or ornaments, has grown dramatically. Tropical butterflies, particularly the birdwing butterflies, are highly prized collectors' items. The Queen Alexandra's birdwing of the forests of New Guinea has been protected since 1966 from unauthorized collecting. Butterfly farms in the region are increasing in number, and recent estimates suggest that controlled breeding of this species and other insects brings in some $50,000 a year. However, in many places habitat loss is now undoing this valuable conservation work.

Trade in birds for the pet trade and collectors' market, both domestic and international, has drastically reduced local populations of certain bird species, particularly cockatoos and parrots.

In the Philippines hundreds of coral reef fish are caught for the aquarium trade. They are doped with cyanide, which damages their livers, so they die within six months of capture, creating further demand for replacements. Stress and disease in the overcrowded conditions of storage and transportation result in a high mortality rate among the fish. Rare corals and mollusk shells are also great collectors' items, and are taken by the thousand as tourist souvenirs.

The Great mormon butterfly, common throughout Southeast Asia, is remarkable for the great variety of form and color of the females; these mimic several other species of swallowtail butterflies.

WETLAND AND MOUNTAIN ANIMALS

A HISTORY OF INVASIONS · RELICS AND INNOVATORS · SLOW AWAKENING TO CONSERVATION

Although relatively small in area, Japan and Korea extend from the subarctic to the subtropics, hosting a wide range of animals. Brown bears, Siberian tigers and the Japanese macaque inhabit the northern forests; waterfowl, storks and rare species of crane overwinter in the lowland wetlands; and corals flourish off the southern coasts. The high mountains of northern Korea and central Japan act as climatic divides, increasing the variety of habitats for wildlife. Many species, such as the Amami woodcock, the Ryukyu robin and the endangered Okinawa woodpecker, have evolved in isolation on the thousands of islands off the coasts of Japan and Korea. On some islands lack of competition has enabled ancient forms to survive – "living fossils" such as the Iriomote cat and the Amami rabbit.

COUNTRIES IN THE REGION

Japan, North Korea, South Korea

ENDEMISM AND DIVERSITY

Diversity Low to medium
Endemism Low to medium

SPECIES

	Total	Threatened	Extinct†
Mammals	150	6	0
Birds	650*	26	4
Others	unknown	9	0

† *species extinct since 1600 - Bonin wood pigeon* (Columba versicolor), *Ryukyu kingfisher* (Halcyon miyakoensis), *Kittlitz's thrush* (Zoothera terrestris), *Bonin grosbeak* (Chaunoproctus ferreorostris)
* *breeding and regular non-breeding species*

NOTABLE THREATENED ENDEMIC SPECIES

Mammals Amami rabbit *(Pentalagus furnessi)*, Iriomote cat *(Felis iriomotensis)*
Birds Short-tailed albatross *(Diomeda albatrus)*, Okinawa rail *(Rallus okinawae)*, Amami thrush *(Zoothera amami)*
Others Japanese giant salamander *(Andrias japonicus)*, Tokyo bitterling *(Tanakia tanago)*

NOTABLE THREATENED NON-ENDEMIC SPECIES

Mammals tiger *(Panthera tigris)*, dhole (*Cuon alpinus*), leopard (*Panthera pardus*), Asiatic black bear (*Ursus thibetanus*)
Birds Red-crowned crane *(Grus japonensis)*, Blakiston's fish owl *(Ketupa blakistoni)*, Fairy pitta *(Pitta nympha)*
Others Greater large blue butterfly *(Maculinea arionides)*, Giant triton *(Charonia tritonis)*

DOMESTICATED ANIMALS (originating in region)

none

A HISTORY OF INVASIONS

Japan separated from mainland Asia only during the last ice age with the result that many of the animals that formerly ranged over northeast continental Asia still inhabit Japan as well as Korea. The region also supports many species found at the European end of the Asian landmass. These include weasels, badgers, the Northern pika, or Mouse hare, the Raccoon dog – so-called because of its resemblance to the raccoon – and the Whooper swan. However, a history of climatic change, culminating in the recent ice age, has fragmented the distributions of some of these species. The Azure-winged magpie, for example, survives only in China, Japan and Korea in Asia; there is also an isolated European population in Spain.

With the end of the last ice age many species gradually invaded the region. The Amur leopard subspecies and numerous birds moved north from the tropics; the Brown bear and the goral (a form of goat antelope) traveled across the Himalayas into the north of the region; and the advancing steppe allowed the passage of animals from the west, such as bustards. Many of the larger mammals such as the leopard and the goral reached Korea but not Japan, although many smaller species are common to both.

Animals of Korea

Hills and mountains cover some 80 percent of Korea. The animals of the northern highlands along the border with China are closely related to those of northern China, Siberia and Hokkaido, the most northerly of Japan's four main islands. Here there are still some Siberian tigers and Brown bears, although both of these species are now scarce.

Many of the lowland animals have affinities with species of central China and the main part of Japan. Extensive coastal wetlands, now rapidly being drained or polluted, lie on a major north–south migration route; they are important overwintering grounds and breeding areas for many migrant birds, including the rare Chinese egret. Some 390 bird species have been recorded in Korea; about 117 of them breed here. Compared with Japan, however, Korea has less diversity of species, as it lacks such extremes of climate and altitude.

Primitive member of the dog family *(above)* The Raccoon dog is a nocturnal predator, hiding by day in a den or in tall vegetation. It prefers to forage for fruit and insects in dense woodland, and it hibernates in winter in the northern part of its range.

Bathing in hot springs *(right)* Japanese macaques are well adapted to the cold weather of Japan's high mountain forests. They have thick, shaggy coats and short tails to minimize heat loss. Their complex social behavior has been studied intensively.

Island diversity

In addition to the four main islands, there are some 4,000 smaller islands off the coast of Japan, and approximately 3,400 little islands off the Korean coast. Hokkaido was separated from the other three large islands of Japan at a relatively early stage and retains many more cold-adapted animals than the rest of the country. For its relatively small area, Japan supports a great diversity of wildlife. There are about 130 mammal species and more than 500 bird species, some 200 of which breed in the region. Japan is also home to 84 species of reptile, including several sea snakes, along with terrestrial snakes such as the habu and mamushi,

which kill a number of people each year. In addition, 58 amphibian species and more than 100,000 insect species are also to be found in Japan.

Japan's northern deciduous and coniferous forests are home to the Brown bear, a distinct Japanese species of serow (a goat antelope) and the Japanese macaque. The Kajika frog and the Japanese giant salamander (the largest living amphibian) are both endemic to Japan, and live in the cold mountain streams.

The subtropical habitats contribute the majority of species, with flying squirrels, flying foxes, Sea cows (dugongs) and, among the birds, Fairy pittas and Emerald doves. Most of the islands off the Korean coast, in contrast to those of Japan, are not volcanic and are nearer the Asian mainland, so they have less diverse wildlife. Only on Korea's warmer, more southerly islands, such as Cheju and Sohukssan, have large numbers of mammal and bird species been recorded.

The marine life of Japan is also rich and varied, with a mixture of tropical Indo-pacific species and forms from more northerly waters. Several species of marine turtle – the Loggerhead sea, Green and Hawksbill – breed in the subtropical Ryukyu Islands. There are also some 1,800 species of freshwater fish, including many varieties of carp; the region is the center of evolution for this group.

Japan's coral reefs contain some 250 kinds of coral, including the rare Blue coral, and numerous species of associated reef fish. There are more than 1,200 varieties of edible marine mollusk, including the spectacular Giant crab; it has a limb span of 3 m (10 ft), and visits Japanese waters from Southeast Asia.

The subtropical marine species extend farther north than the terrestrial animals: the Japan Current carries warmwater species as far north as the Boso Peninsula near Tokyo on Honshu, the second most northerly of Japan's four main islands. Coldwater species, such as the Steller sea lion, do not come farther south than northern Honshu.

RELICS AND INNOVATORS

The mountains of Japan were formed relatively recently, about 20 million years ago. Many of the volcanoes are still active. Complete separation from the continental landmass took place only about a million years ago, and this has enabled certain species to survive: on the mainland they have been eliminated by competition with other species.

"Living fossils"

The Amami rabbit is a so-called "living fossil". It is a member of a family that flourished 20 million years ago and now occurs only in the mountain forests of the southern islands of Amami and Tokuno. Compared with most other rabbits, the Amami rabbit has shorter limbs and ears, smaller eyes and larger claws on the forefeet, which it uses for digging its den. The female is said to shut her young in the den and reexcavate the burrow once a day when she comes to feed them, as does the European rabbit.

The Iriomote cat is another relic from the past, found only on the island of Iriomote, in the Ryukyu archipelago to the far south of Japan. Resembling the Fishing cat, it is thought by some to be ancestral to other present-day cats. It is relatively small, only 60 cm (24 in) from head to tail base, with short legs and tail, round ears and a dark brown coat decorated with bands of spots. Nearly always found near water, the Iriomote cat feeds on waterbirds, rodents, crabs and mudskippers. However, specialization has its costs: today the cat is less isolated from competition and is suffering at the hands of people and the more adaptable species they have introduced.

Enterprising macaques

The myriad volcanic islands of Japan offer the isolation that encourages the evolution of new races and species, especially among groups of animals lacking the ability to swim or fly from one island to another or to the mainland. About two-thirds of Japan's mammals are endemic. Best known of these is the Japanese macaque, a short-tailed monkey that lives farther north than any other primate. Found in the forests from sea level up to about 1,500 m (4,900 ft), it has long thick hair that provides insulation against the cold of winter.

Steller sea lion bulls *(above)*, of the Sea of Japan and northern Pacific waters, compete aggressively for territories on the best beach sites. The bulls are much larger than the females, and each tries to hold on to a harem of females and mate with them.

For species that live at the limits of their natural range, it is advantageous for an individual to evolve new techniques of foraging to improve its chances of survival. It can also benefit by quickly taking advantage of the innovations developed by others. If a series of such innovations are adopted by a population, this can eventually lead to significant increases in range. The Japanese macaque's diet is already broad, including fruits, nuts, grass, insects, seaweed and marine invertebrates. The species has attracted the attention of animal behaviorists studying cultural learning, since it has a great capacity to pick up new habits by imitation. A famous example involved the washing of grain and potatoes fed to the animals by researchers. At one time washing had never been observed, but as

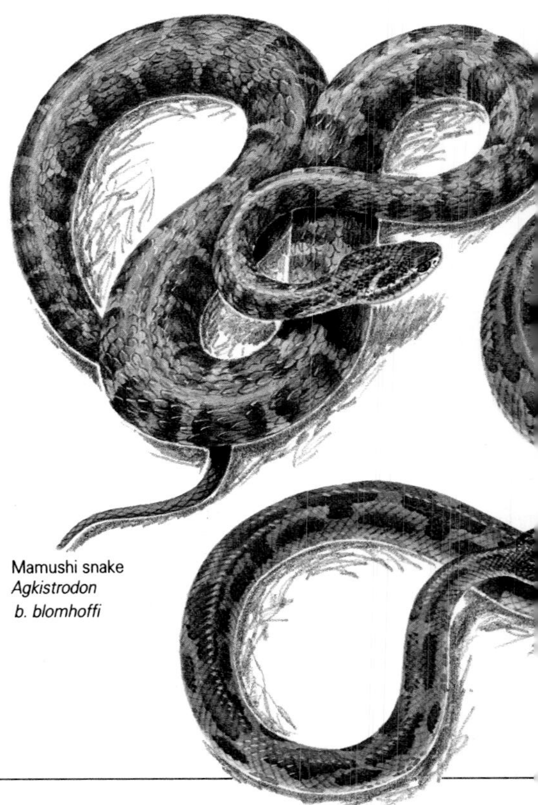

Mamushi snake
Agkistrodon b. blomhoffi

THE JAPANESE GIANT SALAMANDER

Japan is home to the largest amphibian in the world – the Japanese giant salamander. It can reach a length of 1.6 m (5 ft) and lives in the fast-flowing mountain streams of Kyushu and Honshu islands at altitudes of 300–1,000 m (1,000–3,300 ft). It belongs to the family Cryptobranchidae, which includes only two other species, one in China and the other in the United States. The three species of giant salamander all lose their gills as they mature, but they retain other traces of larval characteristics and remain aquatic all their lives. In the fast-flowing, well-aerated water in which they live they absorb sufficient oxygen through their skin. A fold of skin along their flanks increases the surface area through which oxygen can be taken in.

The Japanese giant salamander feeds on a wide variety of invertebrates, fish and amphibians, including smaller salamanders. Its small eyes are set so far to the sides of its head that it has no stereoscopic vision. It frequently hunts by night, relying on its powers of smell and touch rather than sight to locate its prey, which it traps with a rapid sideways snap of the mouth. It is a long-lived animal; one specimen that was kept in captivity is known to have lived for 52 years.

soon as one animal started to wash her food and found it beneficial because it removed sand, the behavior quickly became established in the population.

Tiger of the north

Over on the mainland, in North Korea, lives another animal at the northern end of its range: the Siberian tiger. The tiger is thought to have originated in the far north, in eastern Siberia; the various subspecies evolved as the tiger spread south into areas with warmer climates. The Siberian tiger is the largest of the subspecies and inhabits the Changbai Mountains on the border of China and North Korea, as well as other mountains in the area. It is well equipped to survive in the severe cold and heavy snowfalls of its range; it has thick fur and large pads, which enable it to walk on snow.

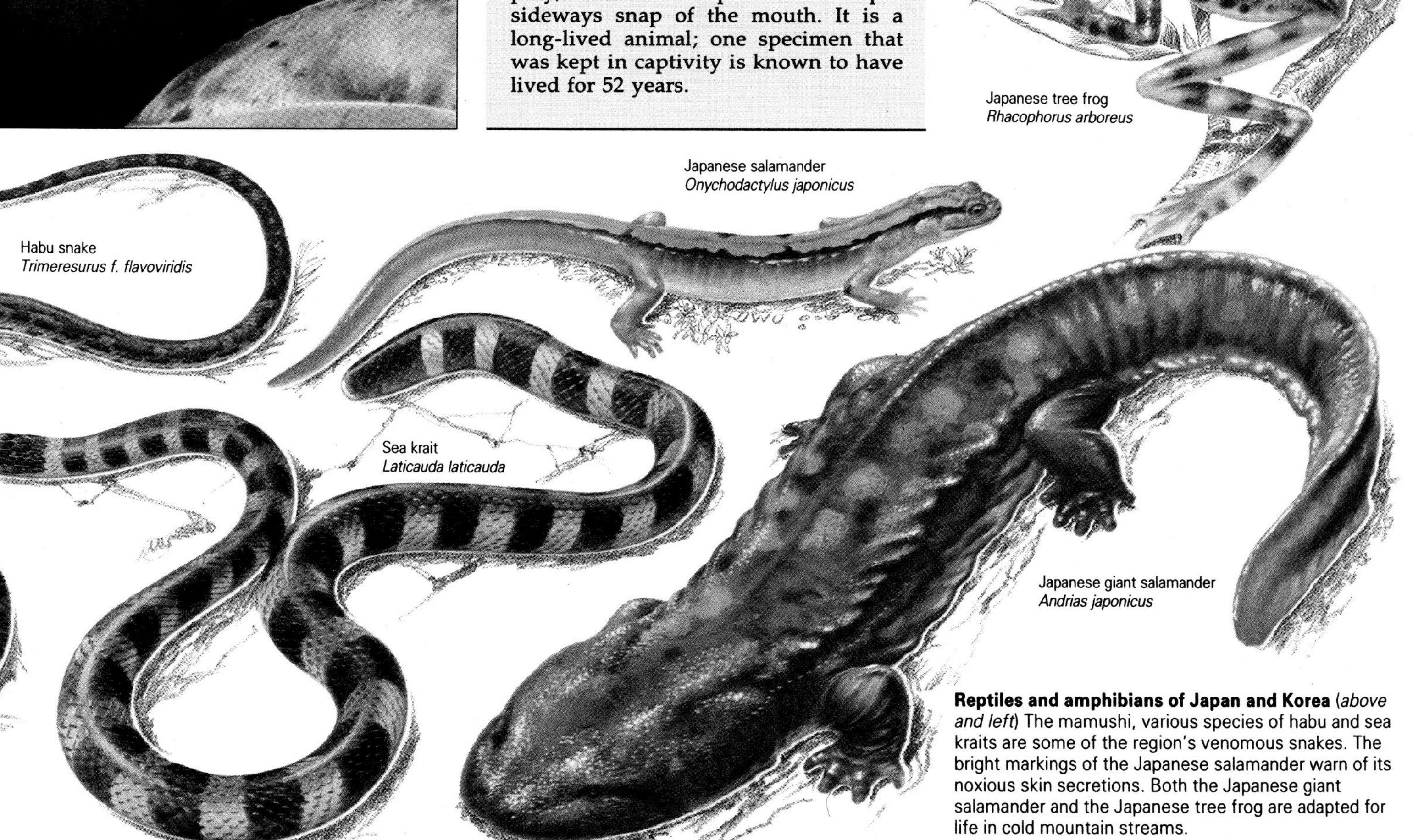

Reptiles and amphibians of Japan and Korea *(above and left)* The mamushi, various species of habu and sea kraits are some of the region's venomous snakes. The bright markings of the Japanese salamander warn of its noxious skin secretions. Both the Japanese giant salamander and the Japanese tree frog are adapted for life in cold mountain streams.

SLOW AWAKENING TO CONSERVATION

In the art and literature of Japan and in the minds of many of the people there is a deep attachment to animals and nature. However, this tradition has not been translated into practical action to conserve wildlife and habitats; there is a fatalistic, apparently indifferent view of the plight of many endangered species. Attitudes appear to be changing slowly now, although Japan remains a major market for the trade in threatened animals and wildlife products. It imports the skins of saltwater crocodiles and, for the leather trade, Yellow, Desert and Bengal monitor lizards; it is also the world's largest importer of turtle products.

Until recently Japan was perhaps the greatest consumer of ivory in the world, much of it used for *hankos*, the prestigious seals that substitute for signatures in contemporary Japan. Japanese scientists are working hard to devise an artificial substitute for ivory. Japan also imports some 80 percent of the world's musk, extracted from the male Musk deer – 1 kg (2 lb) of musk involves the slaughter of some 40 adult deer. The musk is used in perfumes, as a remedy for heart disease and as an aphrodisiac.

Whalers and fishermen

The Japanese whaling fleet still sails in spite of a storm of controversy over the sustainability of its harvest of the great whales. Whaling has a long history in Japan. Initially it was a local industry that harvested various whale species for their meat, until foreign whaling vessels arrived in Japanese waters in the early 19th century and hunted the large whales almost to extinction for their baleen and oil. Japan then turned its attention to deep-sea whaling. The International Whaling Commission agreed to suspend the commercial whaling of all great whale species in 1986; since then there has been increased criticism of Japan's continued killing of the great whales and locally caught Pilot whales and dolphins, which have been taken in growing numbers in recent years.

Of the many species of small whale in Japanese waters, some are harvested commercially for their meat, while others are killed because they destroy or become trapped in fishing nets. Perhaps as many as 30,000 animals, including Short-finned pilot whales and Blue-white dolphins, are killed each year – yet little is known about the effect of this on the various marine mammal populations. The issue is seen by many in Japan not as an ecological one, but as a question of national pride and harsh, short-term economics, with little thought for the future.

The Japanese have traditionally lived on rice and the products of the sea, so there were fewer domesticated animals in Japan than in neighboring lands. One species, however, served people in their harvest of the waters: Temminck's cormorant was used as far back as the 6th century for catching fish, and is still used to this day on the Nagara river on Honshu. Japan's cormorant fishing season is in the summer. The boats go out at night with lamps to attract the fish, and the cormorants sit on the gunwales with rope collars around their necks to stop them swallowing all but the smallest of the fish they catch.

Respect versus exploitation

Certain animals, such as cranes, are revered by the Japanese; many species were protected by the aboriginal Ainu of Hokkaido. Across the Sea of Japan, the Koreans also respect the crane as a symbol of good luck. The tiger, the Common magpie and other species play important roles in Korean folk art and superstition.

Many animals, however, are hunted by the Koreans either for meat or for the medicinal properties of various parts of their bodies. The trade in domestic dogs for the table in South Korea has recently been the subject of protest by various animal welfare organizations. There is also concern at the growing number of captive Asiatic black bears kept in North Korea and China for the bile that can be collected from the gallbladder, and used

Blakiston's fish owl (*above*) is one of seven specialized fish-eating owls in the world. It was revered by the Ainu, the indigenous people of Hokkaido, but habitat depletion and the contamination of its fish prey by pollutants have severely reduced its numbers.

BLAKISTON'S FISH OWL

Blakiston's fish owl is the most northerly of the fish owls and, at 71 cm (28 in) in length and 4 kg (9 lb) in weight, is one of the largest owls in the world. It inhabits wooded valleys throughout eastern Siberia, northeastern China, Hokkaido, Sakhalin and the Kuril Islands.The male and female are noted for the distinctive dueting calls that they make to each other.

The owl hunts not on land but on water. It feeds largely on fish and crayfish – snatching the fish from the surface like the osprey, and catching the crayfish by wading in the shallows. Small mammals, frogs and insects are also part of its diet.

Blakiston's fish owl was traditionally revered by the aboriginal Ainu of Japan as *kamui*, or the protector of the village. Ironically, through lack of protection to its habitat, it is now extremely rare in Japan; it is estimated that perhaps just 20 Blakiston's fish owl pairs remain. Its history exemplifies some of the main problems facing both wetland birds and birds of prey in the region. Its habitat has been destroyed by intensive farming and forestry methods, and its fish prey have been poisoned and contaminated by agricultural chemicals. Buildup of such pollution has led to the birds laying infertile eggs. The fish owl now appears to face extinction here, although dedicated conservationists are trying to protect the remaining pairs. Little is known of its prospects elsewhere in its range.

to treat digestive problems. Hundreds of these animals are reported to be kept in cages in appalling conditions after they have had operations to fix tubes to their gallbladders; these allow the bile to be drained off regularly.

Conservation of species and habitats in Japan and Korea has been slow to start, though it is gaining momentum. North Korea, which has the advantage of a relatively low population density, now has many protected areas. In Japan and South Korea there has been more emphasis on the preservation of scenery and tourist facilities in the protected areas than on management for conservation.

A winter refuge (*above*) Large flocks of Whooper swans spend the long cold months of winter on Japanese lakes and sea coasts. These enormous birds often roost on the snow-covered ice, their beaks tucked into their backfeathers. As their sources of food gradually freeze up, the swans move south to slightly warmer climates.

A local bird (*left*) This distinct East Asian race of the Yellow wagtail is just one example of the region's many races of common bird species. Each race differs slightly from its relatives in plumage, song or breeding time. This diversity has been produced by the variety of climates and habitats.

The Red-Crowned Crane

Display of the Red-crowned crane (*above*) Visual displays or dances reinforce bonding between crane pairs, and between members of a flock. Crane dances have been celebrated for centuries in Japanese art and literature. The young birds often join in, some even dancing with nearby crows.

The crane dance (*below*) The ritual dance consists of head bobbing and deep bows, leaping and grasping objects such as feathers or stones with the bill and tossing them in the air, and running with wings flapping in short, low flights. The bows and leaps usually alternate between pair members, one bird bowing while the other leaps. The "unison call", in which both birds utter distinctive calls according to their sex, occurs simultaneously. If the two cranes are upright at the same time and facing each other, they immediately shift into threat postures – flapping, stamping and arching – before returning to their dance.

Precarious migration (*right*) The region's wetlands are important wintering and stopover sites for several species of crane. Drainage for cultivation and construction is drastically limiting the cranes' opportunities for resting and feeding; widespread use of pesticides and chemicals is a further danger.

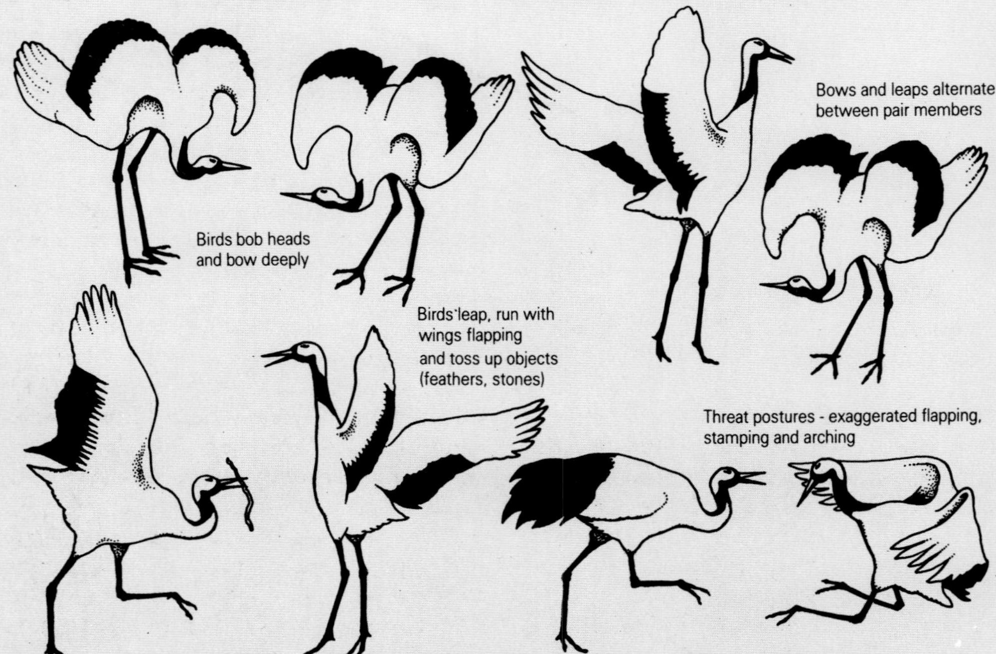
Birds bob heads
and bow deeply
Birds leap, run with
wings flapping
and toss up objects
(feathers, stones)
Bows and leaps alternate
between pair members
Threat postures - exaggerated flapping,
stamping and arching

Spring comes late to eastern Hokkaido, and the marshes of Kushiro, near the coast, are a bleak place to live for most of the year. They are covered in snow and ice throughout the winter, and even in summer there are cold foggy periods and heavy rains that last for several days, interspersed with days of hot sun. It is here that perhaps a third of the world's population of Red-crowned cranes come each spring to nest in the reed beds. Although much reduced in range and

numbers from earlier times, the cranes retain an important place in the minds of the Japanese people, and their fortunes are keenly followed from year to year by the public.

Red-crowned cranes were thought to be extinct at the turn of the century, but a few survived in the marshes of Kushiro and numbers slowly built up again. Other populations breed in northeastern China, and some of these migrate south to Japan and Korea in the winter, but altogether, the total world population is probably not more than 1,500 birds.

Faithful for life

The cranes arrive and settle at the breeding grounds in March, when the snow is still on the ground. They come in pairs, and stay paired for life. They will have already started the annual courtship that serves to bond each pair on the wintering grounds up to 50 km (30 mi) away, but their unusual ritual courtship dancing starts in earnest once they arrive on the marshes.

The timing of nesting is important: if too early there may not be enough food for the chicks; too late and the chicks may not grow large enough to survive the following winter. A nest is built as a platform in the reeds, and generally two eggs are laid by late May. Both sexes share the incubation and feed the young. The adults' diet consists of eels, frogs, fish and sometimes ducklings and other baby birds, while the chicks are fed on insects, tadpoles, snails and small fish. If the eggs and the chicks survive the hazards of the weather, floods, fires, egg thieves and predators, the family flies inland in early September to the farmland wintering grounds, where they feed on small mammals, insects, seeds and bulbs.

Cranes in danger

In former times the cranes were hunted as they flew south in the winter; more recently collision with electricity pylons has been a common cause of death, prompting the authorities to hang marker buoys on the wires. The numbers of adults have been steadily increasing in recent years, but the number of young fails to show a steady increase: the same number of young are fledged each year now as when the adult population was half the size. There is concern that the population is restricted by the lack of suitable breeding territories. At first the birds started to move out to surrounding marshes, but many of these have been, or are being, drained for agriculture or for building. The luxuriant reed plains are fast disappearing under rice and concrete.

Besides affecting the cranes, loss of wetlands threatens egrets, ibises and other waterbirds throughout the region. Many of these bird populations have already been seriously reduced by heavy pesticide use in the 1950s and 1960s.

Japan nevertheless retains some important feeding grounds for other species of crane. Kyushu is a traditional wintering ground for thousands of Hooded and White-naped cranes, and Demoiselle, Sandhill, Common and Siberian white cranes often accompany these large flocks. Kushiro is now one of 72 designated wetlands of international importance, and it is hoped that the growing environmental awareness in Japan will lead to the protection of other wetland areas to preserve these magnificent birds.

WILDLIFE IN ISLAND SECLUSION

ISOLATION AND IMMIGRATION · COPING WITH ARIDITY · UNEASY COEXISTENCE

Throughout the millions of years of Australasia's isolation, animal life developed along different lines from the rest of the world. Australia became isolated from the ancient supercontinent, Gondwanaland, before the rise of advanced placental mammals such as antelopes, deer and monkeys. Instead, it is inhabited by a unique range of animals dominated by pouched mammals – the marsupials – which diversified to fill the evolutionary niches that were elsewhere occupied by the placental mammals. New Zealand became isolated even earlier, and in the absence of predators many species of flightless birds evolved. Bats are the only mammals to have arrived here unaided. Antarctica is virtually barren, but the nutrient-rich seas support large numbers of seabirds and marine mammals.

COUNTRIES IN THE REGION

Australia, Fiji, Kiribati, Nauru, New Zealand, Papua New Guinea, Solomon Islands, Tonga, Tuvalu, Vanuatu, Western Samoa

ENDEMISM AND DIVERSITY

Diversity Low (many coral islands) to high (parts of Australia, New Guinea)
Endemism High to very high

SPECIES

	Total	Threatened	Extinct†
Mammals	450	66	23
Birds	1,500*	135	34
Others	unknown	199	3

† species extinct since 1600
** breeding and regular non-breeding species*

NOTABLE THREATENED ENDEMIC SPECIES

Mammals Northern hairy-nosed wombat *(Lasiorhinus krefftii)*, Woodlark Island cuscus *(Phalanger lullulae)*, numbat *(Myrmecobius fasciatus)*, Greater bilby *(Macrotis lagotis)*, Pohnpei flying fox *(Pteropus molossinus)*, Samoan flying fox *(Pteropus samoensis)*
Birds kagu *(Rhynochetos jubatus)*, Black stilt *(Himantopus novaezelandiae)*, Rapa fruit-dove *(Ptilinopus huttoni)*, kakapo *(Strigops habroptilus)*, Paradise parrot *(Psephotus pulcherrimus)*, Guam flycatcher *(Myiagra freycineti)*
Others Western swamp turtle *(Pseudemydura umbrina)*, tuatara *(Sphenodon punctatus)*, Baw baw frog *(Philoria frosti)*, Short Samoan tree snail *(Samoana abbreviata)*, Stephens Island weta beetle *(Deinacrida rugosa)*, Queen Alexandra's birdwing butterfly *(Ornithoptera alexandrae)*

NOTABLE THREATENED NON-ENDEMIC SPECIES

Mammals dugong *(Dugong dugon)*, Humpback whale *(Megaptera novaeangliae)*
Birds Nicobar pigeon *(Caloenas nicobarica)*
Others Estuarine crocodile *(Crocodylus porosus)*, Whale shark *(Rhincodon typus)*, Giant clam *(Tridacna gigas)*

DOMESTICATED ANIMALS (originating in region)

budgerigar *(Melopsittacus undulatus)*

The most feared of fish The Great white shark – or maneater – has an undeserved reputation: very few swimmers are killed by it, and even fewer eaten. It feeds on large fish and other marine animals, and can reach a length of 6 m (20 ft).

ISOLATION AND IMMIGRATION

Some of the animals of Australia have affinities with those of Africa, India and South America. These are remnants from the time 140 million years ago when the four landmasses – and Antarctica – were fused together in a giant southern supercontinent, covered in tropical and subtropical forests, called Gondwanaland. This broke up over the next 95 million years; India and Africa separated first, followed by New Zealand about 80 million years ago and Australia, which broke away from Antarctica, some 35 million years later. This explains why some of Australia's wildlife, such as the possums, show similarities to animals found in South America. Lungfish are found in Africa and South America, as well as in Australia; and large flightless "ratites" (birds that do not have a keeled breastbone), such as the South American rheas and African ostriches, are related to the emus and cassowaries of Australia and New Guinea. Other ancient Gondwanan animals include turtles of the family Chelidae and several families of insects, spiders and terrestrial mollusks.

After breaking away, Australia drifted north, its climate and vegetation changing as it went. This was the stimulus for the marsupials to evolve into many different forms as they adapted to the new ecological niches. On other continents the marsupials were outcompeted by the rapidly evolving placental mammals, but in Australia they flourished.

Contact with Asia

The Australian continental plate gradually drifted northward and came up against the Southeast Asian continental plate, pushing up the mountains of New Guinea. Marsupials invaded the new territory. Tree kangaroos and other descendants of the occupants of the old Gondwanan forests still survive in the tropical forests of New Guinea and northeastern Australia. Many spectacular bird species have evolved in the isolation of these mountain forests: the birds of paradise, with their splendid courtship plumage; the lyrebirds, with their lacy fantail shaped like a lyre; and the bowerbirds, whose drab males build elaborate bowers which they decorate with flowers, leaves and even human artefacts and paint them with plant extracts applied with a frayed plantstem brush. Australia also supports numerous parrot species; the family has greatly diversified, and includes the cockatoos and the nomadic budgerigars, which wander the deserts in large flocks.

Animals from Southeast Asia were able to reach Australia by island hopping from New Guinea. Bats arrived about 30 million years ago, and ratlike rodents about 7 million years ago. Many species of birds, together with most of the continent's families of lizards and snakes, were also colonizers. Some animals came across the oceans from South America (New Zealand's bats, for example) and India (certain Australian worms). New Zealand and the Pacific islands – because of their isolation – also evolved unique animal species that are specially adapted to the local conditions.

The collision with Southeast Asia interrupted the equatorial ocean current, and warmwater currents now flow down the west and east coasts of Australia, bringing with them an abundant array of tropical marine creatures. The seagrass beds of the north are grazed by Sea cows (dugongs), while the southern coast's kelp forests provide a habitat for a rich variety of fish and marine mammals.

Off the east coast of Australia lies the largest coral reef in the world: the Great Barrier Reef. More than 18 million years old, it contains some 300 species of coral, a third of the world's total. This reef is perhaps the most species-rich habitat on Earth; its 2,000 fish species range from tiny but well-camouflaged gobies to the streamlined 5 m (16.5 ft) long Tiger shark.

Paradise on earth A male Emperor of Germany bird of paradise puts on a dazzling courtship display. Birds of paradise are found only in the upland forests of New Guinea, eastern Australia and the Moluccas. The females are drab, but better camouflaged.

Marsupial diversity

The very earliest marsupials were tree-dwelling animals, very similar in form to today's tree shrews: small, gray-brown insect-eaters with sharp teeth and claws, and long whiskers. The descendants of the early marsupials supplemented their insect diet with fruit and seeds. Similar marsupials still survive; for example the Eastern pygmy possum of Tasmania and the rat opossums of South America.

Terrestrial carnivorous marsupials – the dasyurids – include the "native cats" or quolls, and the related thylacine or Tasmanian wolf, which is believed to have become extinct in the 1930s. Another dasyurid, the Tasmanian devil, will take prey, but prefers carrion. The numbat specializes in eating ants and termites. Like the placental anteaters it has a pointed nose for poking into rotting wood, and a long tongue that can extend to half its body length, enabling it to lick up several hundred termites in a second.

Many marsupials have adapted to a tough diet of leaves and fruit. These include the possums, which evolved grinding teeth and a longer gut for the purpose. Possums have many similarities with the placental primates, such as good stereoscopic vision and opposable thumbs on their hands and feet; some also have prehensile tails – adaptations to an arboreal life. The Striped possum of the Queensland rainforest shows a remarkable similarity to the aye-aye of Madagascar. Both use the nail of their long fourth finger to extract insects from crevices in bark. The Honey possum specializes in feeding on nectar and has a long, brush-tipped tongue. In the open eucalyptus forests, the gliders, relatives of the possums, move from tree to tree using the folds of skin between their outstretched forelimbs and hindlimbs to catch the air like a parachute.

Many leaf-eating animals use symbiotic bacteria in their gut to help them digest tough plant material. One of the marsupial herbivores to do this is the koala. The young koala is fed on a pap of soft green feces, presumably to infect its gut with the right organisms.

Some of the possums and the related bandicoots, harelike marsupials, adapted to life on the ground and colonized the grasslands that were beginning to form. The underground expert is the Marsupial mole, which lives in deserts and semi-deserts. This mole is well adapted for life below ground, with large flat claws for digging, a horny shield over its naked nose and only rudimentary eyes and ears.

As Australia drifted toward the tropics the climate became hotter and drier, and desert grassland increased in area. This permitted kangaroos and wallabies – descendents of the small, possumlike, tree-dwelling marsupials – to extend their range right across the vast desert interior with its rocky outcrops.

COPING WITH ARIDITY

Unusual among the mammals are the monotremes, or egg-laying mammals. This trait is shared only by the Short-beaked echidna of Australia, the Long-beaked echidna of New Guinea and the Australian platypus. The Short-beaked echidna is one of the most widespread and successful animals of the Australian bush, living in both temperate and arid areas. It has a specialized diet of ants and termites, and it occupies a wide range of terrestrial habitats in both Australia and New Guinea.

Echidnas are widespread throughout Australia's hot, arid interior. Yet, lacking sweat glands, they cope poorly with heat, dying quickly when exposed to temperatures above 35°C (95°F). Echidnas avoid high temperatures by digging deep burrows with their powerful, clawed limbs and emerging only at night. To maintain water balance in the arid zone, they select termites in preference to ants because of their higher water content. Prey selection is made partly by smell but probably also by a remarkable battery of electroreceptors located on the tip of the snout. In colder environments echidnas feed mainly on ants and become more active by day. By storing up to 30 percent of their body weight as fat, and by dropping their body temperature from the usual 33°C (91.4°F) to just 6°C (42.8°F), echidnas can enter a state of torpor for up to 10 days; they thus survive blizzards and bouts of extreme cold.

Tree-dwelling possums These marsupials have sharp claws and opposable thumbs for climbing and grasping branches. The Common brushtail possum has large ears and eyes, adaptations to a nocturnal life. The Scaly-tailed possum has a naked prehensile tail, small ears and shorter fur.

Common brushtail possum
Trichosurus vulpecula

Scaly-tailed possum
Wyulda squamicaudata

Refuge underground

Numerous other Australian animals survive the arid zone by burrowing to avoid both loss of moisture and extremes of temperature. Many species of frog burrow to 1 m (3 ft) or more, and emerge to spawn only after periods of heavy rain. The Water-holding frog can absorb water through its skin from the soil of its burrow, and it can store up to 57 percent of its body weight as water in its bladder. In this and several other species, buried individuals retain a cocoon of sloughed-off skin, which reduces water loss by a remarkable 96 percent. Some desert crabs also escape extremes of heat by burrowing, surviving a water loss of up to 32 percent of their wet body weight during extreme drought. Desert-adapted trapdoor and funnel-web spiders are able to take up moisture from the damp soil of their burrows.

Many species of desert gecko exploit the burrows of desert spiders; only a few, such as the Knobtailed gecko, dig their own. Some theridiid spiders (species that spin scaffoldlike webs) occupy the "nose-poke" holes and scratchings of echidnas. Only shaded northfacing holes are used, especially if the adjacent scratchings have created pitfall traps that catch the small invertebrate prey of the spiders.

The ancient lungfish

Lungfish have stayed almost unchanged for more than 110 million years. They live in a few slow-flowing streams on the temperate northeastern coast of Australia. Although they obtain most of their oxygen from water, functional lungs enable them to breathe on land for several days. This ability, coupled with stiffened fin rays and a well-developed pelvic girdle, allows the fish to walk underwater or make short forays overland.

Gliders and possums have a wide variety of diets. The Feathertail possum is omnivorous, while the Yellow-bellied glider feeds mainly on eucalyptus gum and sap, which it obtains by cutting large V-shaped notches in the bark. The Greater glider feeds principally on leaves. These gliders supplement their diet with protein from pollen and insects, especially during the breeding season. The Mountain pygmy possum has a mixed diet of seeds, fruit, leaves, insects and other invertebrates, often foraging on the ground, and even under the snow in winter. It has specialized teeth for cracking open seeds, which it may store as a precaution against winter scarcity.

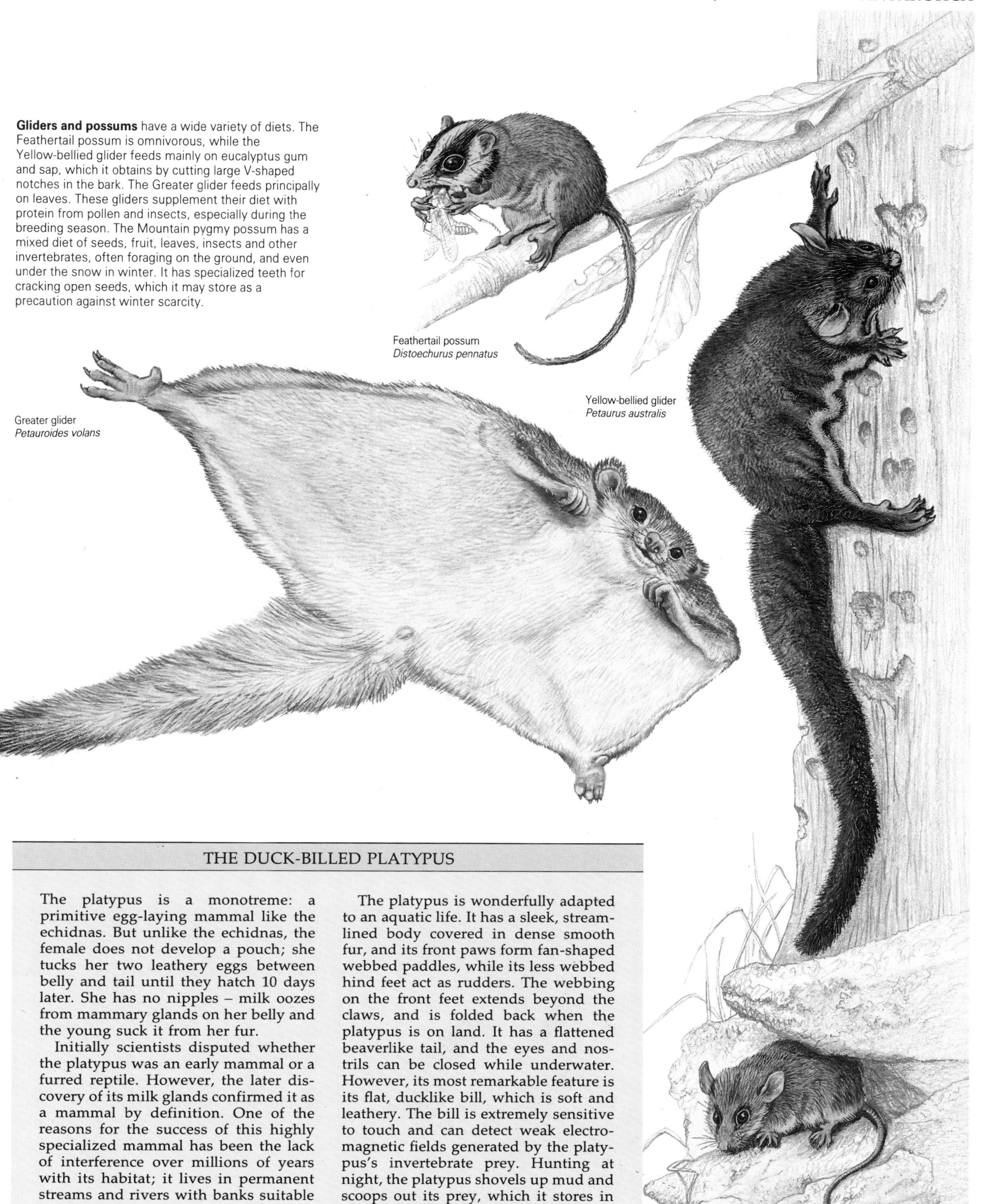

Feathertail possum
Distoechurus pennatus

Yellow-bellied glider
Petaurus australis

Greater glider
Petauroides volans

Mountain pygmy possum
Burramys parvus

THE DUCK-BILLED PLATYPUS

The platypus is a monotreme: a primitive egg-laying mammal like the echidnas. But unlike the echidnas, the female does not develop a pouch; she tucks her two leathery eggs between belly and tail until they hatch 10 days later. She has no nipples – milk oozes from mammary glands on her belly and the young suck it from her fur.

Initially scientists disputed whether the platypus was an early mammal or a furred reptile. However, the later discovery of its milk glands confirmed it as a mammal by definition. One of the reasons for the success of this highly specialized mammal has been the lack of interference over millions of years with its habitat; it lives in permanent streams and rivers with banks suitable for burrowing.

The platypus is wonderfully adapted to an aquatic life. It has a sleek, streamlined body covered in dense smooth fur, and its front paws form fan-shaped webbed paddles, while its less webbed hind feet act as rudders. The webbing on the front feet extends beyond the claws, and is folded back when the platypus is on land. It has a flattened beaverlike tail, and the eyes and nostrils can be closed while underwater. However, its most remarkable feature is its flat, ducklike bill, which is soft and leathery. The bill is extremely sensitive to touch and can detect weak electromagnetic fields generated by the platypus's invertebrate prey. Hunting at night, the platypus shovels up mud and scoops out its prey, which it stores in cheek pouches until it surfaces.

UNEASY COEXISTENCE

Aborigines have occupied Australia for at least 40,000 years and the Pacific islands for up to 20,000 years; European exploitation of the region began only in the 18th century. The European colonizers introduced methods of settled agriculture that imposed rapid and far-reaching changes on the native wildlife. Australia is now a country of 17 million people, 24 million cattle and 138 million sheep. The conflict between maintaining this enormous primary livestock industry, with its annual turnover of $5 billion, and trying to mitigate its appalling impact on the continent's fragile ecosystems, is perhaps the most striking dilemma facing the region.

Use of native species

Although there is an overall agricultural and economic reliance on introduced animals throughout Australasia, much of the native wildlife of the region is also exploited. Ten species of kangaroo are culled for pest control, for the leather and pet food industries and, in Tasmania and southern Australia, for human consumption. Koalas were rare at the time of European settlement due to predation by Aboriginal hunters, but their numbers increased during the 19th century as the indigenous peoples declined. However, the koalas' reprieve was shortlived; in the depressed years of the early 20th century they were slaughtered in great numbers for their fur, with a million being taken in

Eucalyptus eater *(above)* The koala's principal diet of eucalyptus leaves is fibrous and low in protein. The leaves are also poisonous, and must be detoxified in the koala's specialized digestive system. As an adaptation to this poor diet, the koala has low energy requirements: it moves slowly and sleeps for up to 18 hours a day.

Prolific immigrant *(left)* The Cane toad was brought to Australia in the 1930s to combat a sugar cane pest. With no natural enemies, the toads multiplied rapidly, devouring countless native invertebrates, including beneficial insects.

The primitive tuatara *(right)* This is the sole survivor of an order of reptiles that evolved 200 million years ago. Its most famous feature is a "third eye" on top of its head. This has a lens and a retina, and is sensitive to light, but it cannot form visual images.

Queensland alone in 1919. They are now totally protected.

In Tasmania Brushtail possums are trapped for their long silky fur (up to 400,000 animals a year), but with little obvious effect on the population. Tasmanian and Australian brushtails were transported to New Zealand from the mid-19th century to establish a fur industry there but this introduction proved to be a double-edged success. Possum populations reach much higher densities there than in Australia, and their pelts earn several million dollars every year in exports. However, the possums' selective predation of certain tree-dwelling species such as the endemic New Zealand rat has had far-reaching effects on the forests and forest animals of New Zealand.

In New Guinea and on many of the region's smaller islands, land mammals are hunted for food. There are fears for the long-term future of many species, such as cuscuses (relatives of Brushtail possums), due to overhunting; on some islands of Indonesia's Flores Sea (especially Flores, Sumba and Timor) the native rodents have been hunted to extinction. Probably the only endemic mammal to have been domesticated here is the Sulawesi pig. First tamed in the early Holocene epoch – approximately 10,000 years ago – this species was traded extensively among the Lesser and Greater Sunda Islands of southern Indonesia, and persists on Flores, the Moluccas and Timor.

Many other native species are cultivated for food or export throughout Australasia. Prawns, lobsters and shellfish are farmed in New Zealand, New Guinea and Australia, and on some of the Pacific islands. Crocodiles are farmed for their flesh and skins in northern Australia, butterflies for the collecting trade in New Guinea. The illicit export of reptiles, frogs and some insects, especially butterflies, is conducted from Australia and the larger islands of the region, and the trade in colorful birds out of northern Australia has caused alarming declines in the population levels of many species.

FLIGHTLESS BIRDS OF AUSTRALASIA

New Zealand could be called the land of birds. Before the arrival of humans and their exotic animals, the islands boasted only two species of mammals – both bats – and 30 reptiles. The birds had diversified to fill many of the ecological niches usually occupied by mammals, including browsers, grazers, hunters and scavengers. The kiwi is a good example; it not only evolved hairlike feathers, but also a sensitive nasal organ in the head that enables it to detect prey by smell. Many Pacific islands were also colonized by birds that thrive in the absence of mammals.

Many of these birds are flightless or nearly flightless. With no ground-dwelling predators, there is little need for flight, which uses a great deal of energy, and imposes limitations on body size and shape. Many flightless birds are larger than their close relatives: the kakapo, the world's largest parrot, is 63 cm (25 in) tall.

Of all the world's birds known to have become extinct in the last 400 years, more than 90 percent were flightless. They include the moas and some of the rails. Flightless birds were easily captured for food by sailors that passed through the islands; and once ground-dwelling predators such as cats, rats, dogs and pigs had been introduced, the birds were ill-equipped to defend themselves. The only hope for many of the surviving species is to transfer them to islands from which all such predators have been eliminated.

Troublesome newcomers

Few regions have been as devastated as Australasia by the introduction of exotic mammals. Over the last 3,000 years three species of rat have accompanied human settlers as stowaways to most of the Pacific islands. In general, Common rats prey mainly on ground-dwelling birds, whereas Ship and Pacific rats will also raid the nests of tree-dwelling species. Hedgehogs, weasels, ferrets, cats, dogs, foxes and a great many herbivorous mammals have been introduced more recently, particularly to Australia and New Zealand. During the last 200 years, many species of flightless birds throughout the region, and 18 of the 199 species of native Australian mammals, have become extinct. This can be attributed, at least in part, to these unwelcome immigrants.

Among the earliest animal imports to the region were dingoes, or wild dogs, which were brought to mainland Australia by Aborigines some 4,000 years ago. The arrival and spread of dingoes coincided with the decline and mainland extinction of their marsupial counterpart, the thylacine or Tasmanian wolf. Throughout Australasia introduced foxes, along with cats, have wrought havoc with island birds and reptiles – turtles having suffered particularly badly – and small marsupials and rodents. The deadly effects of these predators are sometimes difficult to distinguish from the results of competition from rabbits, goats, deer, domestic stock and other imported species. However, successful poisoning campaigns against these pests in Australia and New Zealand offer some hope for the continued survival of at least some native species.

Kangaroos

Kangaroos are among the most widespread and familiar of the marsupials. Although they are found throughout Australia, New Guinea and nearby islands, in almost every kind of habitat, the greatest numbers are seen on the grasslands. Here they fulfill the same role as the great herds of antelopes in Africa and bison in North America. Ranging from the size of a rat to that of a human, and as fleet-footed as antelopes, kangaroos hop over the grasslands on their powerful hindlegs. Their forelimbs are much weaker, but the front paws are well armed with claws, used for grooming. The kangaroo's long tail serves as a balance when the animal is moving, and in the larger species as a prop when it is moving slowly or resting.

Kangaroos evolved from small possum-like tree-dwelling marsupials, shy and nocturnal animals that fed on plants and insects. A few species of rat kangaroo still live in this way, but many are terrestrial; their tails have remained partly prehensile and are used for carrying nest material, and they have a generalized diet. Rat kangaroos are often regarded as ancestral to other kangaroos, and are placed in a separate family.

As they evolved kangaroos developed enlarged guts that support symbiotic bacteria, which help to break down the tough plant material that the animals feed on. They have extremely sharp incisors for cutting, and broad-ridged molars for grinding their food.

The large Red kangaroos, which inhabit the open grasslands, are the most sociable species. They live in herds of "mobs" of

Meal in a pouch A young Eastern gray kangaroo sucks milk from teats hidden in its mother's pouch. It will continue to suck occasionally, even after the birth of the next young, from a separate teat supplying milk of a different composition than that for the baby.

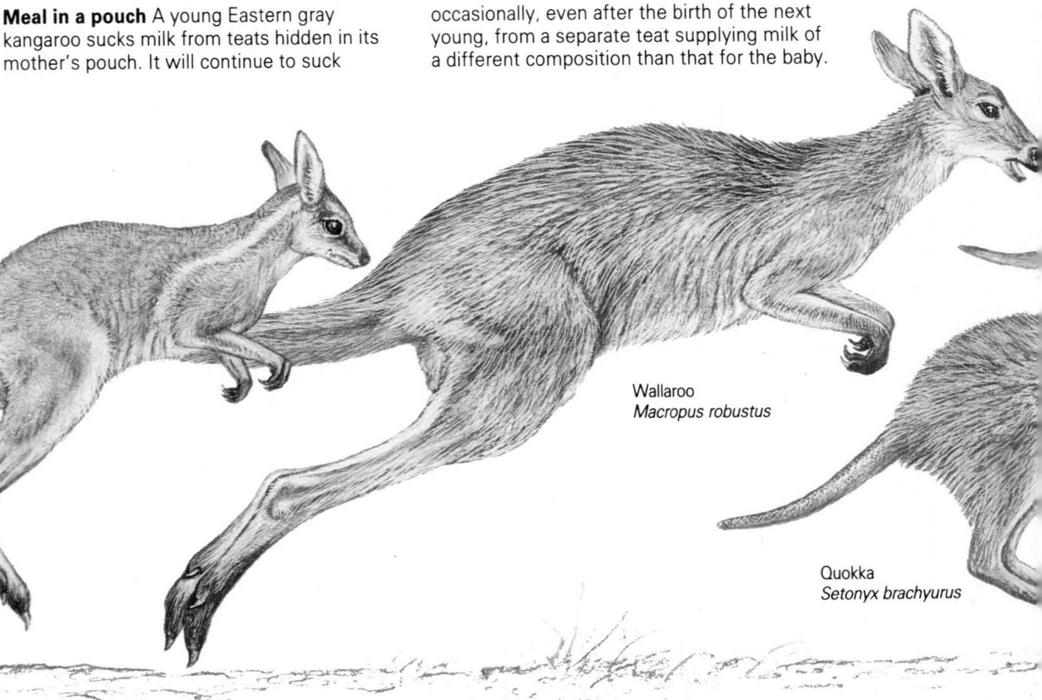

Kangaroo boxing Before a fight, rival males may engage in a stiff-legged walk in the face of the opponent, and in scratching and grooming, standing upright on extended rear legs and using the tail as a prop. The fight is initiated by locking forearms and attempting to push the opponent backward to the ground. Fights usually occur when one male's monopoly of access to an individual or group of females is challenged. There appears to be no defense of territory for its own sake.

up to 10 animals, which increases their safety. A kangaroo's principal defense, however, is flight: a Great gray kangaroo can keep up a speed of 88 kph (55 mph) over a short distance. If cornered, the animal uses its powerful hindlegs and strong claws to defend itself; a kangaroo's kick can disembowel a human. Since the demise of the Tasmanian wolf and local reductions in numbers of dingoes, kangaroos have few enemies apart from the small number of Wedge-tailed eagles that patrol the grasslands, and humans.

Highly adaptable

Kangaroos have adapted well to the hot, arid conditions of the Australian interior. Their thick reflective fur serves to keep out the heat. They pant to cool themselves and also lick their forelimbs and chest; the evaporating mucus has a cooling effect. They can go for weeks and even months without drinking so long as they can eat moist vegetation.

The rock wallabies live in rocky outcrops. Their feet are studded for gripping uneven slopes, and they can leap up to 4 m (13 ft) from rock to rock. The coats of the various species have attractive colors and patterns that camouflage them in the strong light and shade of the rocks. For this reason they have been heavily hunted for their pelts.

The little rabbit-sized hare wallabies are extremely good jumpers. When pursued they skillfully dodge from side to side as they run, sometimes even turning back toward their pursuers and jumping over their heads.

One group of kangaroos has returned to the trees. These are the tree kangaroos of the tropical rainforests. They have strong arms and sharp claws for climbing tree trunks, but on the branches they walk awkwardly on all four legs.

The marsupials' method of reproduction distinguishes them from other mammals. Usually a single, very tiny baby kangaroo or "joey" is born after only a month's gestation. Naked, blind, deaf and with only its forelimbs developed, it finds its way unaided to its mother's pouch; there it fastens itself to one of her four teats, which it probably locates by smell. It stays in the pouch for 5 to 11 months and will still return to be suckled occasionally until it reaches sexual maturity one or two years later.

The kangaroo today

Changes in the composition of the grasslands, together with the loss of shrubs due to livestock farming, have reduced the habitat suitable for many smaller species of kangaroo. This has exposed them to predation by introduced carnivores such as dogs, foxes and cats, as well as to native predators. However, the removal of tall grass is beneficial to the larger species, which increase rapidly, especially where new waterholes have been provided.

Kangaroos are consequently culled as pests. State quotas allow for the annual culling of up to 3 million Red and Gray kangaroos, some 15 percent of the current populations; however, illegal shooting means that perhaps twice this number are taken. Kangaroo meat is highly palatable and suitable for human (and pet) consumption, and the hides can be used for leather. Recent debate has centered on a proposal to farm free-range kangaroos in place of sheep on marginal rangelands. If adopted, this would lower farm income, but would also substantially reduce the damage to vegetation and soil that is currently wrought by stock.

Wallabies and kangaroos The Bridled nailtail wallaby has a horny nail at the tip of its tail. The wallaroo is adapted to desert life. The quokka makes tunnels in dense undergrowth. The Red-legged pademelon lives mainly in forests. The Yellow-footed rock wallaby is a good climber. The Gray forest wallaby has strong arms like those of tree kangaroos.

GLOSSARY

Adaptation
Any feature of a living ORGANISM that enhances its chances of survival in the ENVIRONMENT in which it lives. ADAPTATIONS may be genetic, selected by NATURAL SELECTION, and hence not alterable within the organism's lifetime, or they may be phenotypic, produced by adjustment on the part of the individual and reversible within its lifetime.
Adaptive radiation
A burst of EVOLUTION of a SPECIES or group of species into a number of new forms adapted to a variety of HABITATS or to more specialized modes of life. This may result in the evolution of new species or even of new ORDERS of ORGANISMS.
Adult
A fully developed animal or plant that has reached sexual maturity and is capable of breeding.
Affinities
Similarities between living ORGANISMS indicating EVOLUTION from a common ancestral stock.
Algae
Small photosynthetic (see PHOTOSYNTHESIS) organisms that live in water or damp ground and do not have true stems, leaves or roots.
Alien species
A SPECIES that is not native to (did not evolve in) its present geographical location, but has been introduced by human agency.
Alpine
A treeless ENVIRONMENT found on a mountain above the tree line but beneath the limit of permanent snow.
Ammonites
Members of an ORDER of extinct squidlike MOLLUSKS that had calcareous curved or spiral shells, and are commonly preserved as FOSSILS.
Amphibians
Members of the class Amphibia; VERTEBRATE animals that live on land but whose life cycle requires some time to be spent in water, eg the frog.
Arboreal
Living on or among trees (in the trees as opposed to on the ground).
Arid
Describes an area that is hot and dry. Arid areas are rarely rainless, but rainfall is intermittent and quickly evaporates or sinks into the ground. Little moisture remains in the soil, so vegetation is sparse.
Arthropods
A general term for INVERTEBRATE animals characterized by a segmented body, hard integument (external skeleton) and paired jointed limbs. The group includes a number of PHYLA, namely the CRUSTACEANS, INSECTS and spiders.
Asexual reproduction
Reproduction which does not involve fertilization and is devoid of any sexual process that involves the mixing of genetic material from two different individuals.

Bacteria
Microscopic ORGANISMS, usually single-celled, whose genetic material is not enclosed in a membrane sac.
Baleen
A horny substance, commonly known as whalebone, growing as plates from the upper jaws of FILTERFEEDING whales; it forms a fringelike sieve for filtering PLANKTON from seawater.
Biodiversity
Biological diversity – the genetic diversity of living ORGANISMS.
Biological control
The introduction by humans of one SPECIES (usually a PREDATOR or PARASITE) to control the POPULATION of another species. It is used in agriculture to control weeds and pests.
Biomass
The total mass of all the living ORGANISMS in a defined area or ECOSYSTEM, usually expressed as dry weight per unit area.
Biosphere
The thin layer of the Earth that contains all living ORGANISMS and the ENVIRONMENT that supports them.
Breed
(1) To produce offspring. (2) The crossing of selected individuals to produce offspring with desired characteristics. (3) A race or strain whose members, when crossed, produce offspring with the same characteristics as the parents.
Browser
An HERBIVORE that feeds on the leaves and shoots of trees or shrubs.
Burrowers
Animals that excavate holes in the ground.

Camouflage
The way in which an animal is colored so as to blend with its background and avoid detection. The effectiveness of camouflage coloring is often enhanced by the adoption of specific postures.
Canine distemper
An infectious viral DISEASE that affects members of the dog family. Symptoms include fever, discharge from nose and eyes and, in severe cases, fits.
Carnivore
(1) An animal that eats the flesh of another animal. (2) A member of the MAMMAL ORDER Carnivora, a group that includes cats, dogs, bears, badgers, weasels and other PREDATORS.
Cartilaginous
Made of cartilage rather than of bone. Cartilage is a flexible skeletal tissue that lacks the calcium carbonate that confers rigidity on bone.
Caste
Among social INSECTS, the existence of more than one type of functional individual within the same colony. Each caste has its own specific role in the life of the colony, and is usually distinguished by its morphology, age and/or sex. For example, queens, workers and drones are distinct honeybee castes.
Cells
The basic unit of living matter in the body of an ORGANISM. Each cell is bounded by a protein-lipid membrane.
Cetacean
Member of the MAMMAL ORDER Cetacea, which includes whales, dolphins and porpoises. Cetaceans are streamlined, almost hairless, AQUATIC MAMMALS with forelimbs that are modified to form paddles, no hindlimbs, and a horizontal tail fin.
Circulation system
The structures, usually including heart, arteries and veins, in which and by means of which blood is circulated around the body of an animal.
CITES (Convention on International Trade in Endangered Species)
An international agreement signed by over 90 countries since 1973. SPECIES placed in Appendix I of CITES are considered to be in danger of extinction, and trade is prohibited without an export permit. Appendix II species could be threatened with extinction in the future if trade is not regulated.
Class
A rank in the taxonomic hierarchy coming between PHYLUM and ORDER.
Classification
A system of arranging the different types of living ORGANISMS according to the degree of similarity of their inherited characteristics. The classification system enables organisms to be identified and may also reveal the relationships between different groups. The internationally accepted classification hierarchy groups organisms first into divisions, then PHYLA, CLASSES, ORDERS, GENERA, SPECIES and SUBSPECIES.
Climax vegetation
The final stage in the evolution of an ECOSYSTEM, when the overall composition of the vegetation does not change further (other than regular cyclical changes) and there is no further net growth in biomass. The greatest DIVERSITY of plant and animal life is usually reached at the climax. The ecosystem is then stable and self-sustaining.
Colonization *
The establishment of plants or animals in a new ENVIRONMENT.
Colony
(1) A group of individual animals or plants that are physiologically connected to each other. Examples are colonial CORALS and the Portuguese man-o'-war jellyfish. (2) A distinct localized population of animals, for example termites, seabirds etc.
Commensalism
An association between two or more SPECIES, in which one species benefits from the relationship but the other is not harmed. The term is sometimes used in a broader sense to include relationships in which both species benefit.
Community
A group of POPULATIONS of various SPECIES occupying a common ENVIRONMENT and interacting with each other.
Competition
The struggle between individuals of the same or different SPECIES for food, space, light and nutrients.
Competitor
A living ORGANISM or SPECIES that interacts with another in its attempts to gain a share of a commonly required resource.
Conservation
In relation to living things, the planned management and wise use of living resources in order to maintain their sustainable use and ecological DIVERSITY.
Consumer
An ORGANISM that feeds on living or dead organic material. Consumers may be classified as primary (those that feed on photosynthetic organisms such as green plants), secondary (feeding on herbivorous animals – see HERBIVORE) or tertiary (feeding on other secondary consumers), or as DECOMPOSERS (feeding on dead organic material). Decomposers are sometimes excluded from the definition.
Continental drift
The complex process by which the continents move their positions relative to each other on the plates of the Earth's crust. Also known as plate tectonics.

Convergent evolution
A process whereby two or more groups of ORGANISMS that are not closely related acquire similar ADAPTATIONS independently of each other, through living in similar ENVIRONMENTS, adopting a similar diet, or defending themselves against the same sort of PREDATORS. (When the groups involved are fairly closely related, but still evolved their similar features independently of each other, the phenomenon is known as parallel evolution).
Copepods
A class of small CRUSTACEANS, mostly marine or AQUATIC, characterized by two pairs of antennae, five pairs of oarlike legs and only simple eyes.
Coral reef
A barrier of coral limestone formed by the accumulation of the skeletons of millions of coral POLYPS
Crustaceans
Members of the PHYLUM Crustacea – hard-bodied, mainly AQUATIC animals, usually with five pairs of jointed legs, two pairs of antennae and head and thorax joined, eg crabs, crayfish, shrimps and woodlice.
Culling
The selective removal of animals, normally weak or ageing, from a POPULATION in order to maintain the health of the breeding stock.

DDT
The pesticide dichloro-diphenol-tricholorethane.
Deciduous forest
A forest comprising mainly deciduous trees, ie those that shed their leaves each year during the winter or the dry season.
Decomposers
ORGANISMS such as FUNGI and BACTERIA, which rely upon the dead tissues of other ORGANISMS as an ENERGY source. In order to extract this energy, they break down the organic material, releasing nutrients from those tissues into the ENVIRONMENT.
Desert
A very arid area with less than 25 cm (10 in) precipitation (rainfall, snow, hail, etc.) a year. In hot deserts the rate of evaporation is greater than the rate of precipitation, and there is little vegetation.
Digestive system
The internal organs and glands associated with the breakdown of ingested food into soluble products that can be absorbed by the body tissues.
Dinosaurs
Members of a group of REPTILES that evolved some 175 million years ago and became extinct about 65 million years ago. Their closest living relatives are the crocodiles. Some dinosaurs attained enormous sizes.
Disease
A condition of the body of a living ORGANISM in which its function is impaired (due to the actions of a pathogenic organism or to a genetic disorder). The term is not usually applied to physical injury.
Diversity
A measurement of the variety of SPECIES in a given COMMUNITY or area. The concept sometimes incorporates a measure of the relative abundance of the different species present.
Dormancy
A period during which the metabolic activity of a plant or animal is reduced to such an extent that it can withstand difficult environmental conditions such as cold or drought.
Drought
A long period when rainfall and other precipitation is substantially lower than average.

Echolocation
Orientation by emitting high-pitched sounds and locating the positions of objects by the way in which they reflect the sound. Echolocation is used mainly by bats and dolphins, but also by oilbirds.
Ecological niche
The position in the ecological COMMUNITY occupied by a particular SPECIES defined by the HABITAT it occupies, what it eats and what it is eaten by.
Ecology
The study of ORGANISMS in relation to their physical and living ENVIRONMENT.
Ecosystem
A COMMUNITY of plants and animals that interact with each other, and with the ENVIRONMENT in which they live.
Edentates
Members of the Edentata, an ORDER of MAMMALS in which the teeth are simple and peglike, reduced in number or even absent, as an ADAPTATION to a diet of small INVERTEBRATES. The Edentates include sloths, anteaters and armadillos.
Emigration
The nonseasonal instinctive movement of members of a SPECIES away from the home area in response to overcrowding or food shortages. This is a one-way exodus: the animals do not return.
Endangered species
A SPECIES whose numbers have dropped to such low levels that its continued survival is uncertain.
Endemic
A SPECIES that is native to one specific area, and is therefore often said to be characteristic of that area.
Energy
The ability to do work. Energy may take many forms, such as light, heat and chemical energy; it changes form as it flows through the ECOSYSTEM.
Environment
The living and nonliving surroundings of an ORGANISM.
Evolution
The process by which SPECIES have developed to their current appearance, biology and behavior through the process of NATURAL SELECTION.
Excretory system
The system of organs, ducts and other structures by means of which useless or harmful products of an ORGANISM'S metabolism are eliminated from its body.
Exotic species
A SPECIES that is not native to an area but has become established after being introduced from elsewhere, often for commercial or decorative purposes.
Extinction
The complete elimination of a POPULATION or SPECIES.

Family
A taxonomic term for a group of related ORGANISMS, such as the family Felidae (cat family), which includes the lion, the tiger and all the smaller cats. Most families contain several GENERA, and families are grouped together into ORDERS.
Feces
Wastes discharged from the digestive tract and eliminated from the body.
Filter feeder
An animal that feeds by filtering small items of food out of the water.
Food chain
A succession of living ORGANISMS through which ENERGY and materials may flow in a COMMUNITY. Each is the food for the next one up in the line.
Food web
The complex feeding interactions between SPECIES in a COMMUNITY.
Fossils
Any traces or remains of long-dead ORGANISMS, preserved in a rock or some other material. (Usually only the hard parts, such as the bones of MAMMALS or the shells of MOLLUSKS, are preserved.)
Frugivores
Fruit-eating animals.
Fungi
Nongreen multicellular plantlike ORGANISMS that feed on organic matter.

Game reserve
An area originally set aside for the management and protection of game animals for hunting. Now they are usually areas where all wildlife is protected.
Genera
See GENUS.
Genes
Set of instructions "coded" in chemical form in the nucleus of cells that determine the inherited characteristics of an ORGANISM.
Genetic characteristics
Inherited characteristics of an ORGANISM that are determined by its GENES.
Genetic engineering
The manipulation of genetic material by humans to produce ORGANISMS with new characteristics or new combinations of characteristics.
Genus (pl. genera)
A level of biological CLASSIFICATION of ORGANISMS in which closely related SPECIES are grouped. For example, dogs, wolves, jackals and coyotes are all grouped together in the genus *Canis*.
Gondwanaland
The name given to the ancient southern supercontinent, which was composed of presentday Africa, Australia, Antarctica, India and South America. It began to break up 200 million years ago.
Greenhouse effect
The process in which radiation from the sun passes through the atmosphere, is reflected off the surface of the Earth, and is then trapped by gases in the atmosphere. The build-up of carbon dioxide and other "greenhouse gases" is increasing the effect. There are fears that the temperature of the planet may rise as a result; this global warming is expected to have dire consequences.

Habitat
The external ENVIRONMENT to which an animal or plant is adapted and in which it prefers to live, defined in terms of such factors as vegetation, CLIMATE and altitude.
Harem group
A group of female animals temporarily "possessed" by a particular male during the

breeding season. Males of SPECIES that form harems usually fight for the possession of individual females. (This type of breeding system results in most of the females in a POPULATION being inseminated by just a few males – the "fittest" males.)

Herbicide
A chemical used to kill unwanted vegetation (such as weeds).

Herbivore
An animal that feeds exclusively on living plants. (see also BROWSER, GRAZER).

Hibernation
The state of reduced metabolic activity or DORMANCY in certain MAMMALS during winter.

Hierarchy
The situation in which one animal in a POPULATION dominates another in fights or in access to resources such as food and mates. Hierarchy may be established by means of certain gestures or sounds rather than by fights, and sometimes extends throughout a population, each animal being dominated by those above it and dominating those below it in the hierarchy.

ICBP
The International Council for Bird Preservation. Organization that aims to save the world's birds and their HABITATS from destruction and EXTINCTION.

Insect
A hard-shelled animal with three pairs of jointed legs, one pair of antennae, and a body divided into head, thorax and abdomen.

Introduced species
A SPECIES that has been introduced by humans into an area in which it was not formerly present.

Invertebrate
An animal without a backbone or bones.

IUCN
The International Union for the Conservation of Nature, now called the World Conservation Union. A membership organization bringing together states, government agencies and nongovernmental organizations to promote CONSERVATION and the sustainable use of living resources.

Kingdom
The highest grouping in the CLASSIFICATION of living ORGANISMS. There are five kingdoms: the Monera, including the BACTERIA and blue-green ALGAE); the PROTISTA, including PROTOZOANS and single-celled algae (excepting the blue-green algae); the Fungi; the Animalia – the animals; and the Plantae – the green plants.

Krill
Shrimplike marine CRUSTACEANS, which occur in dense swarms in the oceans, and are the major food of the filter-feeding whales.

Larva (pl. larvae)
An immature stage in the life cycle of many animals. It is the stage in which these animals hatch from the egg, and is usually very different in appearance from the ADULT. An example is the butterfly larva (caterpillar). Larvae are usually incapable of reproduction.

Mammal
A VERTEBRATE animal belonging to the CLASS Mammalia, having a four-chambered heart, fur or hair, and feeding its young on milk secreted by the mammae (nipples). With the exception of the MONOTREMES, mammals do not lay eggs, but give birth to live young.

Marsupial
MAMMALS that have no placenta. The young leaves the womb at a very early stage of development and crawls into a pouch or fold of skin enclosing the nipples, to which it attaches itself.

Migration
The periodic movement, normally seasonal, of animals from one region to another to feed or to breed.

Mollusk
An INVERTEBRATE animal belonging to the PHYLUM Mollusca, a soft-bodied animal usually with a hard shell. Mollusks include snails, slugs, limpets, squid and octopus, oysters and mussels.

Monotreme
An egg-laying MAMMAL. There are only three SPECIES – the Duck-billed platypus and two species of echidnas or spiny anteaters.

Montane
Describes the zone at middle altitudes on the slopes of mountains, below the ALPINE zone.

Morphology
The form and structure of an ORGANISM.

Mutation
An inheritable change in the genetic material (see GENES) of an ORGANISM.

Native
An ORGANISM that lives in a particular locality.

Natural selection
The process by which ORGANISMS not well suited to their ENVIRONMENT are eliminated by predation, parasitism, COMPETITION, etc., and those that are well suited survive to breed and pass on their GENES to the next generation.

Ocean
The great body of salt water that covers more than two thirds of the Earth's surface, particularly that part which lies over what is known as the oceanic crust. Geographers often recognize five distinct oceans – the Atlantic, Pacific, Indian, Arctic and Southern oceans.

Omnivore
An animal that eats a varied diet including both animal and vegetable matter.

Opposable (thumb)
Describes a structure that can be placed opposite another, in this case the thumb can be positioned opposite the other digits, enabling animals to grasp objects.

Organism
Any living thing – animal, plant or microbe.

Parasite
An ORGANISM that lives on or in another organism of a different species and derives nutrients from it, giving nothing beneficial in return.

PCBs (polychlorinated biphenyls)
Highly stable chemicals of low flammability used in a variety of manufactured items, but especially in electrical transformers, fluorescent light bulbs and hydraulic fluid. When they escape into the ECOSYSTEM they are highly persistent and accumulate in the fatty tissues of animals, where they can give rise to cancers and infertility.

Pest
An ORGANISM that causes problems for humans by interfering with their management of ECOSYSTEMS.

Pest control
The reduction of PEST populations by various means, including chemical and BIOLOGICAL CONTROL.

Photosynthesis
The process by which plants make organic compounds, primarily sugars, from carbon dioxide and water, using sunlight as the source of energy and the green pigment chlorophyll, or another related pigment, for trapping the Sun's energy.

Phylum (pl. phyla)
The second largest grouping in the CLASSIFICATION of living ORGANISMS. Each KINGDOM is made up of many phyla, and each phylum contains many CLASSES. An example is the phylum Mollusca the, MOLLUSKS.

Physiology
The functioning of living ORGANISMS and their component parts, particularly internal bodily functions.

Phytoplankton
Microscopic plants that form part of the PLANKTON.

Pioneer
A SPECIES that colonizes a newly exposed or newly altered ENVIRONMENT.

Placental mammals
MAMMALS in which the young develop inside the womb, attached to its wall by a placenta – a fleshy tissue that brings the blood vessels of embryo and mother into close contact, supplying food, oxygen and hormones to the embryo and removing its waste products.

Plankton
The COMMUNITY of MICROSCOPIC plants and animals that float at or near the surface of the sea or a lake.

Polyp
One of the two body forms (polyp and medusa) of jellyfish, sea anemones, corals, etc. Typical polyps are cylindrical animals with a mouth at one end surrounded by a ring of tentacles.

Population
In biological terms, a more or less separate breeding group of animals or plants. The term is also used for the total number of individuals of a particular SPECIES within a given area.

Predator
An animal that feeds on another animal (the PREY).

Prehensile
Describes a structure that is capable of grasping an object by coiling around it, eg the tails of South American monkeys and of chameleons.

Prey
An animal that a PREDATOR hunts and kills for food.

Producers
In an ECOSYSTEM, the plants or other organisms that trap energy (usually from sunlight) and use it to synthesize the organic compounds that form the basis of the FOOD CHAIN.

Productivity
The amount of weight (or energy) gained by an individual, a SPECIES or an ECOSYSTEM per unit area per unit time.

Pronking
A reaction of springbok (small antelope) when excited or threatened, in which the animal leaps into the air with all four feet off the ground, arching its back to display its dorsal ruff of fur.

Protista
A KINGDOM of the living world, whose members include the single-celled protozoans and single-celled algae.

Protozoa
MICROSCOPIC single-celled animals, once considered to comprise a separate kingdom, but now included in the Kingdom PROTISTA).
Purse seine net
A fishing net used to encircle fish, open at one end, with floats at the top and weights at the bottom.

Rabies
A viral DISEASE that affects mainly CARNIVORES, but can also infect humans. An initial stage of heavy production of saliva is followed by depression, fear of water and paralysis. Rabies is transmitted by bites, and is usually fatal.
Race
A taxonomic division subordinate to SUBSPECIES, but linking populations with similar distinct characteristics.
Rainforest
Forest where there is abundant rainfall all year round. The term is normally associated with TROPICAL rainforests, which are rich in plant and animal SPECIES and where growth is lush and very rapid.
Range
The geographical area in which a SPECIES occurs, or over which an individual animal roams.
Refuge (refugia)
A place where a SPECIES of plant or animal has survived after formerly occupying a much larger area. For example, mountaintops are refuges for ARCTIC species left behind as the glaciers retreated at the end of the last period of glaciation.
Related species
SPECIES that belong to the same genus or family, and who share many genetic characteristics.
Reptiles
TERRESTRIAL VERTEBRATES belonging to the CLASS Reptilia. Reptiles are ectothermic ("cold-blooded"), with a body covering of dry, horny scales. There is no distinct larval phase.
Rinderpest
Also known as cattle plague or bovine typhus, a viral DISEASE of wild and domesticated cattle and their relatives, which affects the digestive tract and is often fatal.
Ruminant
Any animal possessing a system of digestion in which newly eaten food is stored in a special stomach compartment, called the rumen, from which it is passed back to the mouth for chewing after eating is completed. Inside the rumen, bacteria help to digest the food. Examples of ruminants are deer, cattle and antelope.
Rut
Stage in the breeding season when male deer herd females to form HAREMS and roar at, and fight with, intruding males.

Salps
Transparent, gelatinous, barrel-shaped, filter-feeding INVERTEBRATES that live in the surface waters of the ocean.
Savanna
A HABITAT of open grassland with scattered trees in TROPICAL areas. Also known as tropical grassland, it covers areas between tropical RAINFOREST and hot deserts. There is a marked dry season each year and too little rain to support large areas of forest.
Scavenger
An animal that feeds on the remains of food killed or collected by other animals.
Scent gland
An organ that produces scented fluids, usually hormones called pheromones, which are used in sexual attraction and in the marking of TERRITORY.
Scent mark
A smear of scented fluid or feces placed by an animal on an object or on the ground in order to inform other members of its SPECIES of its presence. Scent marks are often used by animals to define TERRITORY.
Semiarid
Describes area between ARID deserts and better-watered areas, where there is sufficient moisture to support a little more vegetation than in the DESERT.
Sexual reproduction
Reproduction involving the fusion of sex cells (eggs and sperm) from two animals of opposite sex, thus allowing for the mixing of inherited characteristics in the offspring, an essential part of the mechanism of EVOLUTION.
Siblings
Individuals having one or both parents in common.
Single-celled organism
An ORGANISM that comprises a single cell. In such organisms the cell is usually highly complex, with different regions of the cell specialized for different functions.
Skeleton
Supporting structure of the body. The skeleton may be a bony internal support, as in vertebrates; a hard shell, as in CRUSTACEANS and MOLLUSKS; or, as in the case of worms, may simply take the form of hydrostatic pressure of confined fluids.
Specialization
The evolutionary development of a SPECIES, leading to narrow limits of tolerance and a restricted role (NICHE) in the COMMUNITY.
Species
ORGANISMS that resemble one another and can breed among themselves to produce similar offspring, which can themselves breed with other individuals of their species.
Specimen
An individual plant or animal taken as an example of its SPECIES.
Sponges
AQUATIC multicellular filter-feeding animals without definite tissues or organs. The soft body is supported by calcareous or siliceous spicules. Water is drawn into the hollow interior through holes in the sides. Free-swimming larvae are produced.
Stotting
Behavior of gazelles and certain other antelopes in reaction to attack, or in excitement; the animal bounces up and down on stiff legs.
Subadult
An independent animal that looks very similar to an ADULT, although it may differ in coloring, but has not yet reached sexual maturity and is thus incapable of breeding.
Subfamily
A group of GENERA that are more closely related (genetically similar) to each other than to other members of the same FAMILY.
Suborder
A group of FAMILIES that are more closely related (genetically similar) to each other than to other members of the same ORDER.
Subspecies
A population or group of POPULATIONS of a SPECIES that are distinctly different from other members of the species, and which for various reasons (such as geographical isolation) would not normally mate with other members of the species, but are still capable of doing so.
Subtropical
Describes the area of the Earth's surface between the TROPICS and the TEMPERATE zones. There are marked seasonal changes of temperature, but it is never very cold.
Succession
The development and maturation of an ECOSYSTEM, through changes in the type and abundance of SPECIES. When it reaches maturity it stabilizes as a CLIMAX.

Taxonomy
The scientific CLASSIFICATION of ORGANISMS.
TBT
Tributyl tin, a chemical used as a wood preservative and to prevent the fouling of boats. The substance is toxic in extremely small concentrations and causes serious pollution in estuaries and bays, causing deformity, breeding failure and death of marine organisms. It is also thought to harm the human immune system.
Temperate
Describes the climatic zones in mid latitudes, which cover areas between the warm TROPICAL and cold polar regions, and have a mild climate.
Terrestrial
Describes ORGANISMS whose entire life cycle is spent on the land.
Territory
An area of land occupied by a single animal or group of animals and actively defended.
Third World
Describes the countries of Latin America, Africa, Asia and parts of the Middle East that are not aligned with Western countries, or the Soviet Union, and which have no advanced industrialization.
Trilobites
Primitive marine ARTHROPODS known only from FOSSILS. Their bodies were divided into three parts, and they looked rather like presentday Horseshoe crabs or King crabs.
Trophic levels
The levels at which living organisms are positioned in a FOOD CHAIN.
Tropics
The area lying between the Tropic of Cancer and the Tropic of Capricorn. The tropics mark the latitude farthest from the Equator where the sun is directly overhead at midday in midsummer.

United Nations Man and the Biosphere Programme (MAB)
An attempt to protect ECOSYSTEMS by setting up protected reserves in which the core area of true WILDERNESS is fully protected, but is surrounded by buffer zones allowing some exploitation of natural resources. Local people are involved in the management of these reserves.

Vertebrate
An animal supported by a backbone.

Wetlands
A HABITAT that is waterlogged all or enough of the time to support vegetation adapted to those conditions.
Wilderness
An area of land that has never been modified by human activities and is thus in its natural state.

Further reading

Banister, Keith and Campbell, Andrew (eds.) *The encyclopedia of Underwater Life* (George Allen and Unwin, London, 1985)
Begon, Michael, Harper, John L. and Townsend, Colin R. *Ecology Individuals, Populations and Communities* (Blackwell Scientific Publications, Oxford, 1986)
Berry, R.J. and Hallam, A. (eds.) *The encyclopedia of animal evolution* (Facts On File, New York, 1987)
Collar, N.J. and Andrew, P.C. *Birds to watch. The ICBP World Checklist of Threatened Birds* (ICBP, London, 1988)
Cox, C. and Moore, Peter D. *Biogeography – An Ecological and Evolutionary Approach* (Blackwell Scientific Publications, Oxford, 1973)
Halliday, T., Adler, K. and O'Toole, C. (eds.) *The encyclopedia of Reptiles and Insects* (Grolier International, 1986)
IUCN *Red List of Threatened Animals* (IUCN, Cambridge, 1990)
Little, Colin *The terrestrial invasion – An ecophysiological approach to the origins of land animals* (Cambridge University Press, Cambridge, 1990)
Macdonald, David (ed.) *The encyclopedia of Mammals* (Unwin Hyman, London, 1984)
McNeill Alexander, R. *The Invertebrates* (Cambridge University Press, Cambridge, 1979)
Mason, I.L. (ed.) *Evolution of domesticated animals* (Longman, London, 1984)
Moore, Peter D. (ed.) *The encyclopedia of animal ecology* (Facts On File, New York, 1987)
Perrins, C.M. and Middleton, A.L.A. (eds.) *The encyclopedia of Birds* (George Allen and Unwin, London, 1985)
Sibly, R.M. and Calow, P. *Physiological Ecology of Animals* (Blackwell Scientific Publications, Oxford, 1986)
Slater, Peter J.B. (ed.) *The encyclopedia of animal behavior* (Facts On File, New York, 1987)
Soulé, Michael E. (ed.) *Conservation Biology* (Sinauer Associates, USA, 1986)
Wilson, E.O. *Biodiversity* (National Academy Press, USA, 1988)
Ecology, Evolution and Population Biology Readings from *Scientific American* (W.H. Freeman and Company, USA, 1974)

Acknowledgments

Picture credits

Key to abbreviations: A Ardea, London; **ANT** Australasian Nature Transparencies, Victoria, Australia; **BCL** Bruce Coleman Ltd, Uxbridge, Middlesex; **NHPA** Natural History Photographic Agency, Ardingly, Sussex; **OSF** Oxford Scientific Films, Long Hanborough, Oxford; **PEP** Planet Earth Pictures, London.

b=bottom, l=left, r=right, t=top.

1 Premaphotos/K.G. Preston Mafham **2** NHPA/Anthony Bannister **3** BCL/Mik Dakin **4** A/Martin W. Grosnick **6–7** NHPA/Anthony Bannister **8–9** BCL/Gunter Ziesler **10–11** OSF/Robert A. Tyrrell **11b** George Frame **11t** NHPA/Stephen Dalton **14** OSF/Zig Leszczynski **16–17** OSF/Alan G. Nelson **18** OSF/J.A.L. Cooke **20** PEP/Robert Hessler **22** NHPA/Jany Sauvanet **24** OSF/Okapia **24–25** NHPA/Anthony Bannister **25** NHPA/Jany Sauvanet **26** OSF/Maurice Tibbles **28–29** Jacana/J.P. Ferrero **29** OSF/M. Wendler/Okapia **30–31** OSF **31t** PEP/David Maitland **31b** NHPA/ANT/C. & S. Pollitt **32–33t** Frank Lane Picture Agency/Mark Newman **32–33b** Nature Photographers/ Hugh Miles **34** A/J.L. Mason **35** BCL/Keith Gunnar **36** PEP/Norman Cobley **36–37** PEP/Marty Snyderman **38–39** BCL/Frans Lanting **39t** PEP/Jonathan Scott **39b** PEP/Ford Kristo **40–41t** OSF/Andrew Plumptre **40–41b** OSF/Mike Birkhead **44, 45, 48, 48–49** NHPA/Stephen Krasemann **49** NHPA/Brian Hawkes **50–51** OSF/Ralph A. Reinhold **51** OSF/Ted Levin **52–53** NHPA/Stephen Krasemann **53** BCL/Jeff Foott **55** Premaphotos/K.G. Preston-Mafham **56–57** OSF/Michael Fogden **57** NHPA/John Shaw **58** NHPA/Stephen Krasemann **60t** OSF/Michael Fogden **60b** NHPA/R.J. Erwin **60–61** NHPA/Eric Soder **62** BCL/Jeff Foott **62–63** NHPA/Stephen Krasemann **63** NHPA/John Shaw **65** NHPA/Dave Currey **66** NHPA/Henry Ausloos **66–67** NHPA/ Stephen Krasemann **68–69** OSF/Frank Huber **70–71, 73** OSF/Michael Fogden **71** Premaphotos/K.G. Preston-Mafham **74** OSF/Laurence Gould **74–75** NHPA/Henry Ausloos **75** OSF/Partridge Films Ltd **76, 76–77** NHPA/G.I. Bernard **78–79, 80** Premaphotos/K.G. Preston-Mafham **81** NHPA/G.D.T. Silvestris **84** NHPA/Haroldo Palo **85** NHPA/Martin Wendler **86–87** BCL/Frans Lanting **87** BCL/Francisco Erize **88–89** Survival Anglia/Annie Price **90–91** Naturfotograferna/Janos Jurka **91** OSF/ David Curl **94** BCL/Wedigo Ferchland **94–95t** BCL/Brian and Cherry Alexander **94–95b** BCL/Eckhart Pott **96** BCL/Ronald Thompson **96–97** NHPA/G.I. Bernard **98–99** NHPA/Manfred Danegger **99** NHPA/John Hayward **101** Survival Anglia/Dennis Green **102–103** NHPA/Melvin Grey **103** NHPA/E.A. Janes **104t** BCL/George McCarthy **104b** NHPA/Michael Leach **105** NHPA/Stephen Krasemann **106–107** OSF/David Wright **108–109** NHPA/Henry Ausloos **109** Jacana/Francois Gohier **112–113** NHPA/Pierre Petit **113** Jacana/C. D'Hotel **114** NHPA/George Bernard **114–115** Okapia/St. Meyers **115** BCL/Ernest Duscher **116–117** NHPA/David Woodfall **117** NHPA/Manfred Danegger **118** BCL/Jane Burton **120–121** NHPA/E.A. Janes **121** Okapia/M. Gruber **123** BCL/George McCarthy **124–125** Okapia/M. Gruber **126** NHPA/Gerard Lacz **127** BCL/J.L.G. Grande **130–131** NHPA/Gerard Lacz **131t** NHPA/Laurie Campbell **131b** BCL/Udo Hirsch **132–133** Jacana/Jean-Philippe Varin **134–135** NHPA/Roger Tidman **135** NHPA/Martin Wendler **138** A/Ian Beames **138–139** Premaphotos/R.A. Preston-Mafham **139** NHPA/E.A. Janes **140** Panda Photo/A. Petretti **140–141** Panda Photo/A. Bardi **142–143** Okapia/Hans Reinhard **143** BCL/Bernd Thies **144** BCL/J.L.G. Grande **146–147** OSF/St. Meyers **147t** NHPA/Henry Ausloos **147b** NHPA/Stephen Dalton **148** Survival Anglia/Bomford & Borkowski **148–149** A/Bomford & Borkowski **150** A/John Daniels **151** Okapia/Klaus Schneider **152** NHPA/Hellio & Van Ingen **154** Okapia/D. Hanff **154–155** Okapia/Hans Reinhard **156–157** BCL **158–159** Okapia/Fred Bruemmer **159** Vadim Gippenreiter **162** A/M. Iijima **162–163** Okapia/Ingo Gerlach **164** BCL/Pekka Helo **164–165** Frank Lane Picture Agency/Hans Dieter Brandt **166–167** BCL/G.D. Plage **167, 168** NHPA/Stephen Dalton **170** BCL/Steve Kaufman **170–171** NHPA/Phillipa Scott **172t** BCL/Mark Boulton **172b** Hutchison Library/Bernard Gerard **173** Hutchison Library **174–175** PEP/Peter Scoones **176–177** NHPA/Hellio & Van Ingen **177** OSF/Eyal Bartov **179** NHPA/Peter Johnson **180–181** BCL/Erwin & Peggy Bauer **181t** BCL/P. Evans **181b** NHPA/A. Papazian **182, 182–183** Estate of Wilma George **184–185** BCL/Gunter Ziesler **186** BCL/Hans Reinhard **187** NHPA/Nigel Dennis **188** OSF/Andrew Plumptre **190–191** Premaphotos/K.G. Preston-Mafham **191** NHPA/Anthony Bannister **192** OSF/David Cayless **192–193** BCL/M.P. Kahl **194–195** NHPA/Agence Nature **196** NHPA/Anthony Bannister **197** OSF/G.I. Bernard **199** Anthony Bannister Photo Library **200–201** BCL/O. Langrand **201t** NHPA/Anthony Bannister **201b** PEP/Peter Scoones **203** Premaphotos/K.G. Preston-Mafham **204–205** NHPA/Anthony Bannister **206–207** NHPA/E. Hanumantha Rao **207** BCL/G.D. Plage **208–209** NHPA/Gerard Lacz **210t** NHPA/Hanumantha Rao **210b** Robert & Linda Mitchell **212** NHPA/E. Hanumantha Rao **213** BCL/Gunter Ziesler **214–215** NHPA/E. Hanumantha Rao **215** BCL/Ross Wilmshurst **216** OSF/Harry Fox **218** BCL/Jeff Foott **219t** BCL/Dieter & Mary Plage **219b** World Wildlife Fund, Gland/Sture Karlsson **220** OSF/Zig Leszczynski **220–221** Jacana **222** OSF/Michael Dick **223** NHPA/Morten Strange **226, 227b** Robert & Linda Mitchell **227t** BCL/Gerald Cubitt **228–229** PEP/Pete Atkinson **229** PEP/James D. Watt **230–231** Robert & Linda Mitchell **232, 233** NHPA/Orion Press **234–235** NHPA/Henry Ausloos **236** BCL/Steven Kaufman **237t** NHPA/Phillipa Scott **237b, 238, 238–239** NHPA/Orion Press **240** NHPA/ANT/Kelvin Aitken **241** Frithfoto **244t** NHPA/ANT/ C. & S. Pollitt **244b** BCL/C. & D. Frith **245** BCL/John Markham **246** NHPA/ANT/Dave Watts

Editorial, research and administrative assistance
Jill Bailey, Lionel Bender, Martin Bramwell, Helen Burridge, Shirley Jamieson, Martin Jenkyns, Miles Litvinoff, Hilary McGlynn, Madeleine Samuel, John Stidworthy

Artists
Graham Allen, Rob van Assen, Priscilla Barrett, Trevor Boyer, Robert Gillmor, Roger Gorringe, Richard Lewington, Malcolm McGregor, Sean Milne, Denis Ovenden, The Maltings Partnership, Ian Willis

Page 132/133: Denis Ovenden's artwork from the Collins *Field Guide to Reptiles and Amphibians of Britain and Europe* appears by arrangement with Harper Collins

Cartography
Map p. 17 drafted by Euromap, Pangbourne

Index
Barbara James

Production
Clive Sparling

Typesetting
Brian Blackmore, Peter MacDonald Associates

Color origination
Scantrans pte Limited, Singapore

INDEX

Page numbers in *italics* refer to captions

E

F

G

H

I

J

K

L